AF771341

TABLE OF CONTENTS

Global Customer Support and Sales Offices

Americas (North/South/Central)
US Customers
Toll Free: +1 (800) 545-2522
Fax: +1 (800) 535-9935

Customers Outside US
Toll Free: +1 (800) 460-3110
Tel: +1 (314) 453-7010
Fax: +1(314) 453-7095
E-mail: usbkinfo@elsevier.com

Asia Pacific
Tel: +65 6349 0222
Fax: +65 6733 1817
E-mail: asiabkinfo@elsevier.com

Australia, New Zealand & Fiji
Tel: +61 (02) 9517 8999
Fax: +61 (02) 9517 2249
E-mail: service@elsevier.com.au

Europe, Middle-East and Africa
UK Customers
Tel: +44 (0) 1865 474000
Fax: +44 (0) 1865 474001

Customers Outside UK
Tel: +44 (0) 1865 474110
Fax: +44 (0) 1865 474111
E-mail: eurobkinfo@elsevier.com

Japan
Tel: +81 3 3589 6370
Fax: +81 3 3589 6371
E-mail: books@elsevierjapan.com

ELSEVIER

Bookmark our site at:
www.elsevier.com

ISBN: 0-444-52049X

THESAURUS ALPHABETICAL LISTING

abaca fiber
UF: abaca fibre
BT: vegetable fiber

abaca fibre
USE: abaca fiber

abdominal support belt
BT: belt
support garment

abrading
BT: finishing

abrading machine
BT: finishing machine

abrasion
BT: wear

abrasion resistance
BT: wear resistance

abrasive agent
BT: finishing agent

absorbency
BT: chemical property

absorber
BT: additive

absorption
BT: sorption

absorption isotherm
BT: mathematical expression
thermal property

accelerating agent
BT: dyeing assistant

accessories
BT: product by item type
NT: bag
garment accessory
luggage
umbrella

acetate
BT: organic compound

acetate fabric
BT: regenerated cellulose fabric

acetate fiber
UF: acetate fibre
BT: regenerated cellulose fiber

acetate fibre
USE: acetate fiber

acetate yarn
BT: regenerated cellulose yarn

acetylation
BT: chemical reaction

acid
BT: chemical by property or purpose

acid dye
BT: dye by application or reaction

acid treatment
BT: finishing
NT: carbonizing

acoustic property
BT: physical property

acquisition
BT: business strategy

acrylic
BT: plastic

acrylic acid
BT: organic acid

acrylic blend
BT: manufactured fiber blend
NT: acrylic-cotton blend
acrylic-polyester blend
acrylic-viscose-rayon blend
acrylic-wool blend

acrylic cotton blend
USE: acrylic-cotton blend

acrylic fabric
BT: manufactured fiber fabric

acrylic fiber
UF: acrylic fibre
BT: manufactured fiber

acrylic fibre
USE: acrylic fiber

acrylic polyester blend
USE: acrylic-polyester blend

acrylic polymer
BT: synthetic organic polymer
NT: polyacrylamide
polyacrylate
polyacrylic acid
polyacrylonitrile
polymethyl methacrylate

acrylic viscose rayon blend

USE: acrylic-viscose-rayon blend

acrylic wool blend
USE: acrylic-wool blend

acrylic yarn
BT: manufactured fiber yarn

acrylic-cotton blend
UF: acrylic cotton blend
BT: acrylic blend
cotton blend

acrylic-polyester blend
UF: acrylic polyester blend
BT: acrylic blend
polyester blend

acrylic-viscose-rayon blend
UF: acrylic viscose rayon blend
BT: acrylic blend
viscose-rayon blend

acrylic-wool blend
UF: acrylic wool blend
BT: acrylic blend
wool blend

acrylonitrile
BT: organic compound

activated carbon
BT: adsorber

activator
BT: additive

activewear
BT: sportswear

add on
USE: add-on

add-on
UF: add on
BT: mathematical expression
weight

addition polymer
BT: polymer by polymerization reaction

additive
BT: chemical by property or purpose
NT: absorber
activator
adsorber
afterclearing agent
antifibrillating agent
antifoaming agent
antioxidant
binding agent
bonding agent
buffering agent
catalyst
cationizing agent
complexing agent
cross linking agent
dispersant
dyeing assistant
finishing agent
inhibitor
plasticizing agent
printing assistant
reducing agent
stabilizing agent
surface-active agent
swelling agent
thickening agent

adhesion
BT: physical phenomena
NT: fiber-matrix adhesion

adhesive
USE: adhesive agent

adhesive agent
UF: adhesive
BT: bonding agent

adhesive property
BT: physical property
NT: stickiness

adsorbency
BT: chemical property

adsorber
BT: additive
NT: activated carbon

adsorption
BT: sorption

adsorption isotherm
BT: mathematical expression
thermal property

advertising
BT: marketing

aerial
BT: product by item type

aerobic treatment
BT: environmental treatment

aerodynamics
BT: fluid dynamics

aerofoil
USE: airfoil

aerosol
BT: colloid

aerospace textile
BT: product by property or purpose
NT: hot-air balloon
parachute

aesthetic property
USE: esthetic property

afterclearing
BT: aftertreatment
finishing
removal

afterclearing agent
BT: additive

aftertreatment
BT: general manufacturing process
NT: afterclearing
afterwashing

afterwashing
BT: aftertreatment
washing

agavaceae fiber
UF: agavaceae fibre
BT: vegetable fiber
NT: furcraea fiber

agavaceae fibre
USE: agavaceae fiber

age determination
BT: methodology

aging
BT: damage
finishing
NT: flash aging

agitation
BT: materials handling

agricultural industry
BT: industry
NT: agricultural production
animal husbandry
cultivation
farming
harvesting
pest and disease management
selective breeding
sericulture

agricultural production
BT: agricultural industry
production
NT: crop production
livestock production
yield

agricultural textile
BT: product by property or purpose

agriculture
BT: applied science
NT: agronomy

agronomy
BT: agriculture

air
BT: gas

air bag
BT: automotive textile
vehicle component

air conditioning
BT: Facet B: operations

air filter
BT: gas filter

air filtration
BT: gas filtration

air flow
USE: airflow

air jet
BT: jet

air jet loom
USE: air-jet loom

air jet spinning
USE: air-jet spinning

air laying
BT: dry laying

air pocket
BT: pocket

air pollution
BT: pollution

air pressure
BT: pressure

air-jet loom
UF: air jet loom
BT: shuttleless loom

air-jet spinning
UF: air jet spinning
BT: spinning

aircraft
BT: transportation vehicle

airflow
UF: air flow
BT: gas flow

NT: ventilation

airfoil
UF: aerofoil
BT: vehicle component

airtight
BT: physical property

albumin
BT: protein

alcohol
BT: organic compound

alginate
BT: polysaccharide

alginate fabric
BT: manufactured fiber fabric

alginate fiber
UF: alginate fibre
BT: manufactured fiber

alginate fibre
USE: alginate fiber

alginate yarn
BT: manufactured fiber yarn

alginic acid
BT: polysaccharide

alignment mechanism
BT: machine component

alkali
BT: chemical by property or purpose

alkali treatment
BT: finishing

alkoxylate
BT: organic compound

allergy
BT: disease

alloy
BT: metal

alpaca
BT: animal

alpaca fabric
BT: keratin fiber fabric

alpaca fiber
UF: alpaca fibre
BT: keratin fiber

alpaca fibre
USE: alpaca fiber

alpaca yarn
BT: keratin fiber yarn

alumina fiber
UF: alumina fibre
BT: mineral fiber

alumina fibre
USE: alumina fiber

aluminosilicate fiber
UF: aluminosilicate fibre
BT: mineral fiber

aluminosilicate fibre
USE: aluminosilicate fiber

aluminum fiber
UF: aluminum fibre
BT: metal fiber

aluminum fibre
USE: aluminum fiber

amination
BT: chemical reaction

amine
BT: organic compound

amino acid
BT: organic acid

amino acid analysis
BT: chemical analysis

aminolysis
BT: chemical reaction

ammonia
BT: inorganic compound

ammonium carbonate
BT: inorganic compound

anaerobic treatment
BT: environmental treatment

analysis
BT: methodology
NT: chemical analysis
failure analysis
image analysis
mathematical analysis
microscopy
qualitative analysis
quantitative analysis
spectroscopy
structural analysis
theoretical analysis
thermal analysis

angora
BT: rabbit

angora fabric
BT: keratin fiber fabric

angora fiber
UF: angora fibre
BT: keratin fiber

angora fibre
USE: angora fiber

angora yarn
BT: keratin fiber yarn

animal
BT: organism
NT: alpaca
bollworm
maggot
rabbit
sheep
silkworm

animal fiber
UF: animal fibre
BT: natural fiber
NT: keratin fiber
silk fiber

animal fiber blend
BT: natural fiber blend
NT: silk blend
wool blend

animal fiber fabric
UF: animal fibre fabric
BT: natural fiber fabric
NT: keratin fiber fabric
silk fabric

animal fiber industry
BT: natural fiber industry
NT: cashmere industry
silk industry
wool industry

animal fiber yarn
UF: animal fibre yarn
BT: natural fiber yarn
NT: keratin fiber yarn
silk yarn

animal fibre
USE: animal fiber

animal fibre fabric
USE: animal fiber fabric

animal fibre yarn
USE: animal fiber yarn

animal husbandry
BT: agricultural industry

animal material
BT: general materials by origin or composition
NT: feather
fiber medulla
hair
sericin
silk
skin
wool
wool grease

anionic dye
BT: dye by application or reaction

anisotropy
BT: physical property

annealing
BT: heat treatment

annual report
BT: publication

anthraquinone dye
BT: synthetic dye

anthropometry
BT: measuring
NT: body scanning

antifelting property
BT: resistant property

antifibrillating agent
BT: additive

antifoaming agent
BT: additive

antifungal property
BT: biological resistance

antimicrobial property
BT: biological resistance

antioxidant
BT: additive

antistatic property
BT: electrical resistance

antitwist device
BT: machine component

antiwrap device
BT: machine component

antiyellowing property
BT: resistant property

apparel fabric
BT: fabric by property or purpose

appearance
BT: physical property
NT: color
esthetic property
evenness
fabric appearance
garment appearance
hairiness
pattern (configuration)
streakiness

application (finishing)
UF: chemical application
dye application
BT: finishing
NT: spray application

applicator
BT: equipment by item type
NT: squeegee

applied science
BT: Facet G: art, science and technology
NT: agriculture
engineering
ergonomics

applique
BT: pattern (configuration)

apron
BT: machine component

aramid
BT: synthetic polyamide

aramid blend
BT: manufactured fiber blend
NT: aramid-nylon blend
aramid-polyester blend

aramid fabric
BT: polyamide fabric

aramid fiber
UF: aramid fibre
BT: polyamide fiber

aramid fibre
USE: aramid fiber

aramid nylon blend
USE: aramid-nylon blend

aramid polyester blend
USE: aramid-polyester blend

aramid yarn
BT: polyamide yarn

aramid-nylon blend
UF: aramid nylon blend
BT: aramid blend
nylon blend

aramid-polyester blend
UF: aramid polyester blend
BT: aramid blend
polyester blend

architectural textile
BT: product by property or purpose
NT: awning
canopy
emergency shelter
roofing
tarpaulin
tent

aromatic amine
BT: organic compound

art
USE: textile art

art, design and culture
BT: Facet G: art, science and technology
NT: costume
design
fashion
handicraft
textile art

artificial fur
BT: artificial textile

artificial hair
BT: artificial textile

artificial leather
BT: artificial textile

artificial textile
BT: product by property or purpose
NT: artificial fur
artificial hair
artificial leather
artificial turf

artificial turf
BT: artificial textile
sports surface

asbestos
BT: geological material

asbestos fabric
BT: mineral fiber fabric

asbestos fiber
UF: asbestos fibre
BT: mineral fiber

asbestos fibre
USE: asbestos fiber

asbestos replacement
BT: use

asbestos yarn
BT: mineral fiber yarn

assessment
BT: methodology
NT: objective assessment
performance assessment
risk assessment

atactic polymer
BT: polymer by stereochemical configuration

autoclave
BT: equipment by item type

autoleveling
BT: carding
drawing

automation
BT: process control

automotive textile
BT: product by property or purpose
NT: air bag
seat belt
tire cord

awning
BT: architectural textile

axminster carpet
BT: carpet

azo dye
BT: dye by application or reaction
synthetic dye

babieswear
BT: garment by user

back coating
USE: back-coating

back winding
USE: back-winding

back-coating
UF: back coating
BT: coating

back-winding
UF: back winding
backwinding
BT: winding

backing
BT: product component
NT: backing fabric
carpet backing

backing fabric
BT: backing
fabric by property or purpose

backstaining
BT: dyeing

backwinding
USE: back-winding

bacterium
BT: microorganism

bag
BT: accessories

balance tube
BT: machine component

bale
BT: handling pack

bale opening machine
BT: opening machine

bale plucking
BT: materials handling

bale plucking machine
BT: machine

baling machine
BT: machine

ballistic resistance
BT: resistant property

balloon (yarn)
BT: shape

balloon limiter
BT: machine component

bamboo fiber
UF: bamboo fibre
BT: vegetable fiber

bamboo fibre
USE: bamboo fiber

banana fiber
UF: banana fibre
BT: vegetable fiber

banana fibre
USE: banana fiber

bandage

BT: medical textile

banner
BT: product by item type

bar
BT: defect

barre
SN: woven fabric fault
BT: defect

barrier fabric
BT: fabric by property or purpose

barrier property
BT: resistant property

basalt
BT: geological material

basalt fabric
BT: mineral fiber fabric

basalt fiber
UF: basalt fibre
BT: manufactured fiber

basalt fibre
USE: basalt fiber

basalt yarn
BT: manufactured fiber yarn

basic dye
BT: dye by application or reaction

bast fiber
UF: bast fibre
BT: vegetable fiber
NT: flax fiber
hemp fiber
hibiscus fiber
jute fiber
kenaf fiber
ramie fiber

bast fiber fabric
UF: bast fibre fabric
BT: vegetable fiber fabric
NT: flax fiber fabric
hemp fabric
jute fabric
kenaf fabric
ramie fabric

bast fiber yarn
UF: bast fibre yarn
BT: vegetable fiber yarn
NT: flax fiber yarn
hemp yarn
jute yarn
kenaf yarn
ramie yarn

bast fibre
USE: bast fiber

bast fibre fabric
USE: bast fiber fabric

bast fibre yarn
USE: bast fiber yarn

batch dyeing
BT: dyeing

batching
BT: materials handling

batching machine
BT: machine

bath
BT: equipment by item type
NT: dye bath

batik
BT: dyed fabric
dyeing

battery separator
BT: product by item type

batting
BT: nonwoven fabric

BCF yarn
BT: yarn by form

bead
BT: trim

beadwork
BT: embroidery

beam
SN: machine component
BT: machine component

beam dyeing
BT: dyeing

beaming
BT: warping

bearings
BT: machine component

beating up
USE: beating-up

beating-up
UF: beating up
BT: weaving

beauty product
BT: product by property or purpose

beck dyeing
USE: kettle dyeing

bedding
BT: furnishings
NT: blanket
mattress
quilt
sleeping bag

behavioral or psychological phenomena
BT: Facet D: phenomena
NT: color perception
comfort
decision making
esthetic property
gender identity

belt
BT: garment accessory
NT: abdominal support belt

benchmarking
BT: standard

bending
BT: distortion

bending property
BT: mechanical property

bias
BT: fabric construction

biaxial loading
BT: loading (testing)

bib
BT: outerwear

bicomponent fiber
UF: bicomponent fibre
BT: fiber by form

bicomponent fibre
USE: bicomponent fiber

bicycle frame
BT: sports goods

binder fiber
BT: fiber by property or purpose

binding
BT: product component

binding agent
BT: additive

bioactive property
BT: biological property

biocompatibility
BT: biological property

biodegradability
BT: biological property

biofinishing
BT: finishing
NT: biopolishing
enzymatic treatment

biohazard
BT: health and disease

biological characteristic
BT: biological phenomena
NT: body shape
body size
breed
maturity

biological phenomena
BT: Facet D: phenomena
NT: biological characteristic
biological property
fouling
health and disease

biological property
BT: biological phenomena
NT: bioactive property
biocompatibility
biodegradability

biological resistance
BT: resistant property
NT: antifungal property
antimicrobial property
insect resistance
virus resistance

biological structure or form
BT: Facet F: organisms and biological structures
NT: cocoon
seed

biological treatment
BT: environmental treatment

biomimetic material
BT: general materials by property or purpose

biopolishing
BT: biofinishing

biopolymer
BT: organic polymer

biotechnology
BT: technology

NT: genetic modification

birefringence
BT: fiber property
mathematical expression

blade
BT: machine component

blanket
BT: bedding

bleaching
BT: wet processing
NT: brightening
whitening

bleaching agent
BT: finishing agent

bleaching assistant
BT: finishing agent

bleaching machine
BT: wet processing machine

blend
BT: materials
NT: manufactured fiber blend
natural fiber blend
polymer blend

blend ratio
BT: mathematical expression

blended fabric
BT: fabric by origin or composition

blended fiber
UF: blended fibre
BT: fiber by origin or composition

blended fibre
USE: blended fiber

blended yarn
BT: yarn by origin or composition

blending
BT: fiber manufacture

blending machine
BT: machine

blinds
BT: furnishings

block copolymer
BT: copolymer by structure

blood filter
BT: fluid filter
medical textile

blood filtration
BT: fluid filtration

boarding
BT: heat setting

boarding machine
BT: heat setting machine

bobbin
BT: machine component

bobbin changing
BT: machine function

bobbin holder
BT: machine component

bobbin lace
BT: lace fabric

body armor
BT: protective clothing

body scanning
BT: anthropometry

body shape
BT: biological characteristic
shape

body size
BT: biological characteristic
size

boiler
BT: equipment by item type

boiling
BT: wet processing
NT: kiering

boiling-off
BT: degumming

bollworm
BT: animal

Bombyx mori
BT: silkworm

bonded fiber fabric
UF: bonded fibre fabric
BT: nonwoven fabric

bonded fibre fabric
USE: bonded fiber fabric

bonding
BT: general manufacturing process
NT: fusing
spunbonding

stitch-bonding
welding

bonding agent
BT: additive
NT: adhesive agent

bonding machine
BT: machine
NT: spunbonding machine
stitch-bonding machine

bone repair
BT: medical process

boom
BT: economic conditions

boric acid
BT: inorganic compound

boron
BT: chemical element

boron fiber
UF: boron fibre
BT: manufactured fiber
NT: boron-nitride fiber

boron fibre
USE: boron fiber

boron nitride fiber
USE: boron-nitride fiber

boron nitride fibre
USE: boron-nitride fiber

boron-nitride fiber
UF: boron nitride fiber
boron nitride fibre
boron-nitride fibre
BT: boron fiber

boron-nitride fibre
USE: boron-nitride fiber

boucle yarn
BT: fancy yarn

brace (medical textile)
BT: medical textile

braid
BT: fabric by form

braiding
BT: fabric manufacture

braiding machine
BT: machine

brake (vehicle)
BT: vehicle component

braking system
BT: machine component
NT: yarn brake

branding
BT: trade and commerce
NT: eco-labeling

brassiere
BT: support garment
underwear

breaking
BT: fracture
NT: end breakage
yarn breakage

breaking elongation
BT: elongation

breaking point
BT: mechanical property

breaking strength
BT: strength

breathability
BT: physical property
transport process

breed
BT: biological characteristic

bridge
BT: facility or structure

brightening
BT: bleaching

brightening agent
BT: finishing agent

brush
BT: machine component
product by item type

buckling
BT: distortion

buffering agent
BT: additive

bulk
BT: measure or dimension

bulking
BT: finishing

bullet resistance
BT: resistant property

buoyancy
BT: physical property

business communication
BT: communication

business merger
BT: business strategy

business plan
BT: business strategy
planning

business strategy
BT: management
NT: acquisition
business merger
business plan
cost reduction
joint venture

buttonhole
BT: fastener

buyer
BT: industrial enterprise
labor and employment

buyers guide
BT: publication

buying
BT: supply chain

cable
BT: rope

cabling
BT: yarn manufacture

cabling machine
BT: knitting machine

calculation
BT: methodology

calender bowl
BT: machine component

calendering
BT: heat setting
NT: embossing

calendering machine
BT: finishing machine

calibration
BT: methodology

calorimetry
BT: measuring

cam
BT: machine component

camel hair
BT: hair

camouflage
BT: military textile

can changing
BT: machine function

canopy
BT: architectural textile

canvas
BT: woven fabric
NT: sailcloth

caprolactam
BT: organic compound

car seat
BT: seat
vehicle component

carbon
BT: chemical element

carbon dioxide
BT: oxide

carbon fiber
UF: carbon fibre
BT: manufactured fiber
NT: graphite fiber

carbon fiber fabric
UF: carbon fibre fabric
BT: manufactured fiber fabric
NT: graphite fiber fabric

carbon fiber yarn
UF: carbon fibre yarn
BT: manufactured fiber yarn
NT: graphite fiber yarn

carbon fibre
USE: carbon fiber

carbon fibre fabric
USE: carbon fiber fabric

carbon fibre yarn
USE: carbon fiber yarn

carbonizing
BT: acid treatment

carcinogenicity
BT: health and disease

card
USE: carding machine

card clothing
BT: machine clothing

carding
BT: yarn preparation
NT: autoleveling

carding machine
UF: card
BT: machine

carpet
BT: floor covering
NT: axminster carpet
carpet backing
carpet tile
saxony carpet

carpet backing
BT: backing
carpet

carpet fiber
UF: carpet fibre
BT: fiber by property or purpose

carpet fibre
USE: carpet fiber

carpet industry
BT: textile-clothing industry

carpet loom
BT: loom

carpet tile
BT: carpet

carpet yarn
BT: yarn by property or purpose

carrier
BT: dyeing assistant
printing assistant

cashmere
BT: wool

cashmere fabric
BT: keratin fiber fabric

cashmere fiber
UF: cashmere fibre
BT: keratin fiber

cashmere fibre
USE: cashmere fiber

cashmere industry
BT: animal fiber industry

cashmere yarn
BT: keratin fiber yarn

casting off
BT: knitting

casualwear
BT: garment by property or purpose

catalyst
BT: additive

cationic dye
BT: dye by application or reaction

cationizing
BT: finishing

cationizing agent
BT: additive

cellulase treatment
BT: enzymatic treatment

cellulose
BT: polysaccharide

cellulose acetate
BT: regenerated cellulose
NT: triacetate

cellulose xanthate
BT: regenerated cellulose

cellulosic fiber
UF: cellulosic fibre
BT: natural fiber
NT: vegetable fiber

cellulosic fiber blend
BT: natural fiber blend
NT: cotton blend
flax blend
hemp blend
jute blend
linen blend

cellulosic fiber fabric
UF: cellulosic fibre fabric
BT: natural fiber fabric
NT: vegetable fiber fabric

cellulosic fiber yarn
UF: cellulosic fibre yarn
BT: natural fiber yarn
NT: vegetable fiber yarn

cellulosic fibre
USE: cellulosic fiber

cellulosic fibre fabric
USE: cellulosic fiber fabric

cellulosic fibre yarn

USE: cellulosic fiber yarn

centrifugal force
BT: force

centrifuge modeling
BT: simulation

ceramic fiber
UF: ceramic fibre
BT: manufactured fiber
NT: silicon carbide fiber

ceramic fiber fabric
UF: ceramic fibre fabric
BT: manufactured fiber fabric

ceramic fiber yarn
UF: ceramic fibre yarn
BT: manufactured fiber yarn

ceramic fibre
USE: ceramic fiber

ceramic fibre fabric
USE: ceramic fiber fabric

ceramic fibre yarn
USE: ceramic fiber yarn

ceramic matrix
BT: matrix

ceramic whisker
BT: whisker

ceramics
BT: general materials by origin or composition

chemical
BT: Facet C: materials, products and resources
NT: chemical by form
chemical by property or purpose
chemical element or compound

chemical analysis
BT: analysis
NT: amino acid analysis
chromatography

chemical application
USE: application (finishing)

chemical by form
BT: chemical
NT: monomer
nanotube
polymer

chemical by property or purpose
BT: chemical
NT: acid
additive
alkali
detergent
dye
fertilizer
finishes
lubricant
pesticide
pigment
precursor
refrigerant
solvent

chemical composition
BT: chemical property
NT: formula

chemical compound
BT: chemical element or compound
NT: fluorochemical
inorganic compound
organic compound

chemical element
BT: chemical element or compound
NT: boron
carbon
nickel
oxygen
silicon
silver
titanium

chemical element or compound
BT: chemical
NT: chemical compound
chemical element

chemical etching
BT: etching

chemical modification
BT: modification

chemical phenomena
BT: Facet D: phenomena
NT: chemical process
chemical property
chemical reaction

chemical process
BT: chemical phenomena
NT: coagulating
controlled release
corrosion
coupling
cross linking
decomposition
degradation
desalination
desorption
dispersion
dissolution
distillation

dye migration
dye penetration
dye uptake
exhaustion
ion exchange
kinetics
nucleation
odor emitting
phase separation
purification
soil release
sorption
wetting
wicking

chemical property
BT: chemical phenomena
NT: absorbency
adsorbency
chemical composition
concentration
crystalline orientation
crystallinity
flammability
hydrophilic property
molecular structure
molecular weight
oleophilic property
pH
polymer property
solubility

chemical reaction
BT: chemical phenomena
NT: acetylation
amination
aminolysis
chlorination
deacetylation
dechlorination
electrolysis
hydrolysis
oxidation
photolysis
polymerization
pyrolysis
synthesis

chemical resistance
BT: resistant property
NT: chlorine resistance

chemical vapor deposition
BT: finishing

chemistry
BT: physical science
NT: electrochemistry
polymer science

chenille fabric
BT: woven fabric

chenille yarn
BT: fancy yarn

child labor
UF: child labour
BT: labor

child labour
USE: child labor

childrenswear
BT: garment by user

chintz
BT: fabric by property or purpose

chitin
BT: polysaccharide

chitin fabric
BT: manufactured fiber fabric

chitin fiber
UF: chitin fibre
BT: manufactured fiber

chitin fibre
USE: chitin fiber

chitin yarn
BT: manufactured fiber yarn

chitosan
BT: polysaccharide

chitosan fabric
BT: manufactured fiber fabric

chitosan fiber
UF: chitosan fibre
BT: manufactured fiber

chitosan fibre
USE: chitosan fiber

chitosan yarn
BT: manufactured fiber yarn

chlorination
BT: chemical reaction
water treatment

chlorine resistance
BT: chemical resistance

chromatography
BT: chemical analysis

chrome dye
BT: dye by application or reaction

chute
BT: machine component

chutki
BT: dyeing

cigarette filter
BT: filter

circular knitting
BT: knitting

civil engineering
BT: engineering

clamping
BT: machine function

classification
BT: information processing

classing
BT: information processing

clay liner
BT: geosynthetics
liner

cleaning
BT: finishing
NT: drycleaning
stain removal
wetcleaning

cleaning machine
BT: finishing machine
NT: drycleaning machine
wetcleaning machine

cleanroom
BT: facility or structure

clip
USE: clip (yield)

clip (yield)
UF: clip
BT: yield

clothing industry
BT: textile-clothing industry

coagulating
BT: chemical process

coarsening
BT: finishing

coated fabric
BT: fabric by property or purpose

coating
BT: finishing
NT: back-coating
edge coating
plating

coating agent
BT: finishing agent

coating machine
BT: finishing machine

coatings
BT: finishes

cockling
BT: distortion

cocoon
BT: biological structure or form

coiling
BT: yarn manufacture

coir
BT: plant material

coir fabric
BT: vegetable fiber fabric

coir fiber
UF: coir fibre
BT: vegetable fiber

coir fibre
USE: coir fiber

coir yarn
BT: vegetable fiber yarn

collagen
BT: protein

collar
BT: garment component

colloid
BT: general materials by form
NT: aerosol
emulsion

color
UF: colour
BT: appearance

color change
UF: colour change
BT: optical phenomena
NT: fading
metamerism
yellowing

color communication
BT: communication

color management
UF: colour management

BT: control
production management
NT: color matching

color matching
UF: colour matching
BT: color management
methodology

color matching system
UF: colour matching system
BT: equipment by property or purpose

color perception
UF: colour perception
BT: behavioral or psychological phenomena

color prediction
UF: colour prediction
BT: prediction

colorfastness
UF: colourfastness
BT: dye property

colorimetry
BT: measuring

colour
USE: color

colour change
USE: color change

colour management
USE: color management

colour matching
USE: color matching

colour matching system
USE: color matching system

colour perception
USE: color perception

colour prediction
USE: color prediction

colourfastness
USE: colorfastness

comb
BT: machine component

combing
BT: yarn preparation

combing machine
BT: machine
NT: hackling machine

comfort
BT: behavioral or psychological phenomena

commingling
BT: yarn manufacture

comminuting
BT: general manufacturing process

communication
BT: Facet H: information and communication
NT: business communication
color communication
communication medium
global communication

communication event
BT: communication medium
NT: conference
exhibition

communication medium
BT: communication
NT: communication event
media
publication

compact spinning
BT: spinning

compacting
BT: general manufacturing process

compacting machine
BT: machine

company information
BT: information

competitiveness
BT: market

complexing agent
BT: additive

composite
BT: general materials by origin or composition
NT: fiber reinforced composite
matrix
nanocomposite

composite property
BT: physicochemical property by material
NT: fiber-matrix adhesion
fiber-matrix interface
interfacial property
interlaminar property

composites industry
BT: industry

compression
BT: stress

compressive property
BT: mechanical property

computational fluid dynamics
BT: fluid dynamics

computer
BT: equipment by item type
NT: hardware
neural network
software

computer aided design
UF: computer assisted design
computer-aided design
BT: design

computer aided production
UF: computer-aided production
BT: production

computer assisted design
USE: computer aided design

computer control
BT: control

computer program
BT: software

computer simulation
BT: simulation

computer-aided design
USE: computer aided design

computer-aided production
USE: computer aided production

concentration
BT: chemical property

concrete
BT: general materials by origin or composition

condensation
BT: dynamic or static phenomena

condensation polymer
BT: polymer by polymerization reaction

condenser
BT: equipment by item type

conditioning
BT: washing

conditioning agent
BT: detergent
finishing agent

conductivity
BT: physical property
NT: electrical conductivity
thermal conductivity

cone
BT: machine component

conference
BT: communication event
NT: conference (AATCC)
conference (ATIRA)
conference (Beltwide Cotton)
conference (BTRA)
conference (CIRCOT)
conference (COTTECH)
conference (DWI)
conference (IAF)
conference (IFVTCC)
conference (International Man-made Fibre)
conference (International Silk Association)
conference (International Wool Research)
conference (ITMF)
conference (Material World)
conference (Medical Textiles)
conference (NITRA)
conference (SAMPE)
conference (SDC)
conference (SITRA)
conference (TIFCON)
conference (TIWC)

conference (AATCC)
BT: conference

conference (ATIRA)
BT: conference

conference (Beltwide Cotton)
BT: conference

conference (BTRA)
BT: conference

conference (CIRCOT)
BT: conference

conference (COTTECH)
BT: conference

conference (DWI)
BT: conference

conference (IAF)
BT: conference

conference (IFVTCC)
BT: conference

conference (International Man-made Fibre)
BT: conference

conference (International Silk Association)
BT: conference

conference (International Wool Research)
BT: conference

conference (ITMF)
BT: conference

conference (Material World)
BT: conference

conference (Medical Textiles)
BT: conference

conference (NITRA)
BT: conference

conference (SAMPE)
BT: conference

conference (SDC)
BT: conference

conference (SITRA)
BT: conference

conference (TIFCON)
BT: conference

conference (TIWC)
BT: conference

constant
BT: mathematical expression
NT: elastic modulus
initial modulus

constructed textile
BT: product by property or purpose
NT: three-dimensional textile
tubular material

construction industry
BT: industry

consumer research
BT: research and development

consumer safety
BT: safety

consumption
BT: trade and commerce
NT: energy consumption
water consumption

contaminant
BT: general materials by property or purpose
NT: dust
foreign body
impurity

contamination
BT: environmental phenomena

continuous dyeing
BT: continuous process
dyeing

continuous process
BT: general manufacturing process
NT: continuous dyeing

control
BT: Facet B: operations
NT: color management
computer control
electronic control
erosion control
moisture management
odor control
perspiration control
pest control
pollution control
process control
production control
quality control
temperature control
vibration damping

control system
BT: equipment by property or purpose

controlled release
BT: chemical process

convection
BT: transport process

conveyor belt
BT: materials handling system

cooling
BT: Facet B: operations
thermal phenomena
NT: quenching

cooling device
BT: equipment by property or purpose

copolymer
BT: polymer by monomer sequence
NT: copolymer by structure

copolymer by structure
BT: copolymer
NT: block copolymer
graft copolymer

copolymerization
BT: polymerization

copper sulfate
BT: inorganic compound

copyright
BT: intellectual property

cord
BT: rope
NT: tire cord

cord construction
BT: structural property

corduroy fabric
BT: pile fabric

core spinning
BT: spinning

cork
BT: plant material

corrosion
BT: chemical process

corrosion resistance
BT: resistant property

corset
BT: support garment
underwear

cost
BT: trade and commerce

cost reduction
BT: business strategy

costing
BT: management

costume
BT: art, design and culture

cotton
BT: plant

cotton blend
BT: cellulosic fiber blend
NT: acrylic-cotton blend
cotton-flax blend
cotton-linen blend
cotton-polyester blend
cotton-polypropylene blend
cotton-silk blend
cotton-viscose-rayon blend
cotton-wool blend

cotton fabric
BT: vegetable fiber fabric

cotton fiber
UF: cotton fibre
BT: vegetable fiber

cotton fibre
USE: cotton fiber

cotton flax blend
USE: cotton-flax blend

cotton industry
BT: vegetable fiber industry

cotton linen blend
USE: cotton-linen blend

cotton picking
BT: harvesting

cotton picking machine
BT: machine

cotton polyester blend
USE: cotton-polyester blend

cotton polypropylene blend
USE: cotton-polypropylene blend

cotton silk blend
USE: cotton-silk blend

cotton viscose rayon blend
USE: cotton-viscose-rayon blend

cotton wool
BT: general materials by origin or composition

cotton wool blend
USE: cotton-wool blend

cotton yarn
BT: vegetable fiber yarn

cotton-flax blend
UF: cotton flax blend
BT: cotton blend
flax blend

cotton-linen blend
UF: cotton linen blend
BT: cotton blend
linen blend

cotton-polyester blend
UF: cotton polyester blend
BT: cotton blend
polyester blend

cotton-polypropylene blend
UF: cotton polypropylene blend
BT: cotton blend
polyolefin blend

cotton-silk blend
UF: cotton silk blend
BT: cotton blend
silk blend

cotton-viscose-rayon blend
UF: cotton viscose rayon blend
BT: cotton blend

viscose-rayon blend

cotton-wool blend
UF: cotton wool blend
BT: cotton blend
wool blend

count
BT: thickness
yarn property

counterfeiting
BT: trade and commerce

coupling
BT: chemical process

coupling agent
BT: dyeing assistant

crabbing
BT: heat setting

crabbing machine
BT: wet processing machine

cracking
BT: fracture

crease recovery
BT: measure or dimension
mechanical property

crease resistance
BT: resistant property

creasing
BT: finishing

creel
BT: machine component

creep
BT: strain

crepe
USE: crepe fabric

crepe fabric
UF: crepe
BT: fabric by property or purpose

creping
BT: wet finishing

crimp
BT: fiber property

crimping
BT: dry finishing

crimping machine
BT: texturing machine

crochet machine
BT: knitting machine

crocheted fabric
BT: knitted fabric

crocheting
BT: knitting

crockfastness
BT: dye property

crop production
BT: agricultural production

cross linking
UF: crosslinking
BT: chemical process

cross linking agent
UF: crosslinking agent
BT: additive

cross section
BT: measure or dimension

cross sectional shape
USE: cross-sectional shape

cross-sectional shape
UF: cross sectional shape
BT: shape

crosslapping
BT: nonwoven fabric manufacture

crosslapping machine
BT: nonwoven fabric machine

crosslinking
USE: cross linking

crosslinking **agent**
USE: cross linking agent

cryogenics
BT: physics

crystalline orientation
UF: orientation
BT: chemical property
NT: degree of orientation

crystallinity
BT: chemical property

cuff
BT: garment component

cultivation
BT: agricultural industry

cultural anthropology
BT: social science

cupro fiber
UF: cupro fibre
BT: rayon fiber

cupro fibre
USE: cupro fiber

curing
BT: finishing

curtain
BT: furnishings
NT: net curtain

cushion
BT: furnishings

customer service
BT: service sector

customization
BT: design
product development

cut resistance
BT: resistant property

cutter
BT: machine component

cutting
BT: general manufacturing process
NT: pattern cutting
slitting

cutting machine
BT: machine

cylinder
BT: machine component
NT: doffer
licker-in

damage
BT: physical phenomena
NT: aging
fracture
wear

data management system
BT: equipment by property or purpose

data processing
BT: information processing
NT: electronic data interchange

database
BT: information

deacetylation
BT: chemical reaction

debonding
BT: disintegration

decatizing
BT: wet finishing

decatizing machine
BT: finishing machine

dechlorination
BT: chemical reaction
removal
water treatment

decision making
BT: behavioral or psychological phenomena
planning

decolorizing
BT: removal
water treatment

decomposition
BT: chemical process

decorticating
BT: fiber preparation
separation

decorticating machine
BT: machine

defect
BT: physical property
NT: bar
barre
streakiness

defoliation
BT: fiber preparation
removal

deformation
BT: mechanical property
NT: elongation
strain

defuzzing
BT: dry finishing
removal

degradation
BT: chemical process
NT: enzymatic degradation
photochemical degradation

degree of orientation
BT: crystalline orientation
fiber orientation
measure or dimension

degree of polymerization
BT: measure or dimension
polymer property

degumming
BT: removal
wet processing
NT: boiling-off

deknitting
BT: texturing

delamination
BT: disintegration

delignifying
BT: fiber preparation
removal

delustering
BT: finishing

delustering agent
BT: finishing agent

demand
BT: market

demineralization
BT: removal

denier
BT. density
yarn property

denim fabric
BT: woven fabric

density
BT: measure or dimension
NT: denier
linear density

dental floss
BT: hygiene product

deodorizing
BT: finishing

desalination
BT: chemical process
removal
water treatment

design
BT: art, design and culture
NT: computer aided design
customization
interior design
pattern (plan)
repeat
urban design

desizing
BT: removal
wet processing

desorption
BT: chemical process

detector
BT: instrument

detergent
BT: chemical by property or purpose
NT: conditioning agent

devore
BT: printing

deweighting
BT: wet processing

dialysis
BT: separation

diameter
BT: measure or dimension
NT: fiber diameter
yarn diameter

diaper
BT: hygiene product

die
BT: machine component

dielectric property
BT: electrical property

diffusion
BT: transport process
NT: reverse osmosis

digital method
BT: methodology

digital printing
BT: printing

dimensional stability
BT: stability

direct dye
BT: dye by application or reaction

direct printing
BT: printing

direct spinning
BT: spinning

directory information
BT: information

discharge printing

BT: printing

disease
BT: health and disease
NT: allergy

disease management
BT: pest and disease management

disintegration
BT: physical phenomena
NT: debonding
delamination
fraying

diskette liner
BT: liner

dispersant
UF: dispersing agent
BT: additive

disperse dye
BT: dye by application or reaction

dispersing agent
USE: dispersant

dispersion
BT: chemical process

disposable product
BT: product by property or purpose

dissolution
BT: chemical process

distillation
BT: chemical process
removal

distortion
BT: physical phenomena
NT: bending
buckling
cockling
necking
puckering
relaxing

distribution
BT: supply chain

diving suit
BT: swimwear

DNA
BT: protein

dobby
BT: loom component

dobby loom
BT: loom

doffer
BT: cylinder

doffing
BT: machine function

dog hair
BT: hair

dosing
BT: process control

dosing machine
BT: machine

dosing system
BT: measuring system

double cloth
BT: woven fabric

down time
BT: production management

drafting
BT: yarn preparation
NT: gilling

drafting machine
BT: machine
NT: gilling machine

drainage
BT: Facet B: operations

drape
BT: mechanical property

draw ratio
BT: mathematical expression

draw texturing
USE: draw-texturing

draw texturing machine
USE: draw-texturing machine

draw-texturing
UF: draw texturing
BT: drawing
texturing

draw-texturing machine
UF: draw texturing machine
BT: texturing machine

drawframe
BT: machine component

drawing
BT: fiber manufacture

NT: autoleveling
draw-texturing
spin-drawing

drawing in
USE: drawing-in

drawing in machine
USE: drawing-in machine

drawing machine
BT: machine

drawing-in
UF: drawing in
BT: weaving

drawing-in machine
UF: drawing in machine
BT: machine

dress
BT: outerwear

dressing
BT: medical textile

drive
BT: machine component

drop wire
BT: machine component

drug delivery
BT: medical process

drum
BT: machine component

dry finishing
BT: finishing
NT: crimping
defuzzing
emerizing
grinding
milling
raising
shearing
singeing
stentering

dry laying
BT: nonwoven fabric manufacture
NT: air laying

drycleaning
BT: cleaning

drycleaning industry
BT: industry

drycleaning machine
BT: cleaning machine

drying
BT: heat treatment

drying machine
BT: finishing machine
NT: mangle

dumping
BT: international trade

durability
BT: mechanical property
NT: toughness

dust
BT: contaminant
NT: mote

dust filter
BT: filter

dust filtration
BT: filtration

dye
BT: chemical by property or purpose
NT: dye by application or reaction
dye by origin or composition
dye by property or form

dye application
USE: application (finishing)

dye bath
BT: bath

dye by application or reaction
BT: dye
NT: acid dye
anionic dye
azo dye
basic dye
cationic dye
chrome dye
direct dye
disperse dye
reactive dye
vat dye

dye by origin or composition
BT: dye
NT: indigo dye
metal-complex dye
natural dye
organic dye
sulfur dye
synthetic dye

dye by property or form
BT: dye
NT: leuco dye
photochromic dye

thermochromic dye
trichromatic dye

dye exhaustion
BT: exhaustion

dye fixing
BT: dyeing

dye levelness
BT: dye property

dye manufacture
UF: dye manufacturing
BT: manufacture

dye manufacturing
USE: dye manufacture

dye manufacturing industry
BT: industry

dye migration
BT: chemical process

dye mixture
BT: general materials by property or purpose

dye penetration
BT: chemical process

dye property
BT: physicochemical property by material
NT: colorfastness
crockfastness
dye levelness
dye recipe
lightfastness
washfastness

dye recipe
BT: dye property

dye removal
BT: removal

dye uptake
BT: chemical process

dyeability
BT: physicochemical property by characteristic

dyed fabric
BT: fabric by property or purpose
NT: batik

dyeing
BT: wet processing
NT: backstaining
batch dyeing
batik
beam dyeing
chutki
continuous dyeing
dye fixing
ikat dyeing
jet dyeing
jig dyeing
kettle dyeing
leveling
mass dyeing
melt dyeing
mordant dyeing
one step dyeing
open-width dyeing
pad dyeing
pad-batch dyeing
pad-steam dyeing
piece dyeing
resist dyeing
rope dyeing
space dyeing
spin dyeing
spray dyeing
supercritical carbon dioxide dyeing
tie dyeing
top dyeing
trichromatic dyeing
union dyeing
vat dyeing
winch dyeing

dyeing assistant
BT: additive
NT: accelerating agent
carrier
coupling agent
fixing agent
leveling agent
mordant

dyeing finishing machine
USE: dyeing-finishing machine

dyeing industry
BT: industry

dyeing machine
BT: wet processing machine
NT: pad

dyeing-finishing machine
UF: dyeing finishing machine
BT: finishing machine

dynamic or static phenomena
BT: physical phenomena
NT: condensation
elongation
energy
expansion
flotation
flow
fluid dynamics
force
inflation

kinetics
shrinkage
speed
stability
thermodynamics
transport process
vibration

e-commerce
BT: market

easy care property
BT: physicochemical property by characteristic

eco-labeling
UF: eco-labelling
BT: branding
labeling

eco-labelling
USE: eco-labeling

ecology
BT: life science

economic conditions
BT: society and economics
NT: boom
globalization

economic data
BT: economic information

economic information
BT: information
NT: economic data

economic strategy
BT: society and economics

edge coating
BT: coating

education
BT: society and economics

effluent
BT: waste

effluent treatment
BT: environmental treatment

elastane
BT: polyurethane
NT: spandex

elastane fabric
BT: manufactured fiber fabric
NT: spandex fabric

elastane fiber
UF: elastane fibre
BT: manufactured fiber
NT: spandex fiber

elastane fibre
USE: elastane fiber

elastane yarn
BT: manufactured fiber yarn
NT: spandex yarn

elastic modulus
BT: constant
elastic property
NT: Young modulus

elastic property
BT: mechanical property
NT: elastic modulus
elastic recovery
elasticity

elastic recovery
BT: elastic property

elasticity
BT: elastic property
NT: viscoelasticity

elastomer
BT: polymer by physical property

elastomeric fabric
BT: manufactured fiber fabric

elastomeric fiber
UF: elastomeric fibre
BT: manufactured fiber

elastomeric fibre
USE: elastomeric fiber

elastomeric yarn
BT: manufactured fiber yarn

elastoplasticity
BT: mechanical property

electrect
USE: electret

electret
UF: electrect
BT: general materials by property or purpose

electrical conductivity
BT: conductivity
electrical property

electrical phenomena
BT: physical phenomena
NT: electrical property
electricity

electrical property

BT: electrical phenomena
physical property
NT: dielectric property
electrical conductivity

electrical resistance
BT: resistant property
NT: antistatic property

electricity
BT: electrical phenomena
NT: static electricity

electrochemistry
BT: chemistry

electroless plating
BT: plating

electrolysis
BT: chemical reaction
finishing

electromagnetic device
BT: equipment by property or purpose

electromagnetic resistance
BT: resistant property
NT: infrared resistance
ultraviolet resistance

electron microscopy
BT: microscopy
NT: scanning electron microscopy
transmission electron microscopy

electron spin resonance
BT: spectroscopy

electronic control
BT: control

electronic data interchange
BT: data processing

electronics
BT: technology

electroplating
BT: plating

electrospinning
BT: fiber manufacture

elongation
BT: deformation
dynamic or static phenomena
NT: breaking elongation

embossing
BT: calendering

embroidering
BT: sewing

embroidery
BT: pattern (configuration)
NT: beadwork
hand embroidery

embroidery frame
BT: machine component

embroidery industry
BT: textile-clothing industry

embroidery machine
BT: sewing machine

emergency shelter
BT: architectural textile

emerizing
BT: dry finishing

emerizing machine
BT: finishing machine

emission reduction
BT: environmental management

employment
BT: labor and employment
NT: womens employment

emulsion
BT: colloid

encapsulation
BT: general manufacturing process

end breakage
BT: breaking

end use
BT: use

energy
BT: dynamic or static phenomena
NT: radiation

energy conservation
BT: environmental management

energy consumption
BT: consumption
NT: power consumption

engineering
BT: applied science
NT: civil engineering
process engineering
tissue engineering

environmental impact
BT: environmental phenomena

environmental management
BT: management
NT: emission reduction
energy conservation
environmental protection
pollution control
waste management
water conservation

environmental phenomena
BT: Facet D: phenomena
NT: contamination
environmental impact
indoor air quality
pollution
sustainability

environmental protection
BT: environmental management

environmental treatment
BT: Facet B: operations
NT: aerobic treatment
anaerobic treatment
biological treatment
effluent treatment
ozone treatment
water treatment

enzymatic degradation
BT: degradation

enzymatic treatment
BT: biofinishing
NT: cellulase treatment

enzyme
BT: protein
NT: protease

epoxy resin
BT: synthetic resin

equipment
BT: Facet E: equipment, facilities and structures
NT: equipment by item type
equipment by property or purpose

equipment by item type
BT: equipment
NT: applicator
autoclave
bath
boiler
computer
condenser
fixed bed
fluidized bed
instrument
laser
machine
pump
reactor
stencil
turbine

equipment by property or purpose
BT: equipment
NT: color matching system
control system
cooling device
data management system
electromagnetic device
materials handling system
measuring system
monitoring system
pattern preparation system
piezoelectric device
pneumatic device

ergonomics
BT: applied science

eri silk
BT: silk

erosion control
BT: control

esthetic property
UF: aesthetic property
BT: appearance
behavioral or psychological phenomena

etching
BT: finishing
NT: chemical etching
sputter etching

ethics
BT: social science

ethnicity
BT: society and economics

evenness
BT: appearance

exhaustion
BT: chemical process
NT: dye exhaustion

exhibition
BT: communication event
NT: exhibition (ATME-I)
exhibition (BITMEX)
exhibition (Bobbin Show)
exhibition (CAD Expo)
exhibition (CITME)
exhibition (Domotex)
exhibition (Expo Detergo)
exhibition (Expofil)
exhibition (Heimtextil)
exhibition (IDEA)

exhibition (IFAI)
exhibition (IHE)
exhibition (IKME)
exhibition (IMB)
exhibition (INDA)
exhibition (INDA-TEC)
exhibition (INDEX)
exhibition (Interfloor)
exhibition (ITMA)
exhibition (ITME-I)
exhibition (JEC)
exhibition (JIAM)
exhibition (SEAMS)
exhibition (Techtextil)
exhibition (Texitech)

exhibition (ATME-I)
BT: exhibition

exhibition (BITMEX)
BT: exhibition

exhibition (Bobbin Show)
BT: exhibition

exhibition (CAD Expo)
BT: exhibition

exhibition (CITME)
BT: exhibition

exhibition (Domotex)
BT: exhibition

exhibition (Expo Detergo)
BT: exhibition

exhibition (Expofil)
BT: exhibition

exhibition (Heimtextil)
BT: exhibition

exhibition (IDEA)
BT: exhibition

exhibition (IFAI)
BT: exhibition

exhibition (IHE)
BT: exhibition

exhibition (IKME)
BT: exhibition

exhibition (IMB)
BT: exhibition

exhibition (INDA)
BT: exhibition

exhibition (INDA-TEC)
BT: exhibition

exhibition (INDEX)
BT: exhibition

exhibition (Interfloor)
BT: exhibition

exhibition (ITMA)
BT: exhibition

exhibition (ITME-I)
BT: exhibition

exhibition (JEC)
BT: exhibition

exhibition (JIAM)
BT: exhibition

exhibition (SEAMS)
BT: exhibition

exhibition (Techtextil)
BT: exhibition

exhibition (Texitech)
BT: exhibition

exhibition review
BT: publication

expansion
BT: dynamic or static phenomena
NT: hygral expansion
thermal expansion

exports
BT: international trade

extensibility
BT: mechanical property

extraction method
BT: removal

extrusion
BT: fiber manufacture
NT: flash spinning
melt blowing
melt spinning

eyelet
BT: trim

fabric
BT: materials
NT: fabric by form
fabric by origin or composition
fabric by property or purpose
fabric component

fabric appearance
BT: appearance

fabric property

fabric by form
BT: fabric
NT: braid
knitted fabric
lace fabric
laminate
mesh
microfiber fabric
multiaxial fabric
nonwoven fabric
pile fabric
relief fabric
worsted fabric
woven fabric

fabric by origin or composition
BT: fabric
NT: blended fabric
manufactured fiber fabric
natural fiber fabric
rubber fabric

fabric by property or purpose
BT: fabric
NT: apparel fabric
backing fabric
barrier fabric
chintz
coated fabric
crepe fabric
dyed fabric
filter fabric
furnishing fabric
interfacing
interlining
moire fabric
multilayer fabric
narrow fabric
papermakers fabric
printed fabric
protective fabric
reinforced fabric
sailcloth
smart fabric
spacer fabric

fabric component
BT: fabric
NT: openings
pile
selvedge
tuft
warp
weft

fabric construction
BT: fabric property
structural property
NT: bias
honeycomb construction
pitch
seamless
weave

fabric manufacture
BT: manufacture
NT: braiding
knitting
lace making
laminating
layering
nonwoven fabric manufacture
quilting
tufting
weaving

fabric modification
BT: modification

fabric property
BT: physicochemical property by material
NT: fabric appearance
fabric construction
moire
sett

fabrication
BT: general manufacturing process

Facet A: industry and organization
NT: industry
management
society and economics

Facet B: operations
NT: air conditioning
control
cooling
drainage
environmental treatment
heating
inspection
insulation
irradiation
lighting
machine function
maintenance
manufacture
materials handling
medical process
monitoring
recycling
repair
separation
sterilization
use
ventilation

Facet C: materials, products and resources
NT: chemical
materials
product
resource

Facet D: **phenomena**
NT: behavioral or psychological phenomena
biological phenomena
chemical phenomena
environmental phenomena
mathematical phenomena
physical phenomena
physicochemical property

Facet E: **equipment, facilities and structures**
NT: equipment
facility or structure

Facet F: **organisms and biological structures**
NT: biological structure or form
organism

Facet G: **art, science and technology**
NT: applied science
art, design and culture
life science
physical science
science in general
social science
technology

Facet H: **information and communication**
NT: communication
information

Facet I: **research and methodology**
NT: methodology
research (academic)
research and development

facility or structure
BT: Facet E: equipment, facilities and structures
NT: bridge
cleanroom
mill
museum

fading
BT: color change

failure analysis
BT: analysis

false twist
BT: texturing

fancy yarn
BT: yarn by form
NT: boucle yarn
chenille yarn

farming
BT: agricultural industry
NT: organic farming

fashion
BT: art, design and culture

fastener
BT: product component
NT: buttonhole
zipper

fatigue
BT: stress

feather
BT: animal material

feed
BT: guiding system

felt
BT: nonwoven fabric
NT: press felt

felting
BT: nonwoven fabric manufacture

fertilizer
BT: chemical by property or purpose

fiber
UF: fibre
BT: materials
NT: fiber by form
fiber by origin or composition
fiber by property or purpose

fiber arrangement
UF: fibre arrangement
BT: fiber property
NT: fiber orientation

fiber bundle
UF: fibre bundle
BT: handling pack

fiber by form
BT: fiber
NT: bicomponent fiber
fibrid
fibril
filament
hollow fiber
microfiber
nanofiber
noil
sliver
staple fiber
tow
whisker

fiber by origin or composition
BT: fiber
NT: blended fiber
manufactured fiber
natural fiber
organic fiber
rubber fiber

fiber by property or purpose
BT: fiber
NT: binder fiber
carpet fiber
high-modulus fiber
piezoelectric fiber
thermoplastic fiber

fiber curvature
UF: fibre curvature
BT: fiber property
measure or dimension

fiber diameter
UF: fibre diameter
BT: diameter
fiber property

fiber distribution
UF: fibre distribution
BT: fiber property

fiber fineness
UF: fibre fineness
BT: fiber property

fiber flow
UF: fibre flow
BT: flow

fiber length
UF: fibre length
BT: fiber property
length

fiber manufacture
UF: fibre manufacture
BT: manufacture
NT: blending
drawing
electrospinning
extrusion
fiber preparation
fibrillation

fiber matrix adhesion
USE: fiber-matrix adhesion

fiber matrix interface
USE: fiber-matrix interface

fiber medulla
UF: medulla, fiber
BT: animal material

fiber modification
UF: fibre modification
BT: modification

fiber optics
UF: fibre optics
BT: information technology

fiber orientation
UF: fibre orientation
BT: fiber arrangement
NT: degree of orientation

fiber preparation
UF: fibre preparation
BT: fiber manufacture
preparation
NT: decorticating
defoliation
delignifying
ginning
hackling
lint removal
opening
retting

fiber property
UF: fibre property
BT: physicochemical property by material
NT: birefringence
crimp
fiber arrangement
fiber curvature
fiber diameter
fiber distribution
fiber fineness
fiber length
filament property
linear density
micronaire value

fiber reinforced composite
UF: fibre reinforced composite
BT: composite
NT: preform
prepreg

fiber-matrix adhesion
UF: fiber matrix adhesion
fibre matrix adhesion
fibre-matrix adhesion
BT: adhesion
composite property

fiber-matrix interface
UF: fiber matrix interface
fibre matrix interface
fibre-matrix interface
BT: composite property

fiberfill
UF: fibrefill
BT: padding (material)

fibre
USE: fiber

fibre arrangement
USE: fiber arrangement

fibre bundle
USE: fiber bundle

fibre curvature
USE: fiber curvature

fibre diameter
USE: fiber diameter

fibre distribution
USE: fiber distribution

fibre fineness
USE: fiber fineness

fibre flow
USE: fiber flow

fibre length
USE: fiber length

fibre manufacture
USE: fiber manufacture

fibre matrix adhesion
USE: fiber-matrix adhesion

fibre matrix interface
USE: fiber-matrix interface

fibre modification
USE: fiber modification

fibre optics
USE: fiber optics

fibre orientation
USE: fiber orientation

fibre preparation
USE: fiber preparation

fibre property
USE: fiber property

fibre reinforced composite
USE: fiber reinforced composite

fibre-matrix adhesion
USE: fiber-matrix adhesion

fibre-matrix interface
USE: fiber-matrix interface

fibrefill
USE: fiberfill

fibrid
BT: fiber by form

fibril
BT: fiber by form

fibrillation
BT: fiber manufacture

fibroin
BT: protein

filament
BT: fiber by form
NT: hollow filament
microfilament
monofilament
multifilament
plexifilament

filament property
BT: fiber property

film
BT: general materials by form

filter
BT: machine component
product by item type
NT: cigarette filter
dust filter
fluid filter

filter fabric
BT: fabric by property or purpose

filtration
BT: separation
NT: dust filtration
fluid filtration

financial provision
BT: society and economics

financial service
BT: service sector

fine structure
BT: structural property

finishes
BT: chemical by property or purpose
NT: coatings

finishing
BT: general manufacturing process
NT: abrading
acid treatment
afterclearing
aging
alkali treatment
application (finishing)
biofinishing
bulking
cationizing
chemical vapor deposition
cleaning
coarsening
coating

creasing
curing
delustering
deodorizing
dry finishing
electrolysis
etching
flocking
grafting
heat treatment
laundering
loading (finishing)
mercerizing
metallization
padding (finishing)
plasma treatment
press finishing
pressing
pretreatment
printing
proofing
shrinking
stripping
thinning
voluminizing
wet finishing

finishing agent
BT: additive
NT: abrasive agent
bleaching agent
bleaching assistant
brightening agent
coating agent
conditioning agent
delustering agent
oxidizing agent
sizing agent
softening agent

finishing industry
BT: industry

finishing machine
BT: machine
NT: abrading machine
calendering machine
cleaning machine
coating machine
decatizing machine
drying machine
dyeing-finishing machine
emerizing machine
flocking machine
heat setting machine
laundering machine
mercerizing machine
pressing machine
printing machine
raising machine
shearing machine
shrinking machine
singeing machine
spreading machine
steaming machine

finite element analysis
BT: mathematical analysis

fishing line
BT: sports goods

fit
BT: garment property

fixed bed
BT: equipment by item type

fixing agent
BT: dyeing assistant

flag
BT: product by item type

flame resistance
BT: resistant property

flammability
BT: chemical property

flash aging
BT: aging

flash spinning
BT: extrusion

flat knitting
BT: knitting

flax
BT: plant

flax blend
BT: cellulosic fiber blend
NT: cotton-flax blend
flax-hemp blend
flax-jute blend
flax-polyester blend
flax-viscose-rayon blend
flax-wool blend

flax fiber
UF: flax fibre
BT: bast fiber
NT: linen

flax fiber fabric
UF: flax fibre fabric
BT: bast fiber fabric
NT: linen fabric

flax fiber yarn
BT: bast fiber yarn
NT: linen yarn

flax fibre

USE: flax fiber

flax fibre fabric
USE: flax fiber fabric

flax hemp blend
USE: flax-hemp blend

flax industry
BT: vegetable fiber industry

flax jute blend
USE: flax-jute blend

flax polyester blend
USE: flax-polyester blend

flax viscose rayon blend
USE: flax-viscose-rayon blend

flax wool blend
USE: flax-wool blend

flax-hemp blend
UF: flax hemp blend
BT: flax blend
hemp blend

flax-jute blend
UF: flax jute blend
BT: flax blend
jute blend

flax-polyester blend
UF: flax polyester blend
BT: flax blend
polyester blend

flax-viscose-rayon blend
UF: flax viscose rayon blend
BT: flax blend
viscose-rayon blend

flax-wool blend
UF: flax wool blend
BT: flax blend
wool blend

fleece
BT: wool

flexibility
BT: mechanical property

flocculation
BT: separation

flocking
BT: finishing

flocking machine
BT: finishing machine

floor covering
BT: furnishings
NT: carpet
mat
rug

flotation
BT: dynamic or static phenomena

flow
BT: dynamic or static phenomena
NT: fiber flow
flow rate
fluid flow

flow rate
BT: flow
measure or dimension

fluid
BT: general materials by form
NT: gas
liquid

fluid dynamics
BT: dynamic or static phenomena
fluid mechanics
NT: aerodynamics
computational fluid dynamics
hydrodynamics

fluid filter
BT: filter
NT: blood filter
gas filter

fluid filtration
BT: filtration
NT: blood filtration
gas filtration

fluid flow
BT: flow
NT: gas flow
jet

fluid mechanics
BT: mechanics
NT: fluid dynamics
hydraulics

fluidized bed
BT: equipment by item type

fluorescence
BT: optical property

fluorochemical
BT: chemical compound

fluoropolymer
BT: polymer by chemical composition
NT: perfluoropolymer

fly
BT: waste

foam
BT: general materials by form

folding
BT: materials handling

folding machine
BT: machine

footwear
BT: garment by item type
NT: gaiter
sock
wader

force
BT: dynamic or static phenomena
NT: centrifugal force
friction
impact
pressure
stress

forecasting
BT: management

foreign body
BT: contaminant

formaldehyde
BT: organic compound

formula
BT: chemical composition

fouling
BT: biological phenomena

fourdrinier wire
BT: machine component

Fourier transform analysis
BT: mathematical analysis

fractal analysis
BT: mathematical analysis

fracture
BT: damage
NT: breaking
cracking

fraying
BT: disintegration

free trade
BT: international trade

friction
BT: force

friction coefficient
BT: mathematical expression

friction spinning
BT: spinning

frosting
BT: optical phenomena

fuel
BT: general materials by property or purpose

fuel cell
BT: product by item type

fungicide
BT: pesticide

furcraea fiber
UF: furcraea fibre
furcrea fiber
BT: agavaceae fiber

furcraea fibre
USE: furcraea fiber

furcrea fiber
USE: furcraea fiber

furnishing fabric
BT: fabric by property or purpose

furnishings
BT: product by item type
NT: bedding
blinds
curtain
cushion
floor covering
towel
upholstery
wall covering

furniture
BT: product by item type
NT: seat

fusing
BT: bonding

fuzzy logic
BT: mathematical analysis

gaiter
BT: footwear

galloon
BT: woven fabric

garment
BT: product by item type

NT: garment by item type
garment by property or purpose
garment by user
garment component

garment accessory
BT: accessories
garment by item type
NT: belt
glove
mask
tie
wig

garment appearance
BT: appearance
garment property

garment by item type
BT: garment
NT: footwear
garment accessory
headwear
outerwear
underwear

garment by property or purpose
BT: garment
NT: casualwear
knitwear
maternitywear
protective clothing
reversible garment
sleepwear
sportswear
support garment
swimwear
workwear

garment by user
BT: garment
NT: babieswear
childrenswear
menswear
womenswear

garment component
BT: garment
product component
NT: collar
cuff
gusset
insole
interfacing
interlining
shoulder pad
sleeve
waistband

garment construction
BT: garment property
structural property
NT: seamless

garment hanger
BT: product by item type

garment manufacture
BT: manufacture
NT: millinery
toe closing
whole-garment knitting

garment property
BT: physicochemical property by material
NT: fit
garment appearance
garment construction

gas
BT: fluid
NT: air
vapor

gas filter
BT: fluid filter
NT: air filter

gas filtration
BT: fluid filtration
NT: air filtration

gas flow
BT: fluid flow
NT: airflow

gasket
BT: machine component
seal

gathering
BT: sewing

gauge length
BT: length

gauze
BT: mesh
woven fabric

gear
BT: machine component

gel
BT: general materials by form
NT: hydrogel

gender identity
BT: behavioral or psychological phenomena

general manufacturing process
BT: manufacture
NT: aftertreatment
bonding
comminuting
compacting

continuous process
cutting
encapsulation
fabrication
finishing
knotting
lubricating
making-up
materials selection
mixing
modification
molding
one step process
open-width process
preparation
sewing
wet processing

general materials
BT: materials
NT: general materials by form
general materials by origin or composition
general materials by property or purpose

general materials by form
BT: general materials
NT: colloid
film
fluid
foam
gel
liposome
liquor
membrane
particle
powder
wax

general materials by origin or composition
BT: general materials
NT: animal material
ceramics
composite
concrete
cotton wool
geological material
glass
metal
mineral wool
plant material
plastic
raw material

general materials by property or purpose
BT: general materials
NT: biomimetic material
contaminant
dye mixture
electret
fuel
handling pack
ink
insulating material
packaging
padding (material)
polymer melt
polymer solution
printing paste
reinforcement
sample
spinning solution
substrate
waste

genetic modification
BT: biotechnology
modification

geocell
BT: geosynthetics

geogrid
BT: geosynthetics

geological material
BT: general materials by origin or composition
NT: asbestos
basalt

geomembrane
BT: geosynthetics

geonet
BT: geosynthetics

geosynthetics
BT: product by property or purpose
NT: clay liner
geocell
geogrid
geomembrane
geonet
geotextile

geotextile
BT: geosynthetics

gilling
BT: drafting

gilling machine
BT: drafting machine

ginning
BT: fiber preparation

ginning machine
BT: machine

glass
BT: general materials by origin or composition

glass fiber
UF: glass fibre
glassfiber
glassfibre

BT: manufactured fiber

glass fiber fabric
UF: glass fibre fabric
BT: manufactured fiber fabric

glass fiber yarn
UF: glass fibre yarn
BT: manufactured fiber yarn

glass fibre
USE: glass fiber

glass fibre fabric
USE: glass fiber fabric

glass fibre yarn
USE: glass fiber yarn

glass manufacture
BT: manufacture

glass transition temperature
BT: measure or dimension
physicochemical property by characteristic

glassfiber
USE: glass fiber

glassfibre
USE: glass fiber

global communication
BT: communication

globalization
BT: economic conditions

glove
BT: garment accessory

goat hair
BT: hair

godet
BT: machine component

graft copolymer
BT: copolymer by structure

grafting
BT: finishing

graphite fiber
UF: graphite fibre
BT: carbon fiber

graphite fiber fabric
UF: graphite fibre fabric
BT: carbon fiber fabric

graphite fiber yarn
UF: graphite fibre yarn
BT: carbon fiber yarn

graphite fibre
USE: graphite fiber

graphite fibre fabric
USE: graphite fiber fabric

graphite fibre yarn
USE: graphite fiber yarn

greige weaving
BT: weaving

grinding
BT: dry finishing

gripper
BT: loom component

guide bar
BT: guiding system
machine component

guiding system
BT: materials handling system
NT: feed
guide bar
sliver guide

gusset
BT: garment component

hackling
BT: fiber preparation

hackling machine
BT: combing machine

hair
BT: animal material
NT: camel hair
dog hair
goat hair
horsehair
human hair
mohair

hairiness
BT: appearance
handle

hand embroidery
BT: embroidery

hand knitting
BT: knitting

hand sewing
BT: sewing

hand weaving
BT: weaving

handicraft
BT: art, design and culture

handle
BT: physical property
NT: hairiness

handling pack
BT: general materials by property or purpose
NT: bale
fiber bundle
hank
lap
package

handloom
BT: shuttle loom

hank
BT: handling pack

hardware
BT: computer

harness
BT: machine component

harvesting
BT: agricultural industry
NT: cotton picking

headwear
BT: garment by item type
NT: mask
wig

heald
UF: heddle
BT: machine component

heald frame
BT: machine component

health and disease
BT: biological phenomena
NT: biohazard
carcinogenicity
disease
toxicity

health and welfare
BT: society and economics
NT: health care
occupational health
welfare provision

health care
BT: health and welfare

heat bonding
BT: heat treatment

heat recovery
BT: thermal phenomena

heat resistance
BT: resistant property

heat setting
BT: heat treatment
NT: boarding
calendering
crabbing

heat setting machine
BT: finishing machine
NT: boarding machine

heat transfer
BT: thermal phenomena
transport process

heat transfer printing
BT: transfer printing

heat treatment
BT: finishing
NT: annealing
drying
heat bonding
heat setting
sintering
steaming
thermal treatment

heating
BT: Facet B: operations

heddle
USE: heald

hem
BT: product component

hemp
BT: plant

hemp blend
BT: cellulosic fiber blend
NT: flax-hemp blend

hemp fabric
BT: bast fiber fabric

hemp fiber
UF: hemp fibre
BT: bast fiber

hemp fibre
USE: hemp fiber

hemp yarn
BT. bast fiber yarn

herbicide

BT: pesticide

hibiscus fiber
UF: hibiscus fibre
BT: bast fiber

hibiscus fibre
USE: hibiscus fiber

high modulus fiber
USE: high-modulus fiber

high modulus fibre
USE: high-modulus fiber

high performance
BT: physicochemical property by characteristic

high temperature application
BT: use

high volume instrument testing
USE: high-volume-instrument testing

high-modulus fiber
UF: high modulus fiber
high modulus fibre
high-modulus fibre
BT: fiber by property or purpose

high-modulus fibre
USE: high-modulus fiber

high-volume-instrument testing
UF: high volume instrument testing
BT: testing

historical textile
BT: product by property or purpose

history
BT: social science
NT: social history
textile history

hollow fiber
UF: hollow fibre
BT: fiber by form

hollow fibre
USE: hollow fiber

hollow filament
BT: filament

holographic property
BT: optical property

home laundering
BT: laundering

homopolymer
BT: polymer by monomer sequence

honeycomb construction
BT: fabric construction

honeydew
BT: impurity

hook
BT: machine component

horsehair
BT: hair

horticultural textile
BT: product by property or purpose

hosiery
BT: underwear
NT: sock

hot air balloon
USE: hot-air balloon

hot-air balloon
UF: hot air balloon
BT: aerospace textile

household textile
BT: product by property or purpose

household use
BT: use

housing
BT: machine component

human hair
BT: hair

humidity
BT: physical property

hydraulics
BT: fluid mechanics

hydrocarbon
BT: organic compound
NT: perchloroethylene

hydrodynamics
BT: fluid dynamics

hydroentangling
BT: nonwoven fabric manufacture

hydrogel
BT: gel

hydrogen peroxide
BT: inorganic compound

hydrolysis
BT: chemical reaction

hydrophilic property
BT: chemical property

hydrophobic property
BT: resistant property
NT: water resistance

hygiene product
BT: product by property or purpose
NT: dental floss
diaper
incontinence pad
sanitary product

hygral expansion
BT: expansion

identification method
BT: methodology

ikat dyeing
BT: dyeing

image analysis
BT: analysis
NT: magnetic resonance imaging
X-ray analysis

impact
BT: force

impact resistance
BT: resistant property

implant
BT: medical textile

imports
BT: international trade

impregnating
BT: wet processing

impurity
BT: contaminant
NT: honeydew

incontinence pad
BT: hygiene product

indigo
BT: plant

indigo dye
BT: dye by origin or composition

indoor air quality
BT: environmental phenomena

industrial enterprise
BT: industry
NT: buyer
manufacturer
supplier

industrial laundering
BT: laundering

industrial revolution
BT: social history

industrial textile
BT: product by property or purpose

industrial use
BT: use

industry
BT: Facet A: industry and organization
NT: agricultural industry
composites industry
construction industry
drycleaning industry
dye manufacturing industry
dyeing industry
finishing industry
industrial enterprise
laundering industry
manufactured fiber industry
manufacturing
natural fiber industry
printing industry
service sector
textile-clothing industry

inflatable
BT: physical property

inflation
BT: dynamic or static phenomena

information
BT: Facet H: information and communication
NT: company information
database
directory information
economic information
information processing
information technology

information processing
BT: information
NT: classification
classing
data processing

information technology
BT: information
technology
NT: fiber optics

infrared radiation
BT: radiation

infrared resistance

BT: electromagnetic resistance

inhibitor
BT: additive

initial modulus
BT: constant

injection molding
UF: injection moulding
BT: molding

injection moulding
USE: injection molding

ink
BT: general materials by property or purpose
NT: printing ink

ink jet printing
USE: ink-jet printing

ink-jet printing
UF: ink jet printing
inkjet printing
jet printing
BT: printing

inkjet printing
USE: ink-jet printing

inlay
BT: pattern (configuration)

inorganic compound
BT: chemical compound
NT: ammonia
ammonium carbonate
boric acid
copper sulfate
hydrogen peroxide
oxide
titanium chloride
titanium nitride

inorganic polymer
BT: polymer by chemical composition
NT: polysiloxane
silicone

insect resistance
BT: biological resistance
NT: moth resistance

insole
BT: garment component

inspection
BT: Facet B: operations

inspection machine
BT: machine

instrument
BT: equipment by item type
NT: detector
sensor
testing instrument

insulating material
BT: general materials by property or purpose

insulation
BT: Facet B: operations

intellectual property
BT: legislation
NT: copyright
patent
royalty

interfacial property
BT: composite property

interfacing
BT: fabric by property or purpose
garment component

interior design
BT: design

interlaminar property
BT: composite property

interlining
BT: fabric by property or purpose
garment component

intermingling
BT: yarn manufacture

international trade
BT: trade
NT: dumping
exports
free trade
imports
tariff structure
trade agreement

internet
BT: media

investment
BT: society and economics

ion exchange
BT: chemical process

iridescence
BT: optical property

ironing
BT: pressing

ironing machine

BT: pressing machine

irradiation
BT: Facet B: operations

isotactic polymer
BT: polymer by stereochemical configuration

jack
BT: machine component

jacket
BT: outerwear

jacquard
BT: loom component

jacquard knitting
BT: knitting

jacquard loom
BT: loom

jeans
BT: trousers

jet
BT: fluid flow
machine component
NT: air jet
water jet

jet dyeing
BT: dyeing

jet printing
USE: ink-jet printing

jig dyeing
BT: dyeing

joint venture
BT: business strategy

jute
BT: plant

jute blend
BT: cellulosic fiber blend
NT: flax-jute blend

jute fabric
BT: bast fiber fabric

jute fiber
UF: jute fibre
BT: bast fiber

jute fibre
USE: jute fiber

jute industry
BT: vegetable fiber industry

jute yarn
BT: bast fiber yarn

kapok
BT: plant

kenaf
BT: plant

kenaf fabric
BT: bast fiber fabric

kenaf fiber
UF: kenaf fibre
BT: bast fiber

kenaf fibre
USE: kenaf fiber

kenaf yarn
BT: bast fiber yarn

keratin
BT: protein

keratin fiber
UF: keratin fibre
BT: animal fiber
NT: alpaca fiber
angora fiber
cashmere fiber
mohair fiber
vicuna fiber
wool fiber

keratin fiber fabric
BT: animal fiber fabric
NT: alpaca fabric
angora fabric
cashmere fabric
mohair fabric
wool fabric

keratin fiber yarn
BT: animal fiber yarn
NT: alpaca yarn
angora yarn
cashmere yarn
mohair yarn
wool yarn

keratin fibre
USE: keratin fiber

kettle dyeing
UF: beck dyeing
BT: dyeing

kiering
BT: boiling

kilim

BT: mat

kinetics
BT: chemical process
dynamic or static phenomena
mechanics

kite
BT: product by item type

knife
BT: machine component

knittability
BT: yarn property

knitted fabric
BT: fabric by form
NT: crocheted fabric

knitted yarn
BT: yarn by form

knitting
BT: fabric manufacture
NT: casting off
circular knitting
crocheting
flat knitting
hand knitting
jacquard knitting
linking
narrowing
raschel knitting
warp knitting
weft knitting
whole-garment knitting

knitting industry
BT: textile-clothing industry

knitting machine
BT: machine
NT: cabling machine
crochet machine
linking machine
tubular knitting machine
warp knitting machine

knitting needle
BT: needle

knitwear
BT: garment by property or purpose

knot
BT: product component

knotting
BT: general manufacturing process

label
BT: product component

labeling
UF: labelling
BT: making-up
NT: eco-labeling

labeling machine
UF: labelling machine
BT: machine

labelling
USE: labeling

labelling machine
USE: labeling machine

labor
UF: labour
BT: labor and employment
NT: child labor
labor relations

labor and employment
BT: society and economics
NT: buyer
employment
labor
supplier
wage
working conditions

labor relations
UF: labour relations
BT: labor

labour
USE: labor

labour relations
USE: labor relations

lace
BT: lace fabric

lace fabric
BT: fabric by form
NT: bobbin lace
lace

lace machine
BT: machine

lace making
BT: fabric manufacture

lambswool
BT: wool

laminate
BT: fabric by form

laminating
BT: fabric manufacture

laminating machine
BT: machine

lanolin
BT: wool grease

lap
BT: handling pack

lap forming
BT: winding

lariat
UF: lariat (lasso)
BT: rope

lariat (lasso)
USE: lariat

laser
BT: equipment by item type

latex
BT: plant material

launderability
BT: physicochemical property by characteristic

laundering
BT: finishing
NT: home laundering
industrial laundering

laundering industry
BT: industry

laundering machine
BT: finishing machine

layer separation
BT: separation

layering
BT: fabric manufacture

laying up
USE: laying-up

laying-up
UF: laying up
BT: making-up

leather
BT: skin

legislation
BT: management
NT: intellectual property
standard

length
BT: measure or dimension
NT: fiber length
gauge length

leno
BT: weaving

let off motion
USE: let-off motion

let-off motion
UF: let off motion
BT: machine function

leuco dye
BT: dye by property or form

leveling
BT: dyeing

leveling agent
BT: dyeing assistant

licker in
USE: licker-in

licker-in
UF: licker in
BT: cylinder

life prediction
BT: prediction

life science
BT: Facet G: art, science and technology
NT: ecology
psychology

lifting device
BT: machine component

lightfastness
BT: dye property

lighting
BT: Facet B: operations

linear density
BT: density
fiber property
yarn property

linen
BT: flax fiber

linen blend
BT: cellulosic fiber blend
NT: cotton-linen blend
linen-polyester blend
linen-viscose-rayon blend
linen-wool blend

linen fabric
BT: flax fiber fabric

linen industry
BT: vegetable fiber industry

linen polyester blend
USE: linen-polyester blend

linen viscose rayon blend
USE: linen-viscose-rayon blend

linen wool blend
USE: linen-wool blend

linen yarn
BT: flax fiber yarn

linen-polyester blend
UF: linen polyester blend
BT: linen blend
polyester blend

linen-viscose-rayon blend
UF: linen viscose rayon blend
BT: linen blend
viscose-rayon blend

linen-wool blend
UF: linen wool blend
BT: linen blend
wool blend

liner
BT: product by item type
NT: clay liner
diskette liner

lingerie
BT: underwear

lining
BT: making-up

linking
BT: knitting

linking machine
BT: knitting machine

lint
BT: waste

lint removal
BT: fiber preparation
removal

lipid
BT: organic compound

liposome
BT: general materials by form

liquid
BT: fluid
NT: liquid crystal
moisture
oil
water

liquid crystal
BT: liquid

liquor
BT: general materials by form

livestock production
BT: agricultural production

loading
USE: loading (testing)

loading (finishing)
BT: finishing

loading (testing)
UF: loading
BT: testing
NT: biaxial loading

loft
BT: physical property

logistics
BT: management

loom
BT: machine
NT: carpet loom
dobby loom
jacquard loom
loom component
shuttle loom
shuttleless loom
wave-shed loom

loom component
BT: loom
machine component
NT: dobby
gripper
jacquard
rapier
reed
shuttle
temple

loop
BT: structural property

lubricant
BT: chemical by property or purpose
NT: spin finish

lubricating
BT: general manufacturing process

luggage

BT: accessories

luminescence
BT: optical property

luster
BT: optical property

lyocell
USE: lyocell fiber

lyocell fabric
BT: rayon fiber fabric

lyocell fiber
UF: lyocell
lyocell fibre
BT: rayon fiber

lyocell fibre
USE: lyocell fiber

lyocell yarn
BT: rayon fiber yarn

machinability
BT: physical property

machine
BT: equipment by item type
NT: bale plucking machine
baling machine
batching machine
blending machine
bonding machine
braiding machine
carding machine
combing machine
compacting machine
cotton picking machine
cutting machine
decorticating machine
dosing machine
drafting machine
drawing machine
drawing-in machine
finishing machine
folding machine
ginning machine
inspection machine
knitting machine
labeling machine
lace machine
laminating machine
loom
machine component
nonwoven fabric machine
opening machine
papermaking machine
quilting machine
roving machine
sewing machine
sizing machine
spinneret
spinning machine
texturing machine
toe closing machine
tufting machine
twisting machine
wet processing machine
winding machine

machine clothing
BT: machine component
NT: card clothing
press felt

machine component
BT: machine
NT: alignment mechanism
antitwist device
antiwrap device
apron
balance tube
balloon limiter
beam
bearings
blade
bobbin
bobbin holder
braking system
brush
calender bowl
cam
chute
comb
cone
creel
cutter
cylinder
die
drawframe
drive
drop wire
drum
embroidery frame
filter
fourdrinier wire
gasket
gear
godet
guide bar
harness
heald
heald frame
hook
housing
jack
jet
knife
lifting device
loom component
machine clothing
magnet
mandrel
motor

mounting
navel
needle
needle bar
needle bed
nip
nipper
nozzle
pirn
presser foot
pulse jet
ring
roller
rotor
sinker
sliver can
spindle
spool
stop motion
stuffer box
suction device
take-down mechanism
take-up mechanism
tensioner
traveler
traverse mechanism
tube

machine function
BT: Facet B: operations
NT: bobbin changing
can changing
clamping
doffing
let-off motion
needle penetration
needle selection
sewing speed

made to measure
BT: making-up

maggot
BT: animal

magnet
BT: machine component

magnetic resonance imaging
UF: MRI
BT: image analysis

maintenance
BT: Facet B: operations

making up
USE: making-up

making-up
UF: making up
BT: general manufacturing process
NT: labeling
laying-up
lining
made to measure
marking
pattern cutting
piecing
spreading
toe closing
trimming

man made fiber
USE: manufactured fiber

man made fiber fabric
USE: manufactured fiber fabric

man made fiber industry
USE: manufactured fiber industry

man made fiber yarn
USE: manufactured fiber yarn

man-made fiber
USE: manufactured fiber

man-made fiber fabric
USE: manufactured fiber fabric

man-made fiber industry
USE: manufactured fiber industry

man-made fiber yarn
USE: manufactured fiber yarn

management
BT: Facet A: industry and organization
NT: business strategy
costing
environmental management
forecasting
legislation
logistics
optimization
organizational structure
personnel management
planning
production management
professional practise
regulation
safety

mandrel
BT: machine component

mangle
BT: drying machine

manikin
BT: testing instrument

mannequin
BT: product by item type

manufacture

BT: Facet B: operations
NT: dye manufacture
fabric manufacture
fiber manufacture
garment manufacture
general manufacturing process
glass manufacture
papermaking
rope making
yarn manufacture

manufactured fiber
UF: man made fiber
man-made fiber
manufactured fibre
BT: fiber by origin or composition
NT: acrylic fiber
alginate fiber
basalt fiber
boron fiber
carbon fiber
ceramic fiber
chitin fiber
chitosan fiber
elastane fiber
elastomeric fiber
glass fiber
melamine resin fiber
metal fiber
mineral fiber
modacrylic fiber
novolac fiber
polyamide fiber
polybenzazole fiber
polyester fiber
polyether ether ketone fiber
polyimide fiber
polylactic acid fiber
polyolefin fiber
polyphenylene sulfide fiber
polytetrafluoroethylene fiber
polyurethane fiber
polyvinyl alcohol fiber
polyvinyl chloride fiber
regenerated cellulose fiber
synthetic fiber

manufactured fiber blend
BT: blend
NT: acrylic blend
aramid blend
nylon blend
polyester blend
polyolefin blend
viscose-rayon blend

manufactured fiber fabric
UF: man made fiber fabric
man-made fiber fabric
manufactured fibre fabric
BT: fabric by origin or composition
NT: acrylic fabric
alginate fabric
carbon fiber fabric
ceramic fiber fabric
chitin fabric
chitosan fabric
elastane fabric
elastomeric fabric
glass fiber fabric
metal fiber fabric
mineral fiber fabric
modacrylic fabric
polyamide fabric
polyester fabric
polyether ether ketone fabric
polyimide fabric
polyolefin fabric
polytetrafluoroethylene fabric
polyurethane fabric
polyvinyl alcohol fabric
polyvinyl chloride fabric
regenerated cellulose fabric
synthetic fiber fabric

manufactured fiber industry
UF: man made fiber industry
man-made fiber industry
manufactured fibre industry
BT: industry
NT: synthetic fiber industry

manufactured fiber yarn
UF: man made fiber yarn
man-made fiber yarn
manufactured fibre yarn
BT: yarn by origin or composition
NT: acrylic yarn
alginate yarn
basalt yarn
carbon fiber yarn
ceramic fiber yarn
chitin yarn
chitosan yarn
elastane yarn
elastomeric yarn
glass fiber yarn
metal fiber yarn
mineral fiber yarn
modacrylic yarn
polyamide yarn
polyester yarn
polyether ether ketone yarn
polyimide yarn
polyolefin yarn
polytetrafluoroethylene yarn
polyurethane yarn
polyvinyl alcohol yarn
polyvinyl chloride yarn
regenerated cellulose yarn
synthetic fiber yarn

manufactured fibre
USE: manufactured fiber

manufactured fibre fabric

USE: manufactured fiber fabric

manufactured fibre industry
USE: manufactured fiber industry

manufactured fibre yarn
USE: manufactured fiber yarn

manufacturer
BT: industrial enterprise

manufacturing
BT: industry

marine textile
BT: product by property or purpose

market
BT: trade and commerce
NT: competitiveness
demand
e-commerce
niche market
price
supply

marketing
BT: trade and commerce
NT: advertising
merchandizing

marking
BT: making-up

mask
BT: garment accessory
headwear

mass dyeing
BT: dyeing

mass transfer
BT: transport process

mat
BT: floor covering
NT: kilim

materials
BT: Facet C: materials, products and resources
NT: blend
fabric
fiber
general materials
yarn

materials handling
BT: Facet B: operations
NT: agitation
bale plucking
batching
folding
storage
threading
transport

materials handling system
BT: equipment by property or purpose
NT: conveyor belt
guiding system

materials selection
BT: general manufacturing process

maternitywear
BT: garment by property or purpose

mathematical analysis
BT: analysis
mathematical phenomena
NT: finite element analysis
Fourier transform analysis
fractal analysis
fuzzy logic

mathematical expression
BT: mathematical phenomena
NT: absorption isotherm
add-on
adsorption isotherm
birefringence
blend ratio
constant
draw ratio
friction coefficient
micronaire value
moisture regain
pull-out
zeta potential

mathematical model
BT: model

mathematical phenomena
BT: Facet D: phenomena
NT: mathematical analysis
mathematical expression
measure or dimension

matrix
BT: composite
NT: ceramic matrix

mattress
BT: bedding

maturity
BT: biological characteristic

measure or dimension
BT: mathematical phenomena
NT: bulk
crease recovery
cross section
degree of orientation
degree of polymerization

density
diameter
fiber curvature
flow rate
glass transition temperature
length
shape
size
thickness
weight

measuring
BT: methodology
NT: anthropometry
calorimetry
colorimetry
spectrophotometry
thermogravimetry

measuring system
BT: equipment by property or purpose
NT: dosing system

mechanical property
BT. physical property
NT: bending property
breaking point
compressive property
crease recovery
deformation
drape
durability
elastic property
elastoplasticity
extensibility
flexibility
shear property
stiffness
strength
stress-strain property
stretch
tenacity
tensile property
viscosity

mechanics
BT: physics
NT: fluid mechanics
kinetics
rheology
thermodynamics

mechatronics
BT: technology

media
BT: communication medium
NT: internet

medical process
BT: Facet B: operations
NT: bone repair
drug delivery
tissue engineering

medical textile
BT: product by property or purpose
NT: bandage
blood filter
brace (medical textile)
dressing
implant
orthosis
prosthesis
splint
suture
tourniquet

medical use
BT: use

medulla, fiber
USE: fiber medulla

melamine resin fiber
UF: melamine resin fibre
BT: manufactured fiber

melamine resin fibre
USE: melamine resin fiber

melt blowing
BT: extrusion

melt blowing machine
BT: nonwoven fabric machine

melt dyeing
BT: dyeing

melt spinning
BT: extrusion

melting point
BT: thermal property

membrane
BT: general materials by form

menswear
BT: garment by user

mercerizing
BT: finishing

mercerizing machine
BT: finishing machine

merchandizing
BT: marketing

merino sheep
BT: sheep

merino wool
BT: wool

mesh
BT: fabric by form
NT: gauze
net
scrim

metal
BT: general materials by origin or composition
NT: alloy

metal complex dye
USE: metal-complex dye

metal fiber
UF: metal fibre
BT: manufactured fiber
NT: aluminum fiber

metal fiber fabric
UF: metal fibre fabric
BT: manufactured fiber fabric

metal fiber yarn
UF: metal fibre yarn
BT: manufactured fiber yarn

metal fibre
USE: metal fiber

metal fibre fabric
USE: metal fiber fabric

metal fibre yarn
USE: metal fiber yarn

metal-complex dye
UF: metal complex dye
BT: dye by origin or composition

metallization
BT: finishing

metamerism
BT: color change

metering
BT: process control

methodology
BT: Facet I: research and methodology
NT: age determination
analysis
assessment
calculation
calibration
color matching
digital method
identification method
measuring
prediction
sampling
simulation
testing

microfiber
UF: microfibre
BT: fiber by form

microfiber fabric
UF: microfibre fabric
BT: fabric by form

microfiber yarn
UF: microfibre yarn
BT: yarn by form

microfibre
USE: microfiber

microfibre fabric
USE: microfiber fabric

microfibre yarn
USE: microfiber yarn

microfilament
BT: filament

micronaire value
BT: fiber property
mathematical expression

microorganism
BT: organism
NT: bacterium

microscopy
BT: analysis
NT: electron microscopy

microstructural analysis
BT: structural analysis

microstructural property
BT: microstructure

microstructure
BT: structural property
NT: microstructural property

microwave radiation
BT: radiation

military textile
BT: product by property or purpose
NT: camouflage

mill
BT: facility or structure

millinery
BT: garment manufacture

milling
BT: dry finishing

mineral fiber
UF: mineral fibre
BT: manufactured fiber
NT: alumina fiber
aluminosilicate fiber
asbestos fiber

mineral fiber fabric
UF: mineral fibre fabric
BT: manufactured fiber fabric
NT: asbestos fabric
basalt fabric

mineral fiber yarn
UF: mineral fibre yarn
BT: manufactured fiber yarn
NT: asbestos yarn

mineral fibre
USE: mineral fiber

mineral fibre fabric
USE: mineral fiber fabric

mineral fibre yarn
USE: mineral fiber yarn

mineral wool
BT: general materials by origin or composition

mixing
BT: general manufacturing process

modacrylic
BT: synthetic organic polymer

modacrylic fabric
BT: manufactured fiber fabric

modacrylic fiber
UF: modacrylic fibre
BT: manufactured fiber

modacrylic fibre
USE: modacrylic fiber

modacrylic yarn
BT: manufactured fiber yarn

modal
USE: modal fiber

modal fabric
BT: rayon fiber fabric

modal fiber
UF: modal
modal fibre
BT: rayon fiber
NT: polynosic fiber

modal fibre
USE: modal fiber

modal yarn
BT: rayon fiber yarn

model
BT: simulation
NT: mathematical model

modification
BT: general manufacturing process
NT: chemical modification
fabric modification
fiber modification
genetic modification
surface modification
yarn modification

mohair
BT: hair

mohair fabric
BT: keratin fiber fabric

mohair fiber
UF: mohair fibre
BT: keratin fiber

mohair fibre
USE: mohair fiber

mohair yarn
BT: keratin fiber yarn

moire
BT: fabric property

moire fabric
BT: fabric by property or purpose

moisture
BT: liquid

moisture content
BT: physical property

moisture management
BT: control

moisture regain
BT: mathematical expression

moisture release
BT: moisture transfer

moisture transfer
BT: transport process
NT: moisture release

molding
UF: moulding
BT: general manufacturing process
NT: injection molding

pultrusion
resin transfer molding

molecular structure
BT: chemical property

molecular weight
BT: chemical property
weight

monitoring
BT: Facet B: operations
NT: on-line monitoring

monitoring system
BT: equipment by property or purpose

monofilament
BT: filament

monomer
BT: chemical by form

mop
BT: product by item type

mordant
BT: dyeing assistant

mordant dyeing
BT: dyeing

morphology
BT: science in general

mote
BT: dust

moth resistance
BT: insect resistance

motor
BT: machine component

moulding
USE: molding

mounting
BT: machine component

MRI
USE: magnetic resonance imaging

muga silk
BT: silk

mulberry
BT: plant

mulberry silk
BT: silk

multiaxial fabric
BT: fabric by form

multifilament
BT: filament

multilayer fabric
BT: fabric by property or purpose

multiphase weaving
BT: weaving

museum
BT: facility or structure

muslin
BT: woven fabric

nanocomposite
BT: composite

nanofiber
UF: nanofibre
BT: fiber by form

nanofibre
USE: nanofiber

nanotechnology
BT: technology

nanotube
BT: chemical by form

nap
BT: pile

napping
BT: raising

napping machine
BT: raising machine

narrow fabric
BT: fabric by property or purpose
NT: tape

narrowing
BT: knitting

natural dye
BT: dye by origin or composition
NT: vegetable dye

natural fiber
UF: natural fibre
BT: fiber by origin or composition
NT: animal fiber
cellulosic fiber
zein fiber

natural fiber blend
BT: blend
NT: animal fiber blend

cellulosic fiber blend

natural fiber fabric
UF: natural fibre fabric
BT: fabric by origin or composition
NT: animal fiber fabric
cellulosic fiber fabric

natural fiber industry
UF: natural fibre industry
BT: industry
NT: animal fiber industry
vegetable fiber industry

natural fiber yarn
UF: natural fibre yarn
BT: yarn by origin or composition
NT: animal fiber yarn
cellulosic fiber yarn

natural fibre
USE: natural fiber

natural fibre fabric
USE: natural fiber fabric

natural fibre industry
USE: natural fiber industry

natural fibre yarn
USE: natural fiber yarn

natural organic polymer
BT: organic polymer
NT: natural polyamide
natural polyisoprene
polyamine
polysaccharide

natural polyamide
BT: natural organic polymer
NT: polypeptide
protein

natural polyisoprene
UF: polyisoprene
BT: natural organic polymer
NT: natural rubber

natural resin
BT: plant material

natural rubber
BT: natural polyisoprene
plant material

navel
BT: machine component

necking
BT: distortion

needle
BT: machine component
NT: knitting needle

needle bar
BT: machine component

needle bed
BT: machine component

needle penetration
BT: machine function

needle selection
BT: machine function

needled fabric
BT: nonwoven fabric

needling
BT: nonwoven fabric manufacture

needling machine
BT: nonwoven fabric machine

nep
BT: waste

net
BT: mesh

net curtain
BT: curtain

neural network
BT: computer

niche market
BT: market

nickel
BT: chemical element

nip
BT: machine component

nipper
BT: machine component

noil
BT: fiber by form

noise control
BT: pollution control

nonwoven fabric
BT: fabric by form
NT: batting
bonded fiber fabric
felt
needled fabric
web

nonwoven fabric industry

BT: textile-clothing industry

nonwoven fabric machine
BT: machine
NT: crosslapping machine
melt blowing machine
needling machine
spunbonding machine
stitch-bonding machine
web forming machine

nonwoven fabric manufacture
BT: fabric manufacture
NT: crosslapping
dry laying
felting
hydroentangling
needling
spunbonding
stitch-bonding
web forming
wet laying

novolac fiber
UF: novolac fibre
BT: manufactured fiber

novolac fibre
USE: novolac fiber

nozzle
BT: machine component

nuclear application
BT: use

nucleation
BT: chemical process

nylon
BT: synthetic polyamide

nylon 11
BT: synthetic polyamide

nylon 4
BT: synthetic polyamide

nylon 46
BT: synthetic polyamide

nylon 6
BT: synthetic polyamide

nylon 66
BT: synthetic polyamide

nylon blend
BT: manufactured fiber blend
NT: aramid-nylon blend
nylon-polyester blend
nylon-polyethylene blend
nylon-polypropylene blend
nylon-silk blend
nylon-viscose-rayon blend
nylon-wool blend

nylon fabric
BT: polyamide fabric

nylon fiber
UF: nylon fibre
BT: polyamide fiber

nylon fibre
USE: nylon fiber

nylon polyester blend
USE: nylon-polyester blend

nylon polyethylene blend
USE: nylon-polyethylene blend

nylon polypropylene blend
USE: nylon-polypropylene blend

nylon silk blend
USE: nylon-silk blend

nylon viscose rayon blend
USE: nylon-viscose-rayon blend

nylon wool blend
USE: nylon-wool blend

nylon yarn
BT: polyamide yarn

nylon-polyester blend
UF: nylon polyester blend
BT: nylon blend
polyester blend

nylon-polyethylene blend
UF: nylon polyethylene blend
BT: nylon blend
polyolefin blend

nylon-polypropylene blend
UF: nylon polypropylene blend
BT: nylon blend
polyolefin blend

nylon-silk blend
UF: nylon silk blend
BT: nylon blend
silk blend

nylon-viscose-rayon blend
UF: nylon viscose rayon blend
BT: nylon blend
viscose-rayon blend

nylon-wool blend
UF: nylon wool blend
BT: nylon blend

wool blend

objective assessment
BT: assessment

occupational health
BT: health and welfare

odor control
UF: odour control
BT: control

odor emitting
UF: odour emitting
BT: chemical process

odour control
USE: odor control

odour emitting
USE: odor emitting

offshore application
BT: use

oil
BT: liquid

oil removal
BT: removal

oil resistance
BT: resistant property

oleophilic property
BT: chemical property

oligomer
BT: polymer by monomer sequence

on line monitoring
USE: on-line monitoring

on-line monitoring
UF: on line monitoring
BT: monitoring

one step dyeing
UF: one-step dyeing
BT: dyeing
one step process

one step process
UF: one-step process
BT: general manufacturing process
NT: one step dyeing

one-step dyeing
USE: one step dyeing

one-step process
USE: one step process

open end spinning
USE: open-end spinning

open width dyeing
USE: open-width dyeing

open width process
USE: open-width process

open-end spinning
UF: open end spinning
BT: spinning

open-width dyeing
UF: open width dyeing
BT: dyeing
open-width process

open-width process
UF: open width process
BT: general manufacturing process
NT: open-width dyeing

opening
BT: fiber preparation
separation
NT: scutching

opening machine
BT: machine
NT: bale opening machine
scutching machine

openings
BT: fabric component
product component
NT: perforation
shed
slit

optical phenomena
BT: physical phenomena
NT: color change
frosting
optical property
shade variation

optical property
BT: optical phenomena
physical property
NT: fluorescence
holographic property
iridescence
luminescence
luster
reflective property
transparency
visibility

optimization
BT: management

organic acid

BT: organic compound
NT: acrylic acid
amino acid
polycarboxylic acid

organic compound
BT: chemical compound
NT: acetate
acrylonitrile
alcohol
alkoxylate
amine
aromatic amine
caprolactam
formaldehyde
hydrocarbon
lipid
organic acid
volatile organic compound

organic dye
BT: dye by origin or composition
NT: synthetic-organic dye

organic farming
BT: farming

organic fiber
UF: organic fibre
BT: fiber by origin or composition

organic fibre
USE: organic fiber

organic polymer
BT: polymer by chemical composition
NT: biopolymer
natural organic polymer
semisynthetic organic polymer
synthetic organic polymer

organic product
BT: product by property or purpose

organism
BT: Facet F: organisms and biological structures
NT: animal
microorganism
plant

organizational structure
BT: management

orientation
USE: crystalline orientation

orthosis
BT: medical textile

outerwear
BT: garment by item type
NT: bib
dress
jacket
sari
shirt
skirt
suit
sweater
trousers
uniform

outsourcing
BT: supply chain

overlock sewing
BT: sewing

oxidation
BT: chemical reaction

oxide
BT: inorganic compound
NT: carbon dioxide
titanium dioxide
titanium oxide
zinc oxide

oxidizing agent
BT: finishing agent

oxygen
BT: chemical element

ozone treatment
BT: environmental treatment

package
BT: handling pack

packaging
BT: general materials by property or purpose

pad
BT: dyeing machine

pad batch dyeing
USE: pad-batch dyeing

pad dyeing
BT: dyeing

pad steam dyeing
USE: pad-steam dyeing

pad-batch dyeing
UF: pad batch dyeing
BT: dyeing

pad-steam dyeing
UF: pad steam dyeing
BT: dyeing

padding
USE: padding (material)

padding (finishing)
BT: finishing

padding (material)
UF: padding
padding (product)
BT: general materials by property or purpose
NT: fiberfill
stuffing
wadding

padding (product)
USE: padding (material)

papermakers fabric
BT: fabric by property or purpose
NT: press felt

papermaking
BT: manufacture

papermaking machine
BT: machine

parachute
BT: aerospace textile

partially oriented yarn
BT: yarn by form

particle
BT: general materials by form

particle size
BT: size

patchwork
BT: pattern (configuration)

patent
BT: intellectual property

pattern (configuration)
UF: patterns
BT: appearance
NT: applique
embroidery
inlay
patchwork

pattern (plan)
BT: design

pattern cutting
BT: cutting
making-up

pattern preparation system
BT: equipment by property or purpose

patterning
BT: surface modification

patterns
USE: pattern (configuration)

peat fiber
UF: peat fibre
BT: vegetable fiber

peat fibre
USE: peat fiber

perchloroethylene
BT: hydrocarbon

perfluoropolymer
BT: fluoropolymer

perforation
BT: openings

performance assessment
BT: assessment

permeability
BT: physical property

personnel management
BT: management

perspiration control
BT: control

pest and disease management
BT: agricultural industry
NT: disease management
pest control

pest control
BT: control
pest and disease management

pesticide
BT: chemical by property or purpose
NT: fungicide
herbicide

pH
BT: chemical property

phase change material
USE: phase-change material

phase separation
BT: chemical process
separation

phase-change material
UF: phase change material
BT: technical textile

photochemical degradation
BT: degradation

photochromic dye
BT: dye by property or form

photolysis
BT: chemical reaction

physical phenomena
BT: Facet D: phenomena
NT: adhesion
damage
disintegration
distortion
dynamic or static phenomena
electrical phenomena
optical phenomena
physical property
physicomechanical property
thermal phenomena

physical property
BT: physical phenomena
NT: acoustic property
adhesive property
airtight
anisotropy
appearance
breathability
buoyancy
conductivity
defect
electrical property
handle
humidity
inflatable
loft
machinability
mechanical property
moisture content
optical property
permeability
porosity
structural property
thermal property
uniformity

physical science
BT: Facet G: art, science and technology
NT: chemistry
physics

physicochemical property
BT: Facet D: phenomena
NT: physicochemical property by characteristic
physicochemical property by material

physicochemical property by characteristic
BT: physicochemical property
NT: dyeability
easy care property
glass transition temperature
high performance
launderability
printability
reproducibility
resistant property
surface property

physicochemical property by material
BT: physicochemical property
NT: composite property
dye property
fabric property
fiber property
garment property
yarn property

physicomechanical property
BT: physical phenomena

physics
BT: physical science
NT: cryogenics
mechanics
tribology
ultrasonics

pick
BT: weft

picking
BT: weaving

piece dyeing
BT: dyeing

piecing
BT: making-up

piezoelectric device
BT: equipment by property or purpose

piezoelectric fiber
UF: piezoelectric fibre
BT: fiber by property or purpose

piezoelectric fibre
USE: piezoelectric fiber

pigment
BT: chemical by property or purpose

pigment printing
BT: printing

pile
BT: fabric component
NT: nap

pile fabric
BT: fabric by form
NT: corduroy fabric
terry fabric
velour fabric
velvet fabric

pill resistance

BT: resistant property

pilling
BT: wear

pineapple leaf fiber
UF: pineapple leaf fibre
BT: vegetable fiber

pineapple leaf fibre
USE: pineapple leaf fiber

pipe
BT: product by item type

pirn
BT: machine component

pitch
BT: fabric construction

planning
BT: management
NT: business plan
decision making

plant
BT: organism
NT: cotton
flax
hemp
indigo
jute
kapok
kenaf
mulberry
ramie
sisal

plant extract
BT: plant material

plant layout
BT: production management

plant material
BT: general materials by origin or composition
NT: coir
cork
latex
natural resin
natural rubber
plant extract

plasma treatment
BT: finishing

plastic
BT: general materials by origin or composition
NT: acrylic

plasticizing agent
BT: additive

plating
BT: coating
NT: electroless plating
electroplating

pleat
BT: product component

plexifilament
BT: filament

plied yarn
BT: yarn by form

plying
BT: twisting

pneumatic device
BT: equipment by property or purpose

pocket
BT: product component
NT: air pocket

pollutant removal
BT: removal

pollution
BT: environmental phenomena
NT: air pollution
water pollution

pollution control
BT: control
environmental management
NT: noise control

poly(trimethylene terephthalate)
USE: polytrimethylene terephthalate

polyacrylamide
BT: acrylic polymer

polyacrylate
BT: acrylic polymer

polyacrylic acid
BT: acrylic polymer

polyacrylonitrile
BT: acrylic polymer
vinyl polymer

polyamide
USE: synthetic polyamide

polyamide fabric
BT: manufactured fiber fabric
NT: aramid fabric
nylon fabric

polyamide fiber

BT: manufactured fiber
NT: aramid fiber
nylon fiber

polyamide yarn
BT: manufactured fiber yarn
NT: aramid yarn
nylon yarn

polyamine
BT: natural organic polymer

polyaniline
BT: synthetic organic polymer

polyarylene sulfide
UF: polyarylene sulphide
polyarylenesulfide
polyarylenesulphide
BT: polysulfide

polyarylene sulphide
USE: polyarylene sulfide

polyarylenesulfide
USE: polyarylene sulfide

polyarylenesulphide
USE: polyarylene sulfide

polyarylether ketone
UF: polyaryletherketone
BT: polyether
polyketone

polyaryletherketone
USE: polyarylether ketone

polybenzazole fiber
UF: polybenzazole fibre
BT: manufactured fiber

polybenzazole fibre
USE: polybenzazole fiber

polybenzimidazole
BT: synthetic organic polymer

polybenzoxazole
BT: synthetic organic polymer
NT: polyphenylene benzobisoxazole
polyphenylene benzobisthiazole

polybutadiene
BT: synthetic organic polymer

polybutylene
BT: polyolefin

polybutylene terephthalate
BT: polyester

polybutylene terephthalate fiber
UF: polybutylene terephthalate fibre
BT: polyester fiber

polybutylene terephthalate fibre
USE: polybutylene terephthalate fiber

polycaprolactone
BT: polyester

polycarbonate
BT: polyester

polycarboxylic acid
BT: organic acid

polydimethylsiloxane
BT: synthetic organic polymer

polyester
BT: synthetic organic polymer
NT: polybutylene terephthalate
polycaprolactone
polycarbonate
polyethylene naphthalate
polyethylene terephthalate
polypropylene terephthalate
polytrimethylene terephthalate

polyester blend
BT: manufactured fiber blend
NT: acrylic-polyester blend
aramid-polyester blend
cotton-polyester blend
flax-polyester blend
linen-polyester blend
nylon-polyester blend
polyester-polyethylene blend
polyester-polypropylene blend
polyester-silk blend
polyester-viscose-rayon blend
polyester-wool blend

polyester fabric
BT: manufactured fiber fabric
NT: polyethylene terephthalate fabric

polyester fiber
UF: polyester fibre
BT: manufactured fiber
NT: polybutylene terephthalate fiber
polyethylene terephthalate fiber

polyester fibre
USE: polyester fiber

polyester polyethylene blend
USE: polyester-polyethylene blend

polyester polypropylene blend
USE: polyester-polypropylene blend

polyester silk blend
USE: polyester-silk blend

polyester viscose rayon blend
USE: polyester-viscose-rayon blend

polyester wool blend
USE: polyester-wool blend

polyester yarn
BT: manufactured fiber yarn
NT: polyethylene terephthalate yarn

polyester-polyethylene blend
UF: polyester polyethylene blend
BT: polyester blend
polyolefin blend

polyester-polypropylene blend
UF: polyester polypropylene blend
BT: polyester blend
polyolefin blend

polyester-silk blend
UF: polyester silk blend
BT: polyester blend
silk blend

polyester-viscose-rayon blend
UF: polyester viscose rayon blend
BT: polyester blend
viscose-rayon blend

polyester-wool blend
UF: polyester wool blend
BT: polyester blend
wool blend

polyether
BT: synthetic organic polymer
NT: polyarylether ketone
polyether ether ketone
polyether imide
polyether ketone
polyether sulfone
polyethylene glycol
polyphenylene oxide

polyether ether ketone
UF: polyetheretherketone
BT: polyether
polyketone

polyether ether ketone fabric
UF: polyetheretherketone fabric
BT: manufactured fiber fabric

polyether ether ketone fiber
UF: polyetheretherketone fiber
polyetheretherketone fibre
BT: manufactured fiber

polyether ether ketone yarn
UF: polyetheretherketone yarn
BT: manufactured fiber yarn

polyether imide
UF: polyetherimide
BT: polyether
polyimide

polyether ketone
BT: polyether
polyketone

polyether sulfone
UF: polyether sulphone
polyethersulfone
polyethersulphone
BT: polyether
polysulfone

polyether sulphone
USE: polyether sulfone

polyetheretherketone
USE: polyether ether ketone

polyetheretherketone fabric
USE: polyether ether ketone fabric

polyetheretherketone fiber
USE: polyether ether ketone fiber

polyetheretherketone fibre
USE: polyether ether ketone fiber

polyetheretherketone yarn
USE: polyether ether ketone yarn

polyetherimide
USE: polyether imide

polyethersulfone
USE: polyether sulfone

polyethersulphone
USE: polyether sulfone

polyethylene
BT: polyolefin

polyethylene fabric
BT: polyolefin fabric

polyethylene fiber
UF: polyethylene fibre
BT: polyolefin fiber

polyethylene fibre
USE: polyethylene fiber

polyethylene glycol
BT: polyether

polyethylene imine
BT: synthetic organic polymer

polyethylene naphthalate
BT: polyester

polyethylene polypropylene blend
USE: polyethylene-polypropylene blend

polyethylene terephthalate
BT: polyester

polyethylene terephthalate fabric
BT: polyester fabric

polyethylene terephthalate fiber
UF: polyethylene terephthalate fibre
BT: polyester fiber

polyethylene terephthalate fibre
USE: polyethylene terephthalate fiber

polyethylene terephthalate yarn
BT: polyester yarn

polyethylene yarn
BT: polyolefin yarn

polyethylene-polypropylene blend
UF: polyethylene polypropylene blend
BT: polyolefin blend

polyimide
BT: synthetic organic polymer
NT: polyether imide

polyimide fabric
BT: manufactured fiber fabric

polyimide fiber
UF: polyimide fibre
BT: manufactured fiber

polyimide fibre
USE: polyimide fiber

polyimide yarn
BT: manufactured fiber yarn

polyisoprene
USE: natural polyisoprene

polyketone
BT: synthetic organic polymer
NT: polyarylether ketone
polyether ether ketone
polyether ketone

polylactic acid
BT: semisynthetic organic polymer
NT: polylactide

polylactic acid fiber
UF: polylactic acid fibre
BT: manufactured fiber

polylactic acid fibre
USE: polylactic acid fiber

polylactide
BT: polylactic acid

polymer
BT: chemical by form
NT: polymer by chemical composition
polymer by monomer sequence
polymer by physical property
polymer by polymerization reaction
polymer by stereochemical configuration

polymer blend
BT: blend

polymer by chemical composition
BT: polymer
NT: fluoropolymer
inorganic polymer
organic polymer
vinyl polymer

polymer by monomer sequence
BT: polymer
NT: copolymer
homopolymer
oligomer
terpolymer

polymer by physical property
BT: polymer
NT: elastomer
thermoplastic polymer
thermosetting polymer

polymer by polymerization reaction
BT: polymer
NT: addition polymer
condensation polymer

polymer by stereochemical configuration
BT: polymer
NT: atactic polymer
isotactic polymer
syndiotactic polymer

polymer melt
BT: general materials by property or purpose

polymer property
BT: chemical property
NT: degree of polymerization

polymer science
BT: chemistry

polymer solution
BT: general materials by property or purpose

polymerization
BT: chemical reaction

NT: copolymerization

polymethyl methacrylate
BT: acrylic polymer
vinyl polymer

polynosic fiber
UF: polynosic fibre
BT: modal fiber

polynosic fibre
USE: polynosic fiber

polyolefin
BT: synthetic organic polymer
NT: polybutylene
polyethylene
polypropylene

polyolefin blend
BT: manufactured fiber blend
NT: cotton-polypropylene blend
nylon-polyethylene blend
nylon-polypropylene blend
polyester-polyethylene blend
polyester-polypropylene blend
polyethylene-polypropylene blend
polypropylene-viscose-rayon blend
polypropylene-wool blend

polyolefin fabric
BT: manufactured fiber fabric
NT: polyethylene fabric
polypropylene fabric

polyolefin fiber
UF: polyolefin fibre
BT: manufactured fiber
NT: polyethylene fiber
polypropylene fiber

polyolefin fibre
USE: polyolefin fiber

polyolefin yarn
BT: manufactured fiber yarn
NT: polyethylene yarn
polypropylene yarn

polypeptide
BT: natural polyamide

polyphenol
BT: synthetic organic polymer

polyphenylene
BT: synthetic organic polymer

polyphenylene benzobisoxazole
BT: polybenzoxazole

polyphenylene benzobisthiazole
BT: polybenzoxazole

polyphenylene oxide
BT: polyether

polyphenylene sulfide
BT: polysulfide

polyphenylene sulfide fiber
UF: polyphenylene sulfide fibre
polyphenylene sulphide fiber
polyphenylene sulphide fibre
polyphenylenesulfide fiber
polyphenylenesulfide fibre
polyphenylenesulphide fiber
polyphenylenesulphide fibre
BT: manufactured fiber

polyphenylene sulfide fibre
USE: polyphenylene sulfide fiber

polyphenylene sulphide fiber
USE: polyphenylene sulfide fiber

polyphenylene sulphide fibre
USE: polyphenylene sulfide fiber

polyphenylene terephthalamide
BT: synthetic polyamide

polyphenylenesulfide fiber
USE: polyphenylene sulfide fiber

polyphenylenesulfide fibre
USE: polyphenylene sulfide fiber

polyphenylenesulphide fiber
USE: polyphenylene sulfide fiber

polyphenylenesulphide fibre
USE: polyphenylene sulfide fiber

polypropylene
BT: polyolefin

polypropylene fabric
BT: polyolefin fabric

polypropylene fiber
UF: polypropylene fibre
BT: polyolefin fiber

polypropylene fibre
USE: polypropylene fiber

polypropylene terephthalate
BT: polyester

polypropylene viscose rayon blend
USE: polypropylene-viscose-rayon blend

polypropylene wool blend
USE: polypropylene-wool blend

polypropylene yarn
BT: polyolefin yarn

polypropylene-viscose-rayon blend
UF: polypropylene viscose rayon blend
BT: polyolefin blend
viscose-rayon blend

polypropylene-wool blend
UF: polypropylene wool blend
BT: polyolefin blend
wool blend

polypyrrole
BT: synthetic organic polymer

polysaccharide
BT: natural organic polymer
NT: alginate
alginic acid
cellulose
chitin
chitosan
starch

polysiloxane
BT: inorganic polymer

polystyrene
BT: synthetic organic polymer
vinyl polymer

polysuccinimide
BT: synthetic organic polymer

polysulfide
BT: synthetic organic polymer
NT: polyarylene sulfide
polyphenylene sulfide

polysulfone
BT: synthetic organic polymer
NT: polyether sulfone

polytetrafluoroethylene
BT: synthetic organic polymer

polytetrafluoroethylene fabric
BT: manufactured fiber fabric

polytetrafluoroethylene fiber
UF: polytetrafluoroethylene fibre
BT: manufactured fiber

polytetrafluoroethylene fibre
USE: polytetrafluoroethylene fiber

polytetrafluoroethylene yarn
BT: manufactured fiber yarn

polytrimethylene terephthalate
UF: poly(trimethylene terephthalate)
BT: polyester

polyurethane
BT: synthetic organic polymer
NT: elastane

polyurethane fabric
BT: manufactured fiber fabric

polyurethane fiber
UF: polyurethane fibre
BT: manufactured fiber

polyurethane fibre
USE: polyurethane fiber

polyurethane yarn
BT: manufactured fiber yarn

polyvinyl acetate
BT: synthetic organic polymer
vinyl polymer

polyvinyl alcohol
BT: synthetic organic polymer
vinyl polymer

polyvinyl alcohol fabric
BT: manufactured fiber fabric

polyvinyl alcohol fiber
UF: polyvinyl alcohol fibre
BT: manufactured fiber

polyvinyl alcohol fibre
USE: polyvinyl alcohol fiber

polyvinyl alcohol yarn
BT: manufactured fiber yarn

polyvinyl chloride
BT: synthetic organic polymer
vinyl polymer

polyvinyl chloride fabric
BT: manufactured fiber fabric

polyvinyl chloride fiber
UF: polyvinyl chloride fibre
BT: manufactured fiber

polyvinyl chloride fibre
USE: polyvinyl chloride fiber

polyvinyl chloride yarn
BT: manufactured fiber yarn

polyvinylidene fluoride
BT: synthetic organic polymer
vinyl polymer

pore size
BT: size

porosity
BT: physical property

powder
BT: general materials by form

power consumption
BT: energy consumption

precursor
BT: chemical by property or purpose

prediction
BT: methodology
NT: color prediction
life prediction

preform
BT: fiber reinforced composite

preparation
BT: general manufacturing process
NT: fiber preparation
warp preparation
yarn preparation

prepreg
BT: fiber reinforced composite

press felt
BT: felt
machine clothing
papermakers fabric

press finishing
BT: finishing

presser foot
BT: machine component

pressing
BT: finishing
NT: ironing

pressing machine
BT: finishing machine
NT: ironing machine

pressure
BT: force
NT: air pressure
vacuum

pretreatment
BT: finishing

price
BT: market

printability
BT: physicochemical property by characteristic

printed fabric
BT: fabric by property or purpose

printing
BT: finishing
NT: devore
digital printing
direct printing
discharge printing
ink-jet printing
pigment printing
reactive printing
resist printing
rotary printing
screen printing
transfer printing

printing assistant
BT: additive
NT: carrier

printing industry
BT: industry

printing ink
BT: ink

printing machine
BT: finishing machine

printing paste
BT: general materials by property or purpose

process control
BT: control
NT: automation
dosing
metering
tension control

process engineering
BT: engineering

product
BT: Facet C: materials, products and resources
NT: product by item type
product by property or purpose
product component

product by item type
BT: product
NT: accessories
aerial
banner
battery separator
brush
filter
flag
fuel cell
furnishings
furniture
garment
garment hanger
kite

liner
mannequin
mop
pipe
rope
sewing thread
toy
transportation vehicle
wipe

product by property or purpose
BT: product
NT: aerospace textile
agricultural textile
architectural textile
artificial textile
automotive textile
beauty product
constructed textile
disposable product
geosynthetics
historical textile
horticultural textile
household textile
hygiene product
industrial textile
marine textile
medical textile
military textile
organic product
sports goods
sports surface
surgical product
technical textile
traditional textile

product component
BT: product
NT: backing
binding
fastener
garment component
hem
knot
label
openings
pleat
pocket
seal
seam
stitch
trim
vehicle component

product development
BT: research and development
NT: customization

production
BT: society and economics
NT: agricultural production
computer aided production
productivity

production control
BT: control
production management

production management
BT: management
NT: color management
down time
plant layout
production control
quality
quick response
right first time approach

productivity
BT: production

professional association profile
BT: publication

professional practise
BT: management

profitability
BT: society and economics

proofing
BT: finishing
NT: waterproofing

prosthesis
BT: medical textile

protease
BT: enzyme

protectionism
BT: trade

protective clothing
UF: protective garment
BT: garment by property or purpose
NT: body armor

protective fabric
BT: fabric by property or purpose

protective garment
USE: protective clothing

protein
BT: natural polyamide
NT: albumin
collagen
DNA
enzyme
fibroin
keratin
zein

psychology
BT: life science

publication
BT: communication medium
NT: annual report
buyers guide
exhibition review
professional association profile

puckering
BT: distortion

pull out
USE: pull-out

pull-out
UF: pull out
BT: mathematical expression

pulse jet
BT: machine component

pultrusion
BT: molding

pump
BT: equipment by item type

puncture resistance
BT: resistant property

purification
BT: chemical process
removal

pyrolysis
BT: chemical reaction

qualitative analysis
BT: analysis

quality
BT: production management
NT: quality assurance
quality control

quality assurance
BT: quality

quality control
BT: control
quality

quantitative analysis
BT: analysis

quenching
BT: cooling

quick response
BT: production management

quilt
BT: bedding

quilting
BT: fabric manufacture

quilting machine
BT: machine

rabbit
BT: animal
NT: angora

radiation
BT: energy
NT: infrared radiation
microwave radiation
radio frequency
ultraviolet radiation

radio frequency
BT: radiation

raising
BT: dry finishing
NT: napping

raising machine
BT: finishing machine
NT: napping machine

Raman spectroscopy
BT: spectroscopy

ramie
BT: plant

ramie fabric
BT: bast fiber fabric

ramie fiber
UF: ramie fibre
BT: bast fiber

ramie fibre
USE: ramie fiber

ramie yarn
BT: bast fiber yarn

rapier
BT: loom component

rapier loom
BT: shuttleless loom

raschel knitting
BT: knitting

raw material
BT: general materials by origin or composition
NT: raw silk

raw silk
BT: raw material

silk

rayon fiber
UF: rayon fibre
BT: regenerated cellulose fiber
NT: cupro fiber
lyocell fiber
modal fiber
viscose rayon fiber

rayon fiber fabric
BT: regenerated cellulose fabric
NT: lyocell fabric
modal fabric
viscose rayon fabric

rayon fiber yarn
BT: regenerated cellulose yarn
NT: lyocell yarn
modal yarn
viscose rayon yarn

rayon fibre
USE: rayon fiber

reactive dye
UF: remazol dye
BT: dye by application or reaction

reactive printing
BT: printing

reactor
BT: equipment by item type

recycling
BT: Facet B: operations

reducing agent
BT: additive

reed
BT: loom component

reeding
BT: weaving

reeling
BT: winding

reflective property
BT: optical property

refrigerant
BT: chemical by property or purpose

regenerated cellulose
BT: semisynthetic organic polymer
NT: cellulose acetate
cellulose xanthate
viscose rayon

regenerated cellulose fabric
BT: manufactured fiber fabric
NT: acetate fabric
rayon fiber fabric
triacetate fabric

regenerated cellulose fiber
UF: regenerated cellulose fibre
BT: manufactured fiber
NT: acetate fiber
rayon fiber
triacetate fiber

regenerated cellulose fibre
USE: regenerated cellulose fiber

regenerated cellulose yarn
BT: manufactured fiber yarn
NT: acetate yarn
rayon fiber yarn
triacetate yarn

regulation
UF: regulations
BT: management

regulations
USE: regulation

reinforced fabric
BT: fabric by property or purpose

reinforcement
BT: general materials by property or purpose
NT: soil reinforcement

relaxing
BT: distortion

relief fabric
BT: fabric by form

remazol dye
USE: reactive dye

removal
BT: separation
NT: afterclearing
dechlorination
decolorizing
defoliation
defuzzing
degumming
delignifying
demineralization
desalination
desizing
distillation
dye removal
extraction method
lint removal
oil removal
pollutant removal
purification

stain removal
stripping

renewable resource
BT: resource

repair
BT: Facet B: operations
NT: restoration
textile conservation

repeat
BT: design

reproducibility
BT: physicochemical property by characteristic

research (academic)
BT: Facet I: research and methodology

research and development
BT: Facet I: research and methodology
NT: consumer research
product development

resin
USE: synthetic resin

resin transfer molding
UF: resin transfer moulding
BT: molding

resin transfer moulding
USE: resin transfer molding

resist dyeing
BT: dyeing

resist printing
BT: printing

resistant property
BT: physicochemical property by characteristic
NT: antifelting property
antiyellowing property
ballistic resistance
barrier property
biological resistance
bullet resistance
chemical resistance
corrosion resistance
crease resistance
cut resistance
electrical resistance
electromagnetic resistance
flame resistance
heat resistance
hydrophobic property
impact resistance
oil resistance
pill resistance
puncture resistance
shrink resistance
slip resistance
soil resistance
stab resistance
stain resistance
tear resistance
wear resistance
weather resistance

resource
BT: Facet C: materials, products and resources
NT: renewable resource
sustainable resource

restoration
BT: repair

retailing
BT: service sector

retting
BT: fiber preparation

reverse osmosis
BT: diffusion

reversible garment
BT: garment by property or purpose

rewinding
BT: winding

rheology
BT: mechanics

ribbon
BT: woven fabric

right first time approach
BT: production management

ring
BT: machine component

ring spinning
BT: spinning

rinsing
BT: washing

risk assessment
BT: assessment

robotics
BT: technology

roller
BT: machine component

roofing
BT: architectural textile

rope
BT: product by item type

NT: cable
cord
lariat

rope dyeing
BT: dyeing

rope making
BT: manufacture

rotary printing
BT: printing

rotor
BT: machine component

rotor spinning
BT: spinning

roving
BT: yarn preparation

roving machine
BT: machine

royalties
USE: royalty

royalty
UF: royalties
BT: intellectual property

rubber
USE: synthetic rubber

rubber fabric
BT: fabric by origin or composition

rubber fiber
UF: rubber fibre
BT: fiber by origin or composition

rubber fibre
USE: rubber fiber

rubber yarn
BT: yarn by origin or composition

rug
BT: floor covering

s-twist
BT: twist

safety
BT: management
NT: consumer safety

sailcloth
BT: canvas
fabric by property or purpose

sales
BT: trade and commerce
NT: turnover

sample
BT: general materials by property or purpose

sampling
BT: methodology

sanitary napkin
BT: sanitary product

sanitary product
BT: hygiene product
NT: sanitary napkin

sari
BT: outerwear

saxony carpet
BT: carpet

scanning electron microscopy
BT: electron microscopy

scanning technology
BT: technology

science in general
BT: Facet G: art, science and technology
NT: morphology
terminology
theory

scouring
BT: washing

scouring machine
BT: washing machine

screen printing
BT: printing

scrim
BT: mesh

scutching
BT: opening

scutching machine
BT: opening machine

seal
BT: product component
NT: gasket

seam
BT: product component

seamless
BT: fabric construction
garment construction

seat
BT: furniture
NT: car seat

seat belt
BT: automotive textile
vehicle component

seat cover
BT: upholstery

seed
BT: biological structure or form

selective breeding
BT: agricultural industry

selvedge
BT: fabric component

semisynthetic organic polymer
BT: organic polymer
NT: polylactic acid
regenerated cellulose

sensor
BT: instrument

separation
BT: Facet B: operations
NT: decorticating
dialysis
filtration
flocculation
layer separation
opening
phase separation
removal

sequin
BT: trim

sericin
BT: animal material

sericulture
BT: agricultural industry

service sector
BT: industry
NT: customer service
financial service
retailing

sett
BT: fabric property

sewing
BT: general manufacturing process
NT: embroidering
gathering
hand sewing
overlock sewing
toe closing

sewing machine
BT: machine
NT: embroidery machine

sewing speed
BT: machine function
speed

sewing thread
BT: product by item type

shade variation
BT: optical phenomena

shape
BT: measure or dimension
NT: balloon (yarn)
body shape
cross-sectional shape

shape memory material
BT: technical textile

shear property
BT: mechanical property

shearing
BT: dry finishing

shearing machine
BT: finishing machine

sheath core structure
USE: sheath-core structure

sheath-core structure
UF: sheath core structure
BT: structural property

shed
BT: openings

sheep
BT: animal
NT: merino sheep

shingosen fabric
BT: synthetic fiber fabric

shirt
BT: outerwear
NT: T-shirt

shoulder pad
BT: garment component

shrink resistance
BT: resistant property

shrinkage
BT: dynamic or static phenomena

shrinking
BT: finishing

shrinking machine
BT: finishing machine

shuttle
BT: loom component

shuttle loom
BT: loom
NT: handloom

shuttleless loom
BT: loom
NT: air-jet loom
rapier loom
water-jet loom

silicon
BT: chemical element

silicon carbide fiber
UF: silicon carbide fibre
BT: ceramic fiber

silicon carbide fibre
USE: silicon carbide fiber

silicone
BT: inorganic polymer

silk
BT: animal material
NT: eri silk
muga silk
mulberry silk
raw silk
spider silk
tasar silk
tussah silk

silk blend
BT: animal fiber blend
NT: cotton-silk blend
nylon-silk blend
polyester-silk blend
silk-viscose-rayon blend
silk-wool blend

silk fabric
BT: animal fiber fabric

silk fiber
UF: silk fibre
BT: animal fiber

silk fibre
USE: silk fiber

silk industry
BT: animal fiber industry

silk viscose rayon blend
USE: silk-viscose-rayon blend

silk wool blend
USE: silk-wool blend

silk yarn
BT: animal fiber yarn

silk-viscose-rayon blend
UF: silk viscose rayon blend
BT: silk blend
viscose-rayon blend

silk-wool blend
UF: silk wool blend
BT: silk blend
wool blend

silkworm
BT: animal
NT: Bombyx mori

silver
BT: chemical element

simulation
BT: methodology
NT: centrifuge modeling
computer simulation
model

singeing
BT: dry finishing

singeing machine
BT: finishing machine

sinker
BT: machine component

sintering
BT: heat treatment

sisal
BT: plant

sisal fabric
BT: vegetable fiber fabric

sisal fiber
UF: sisal fibre
BT: vegetable fiber

sisal fibre
USE: sisal fiber

sisal yarn
BT: vegetable fiber yarn

size
BT: measure or dimension

NT: body size
particle size
pore size

sizing
UF: slashing
BT: warp preparation

sizing agent
UF: slashing agent
BT: finishing agent

sizing machine
UF: slashing machine
BT: machine

skin
BT: animal material
NT: leather
skin (human)
snakeskin

skin (human)
BT: skin

skirt
BT: outerwear

skiwear
BT: sportswear

slashing
USE: sizing

slashing agent
USE: sizing agent

slashing machine
USE: sizing machine

sleeping bag
BT: bedding

sleepwear
BT: garment by property or purpose

sleeve
BT: garment component

slip resistance
BT: resistant property

slit
BT: openings

slitting
BT: cutting

sliver
BT: fiber by form

sliver can
BT: machine component

sliver guide
BT: guiding system

smart fabric
BT: fabric by property or purpose

snakeskin
BT: skin

social history
BT: history
society and economics
NT: industrial revolution

social impact
BT: society and economics

social science
BT: Facet G: art, science and technology
NT: cultural anthropology
ethics
history

society and economics
BT: Facet A: industry and organization
NT: economic conditions
economic strategy
education
ethnicity
financial provision
health and welfare
investment
labor and employment
production
profitability
social history
social impact
trade and commerce
training

sock
BT: footwear
hosiery

softening agent
BT: finishing agent

software
BT: computer
NT: computer program

soil burial test
UF: soil test
BT: testing

soil reinforcement
BT: reinforcement

soil release
BT: chemical process

soil resistance

BT: resistant property

soil test
USE: soil burial test

solubility
BT: chemical property

solvent
BT: chemical by property or purpose

sorption
BT: chemical process
NT: absorption
adsorption

sourcing
BT: supply chain

space dyeing
BT: dyeing

spacer fabric
BT: fabric by property or purpose

spandex
BT: elastane

spandex fabric
BT: elastane fabric

spandex fiber
UF: spandex fibre
BT: elastane fiber

spandex fibre
USE: spandex fiber

spandex yarn
BT: elastane yarn

Spanish broom fiber
UF: Spanish broom fibre
BT: vegetable fiber

Spanish broom fibre
USE: Spanish broom fiber

spectrophotometry
BT: measuring

spectroscopy
BT: analysis
NT: electron spin resonance
Raman spectroscopy

speed
BT: dynamic or static phenomena
NT: sewing speed

spider silk
BT: silk

spin drawing
USE: spin-drawing

spin dyeing
BT: dyeing

spin finish
BT: lubricant

spin-drawing
UF: spin drawing
BT: drawing
spinning

spindle
BT: machine component

spinnability
BT: yarn property

spinneret
BT: machine

spinning
BT: yarn manufacture
NT: air-jet spinning
compact spinning
core spinning
direct spinning
friction spinning
open-end spinning
ring spinning
rotor spinning
spin-drawing
tow conversion
vortex spinning
wrap spinning

spinning industry
BT: textile-clothing industry

spinning machine
BT: machine

spinning solution
BT: general materials by property or purpose

splicing
BT: yarn manufacture

splint
BT: medical textile

spool
BT: machine component

sports goods
BT: product by property or purpose
NT: bicycle frame
fishing line

sports surface
BT: product by property or purpose

NT: artificial turf

sportswear
BT: garment by property or purpose
NT: activewear
skiwear

spray application
BT: application (finishing)

spray dyeing
BT: dyeing

spreading
BT: making-up

spreading machine
BT: finishing machine

spunbonding
BT: bonding
nonwoven fabric manufacture

spunbonding machine
BT: bonding machine
nonwoven fabric machine

sputter etching
BT: etching

squeegee
BT: applicator

stab resistance
BT: resistant property

stability
BT: dynamic or static phenomena
NT: dimensional stability
thermal stability

stabilizing agent
BT: additive

stain removal
BT: cleaning
removal

stain resistance
BT: resistant property

standard
BT: legislation
NT: benchmarking

staple fiber
UF: staple fibre
BT: fiber by form

staple fibre
USE: staple fiber

starch
BT: polysaccharide

static electricity
BT: electricity

steam
BT: water vapor

steaming
BT: heat treatment

steaming machine
BT: finishing machine

stencil
BT: equipment by item type

stentering
BT: dry finishing

stentering machine
BT: wet processing machine

sterilization
BT: Facet B: operations

stickiness
BT: adhesive property

stiffness
BT: mechanical property

stitch
BT: product component

stitch bonding
USE: stitch-bonding

stitch bonding machine
USE: stitch-bonding machine

stitch-bonding
UF: stitch bonding
BT: bonding
nonwoven fabric manufacture

stitch-bonding machine
UF: stitch bonding machine
BT: bonding machine
nonwoven fabric machine

stonewashing
BT: wet finishing

stop motion
BT: machine component

storage
BT: materials handling

strain
BT: deformation
NT: creep

streakiness
BT: appearance
defect

strength
BT: mechanical property
NT: breaking strength

stress
BT: force
NT: compression
fatigue
tension
torque

stress strain property
USE: stress-strain property

stress-strain property
UF: stress strain property
BT: mechanical property

stretch
BT: mechanical property

stripping
BT: finishing
removal

structural analysis
BT: analysis
NT: microstructural analysis

structural property
BT: physical property
NT: cord construction
fabric construction
fine structure
garment construction
loop
microstructure
sheath-core structure
yarn construction

stuffer box
BT: machine component

stuffing
BT: padding (material)

substrate
BT: general materials by property or purpose

suction device
BT: machine component

suit
BT: outerwear

sulfur dye
UF: sulphur dye
BT: dye by origin or composition

sulphur dye
USE: sulfur dye

superabsorbent material
BT: technical textile

supercritical carbon dioxide dyeing
BT: dyeing

supplier
BT: industrial enterprise
labor and employment

supply
BT: market

supply chain
BT: trade and commerce
NT: buying
distribution
outsourcing
sourcing

support garment
BT: garment by property or purpose
NT: abdominal support belt
brassiere
corset

surface active agent
USE: surface-active agent

surface modification
BT: modification
NT: patterning

surface property
BT: physicochemical property by characteristic

surface-active agent
UF: surface active agent
BT: additive

surgical product
BT: product by property or purpose

sustainability
BT: environmental phenomena

sustainable resource
BT: resource

suture
BT: medical textile

sweater
BT: outerwear

swelling agent
BT: additive

swimwear

BT: garment by property or purpose
NT: diving suit

syndiotactic polymer
BT: polymer by stereochemical configuration

synthesis
BT: chemical reaction

synthetic dye
BT: dye by origin or composition
NT: anthraquinone dye
azo dye
synthetic-organic dye

synthetic fiber
UF: synthetic fibre
BT: manufactured fiber

synthetic fiber fabric
UF: synthetic fibre fabric
BT: manufactured fiber fabric
NT: shingosen fabric

synthetic fiber industry
UF. synthetic fibre industry
BT: manufactured fiber industry

synthetic fiber yarn
UF: synthetic fibre yarn
BT: manufactured fiber yarn

synthetic fibre
USE: synthetic fiber

synthetic fibre fabric
USE: synthetic fiber fabric

synthetic fibre industry
USE: synthetic fiber industry

synthetic fibre yarn
USE: synthetic fiber yarn

synthetic organic dye
USE: synthetic-organic dye

synthetic organic polymer
BT: organic polymer
NT: acrylic polymer
modacrylic
polyaniline
polybenzimidazole
polybenzoxazole
polybutadiene
polydimethylsiloxane
polyester
polyether
polyethylene imine
polyimide
polyketone
polyolefin
polyphenol
polyphenylene
polypyrrole
polystyrene
polysuccinimide
polysulfide
polysulfone
polytetrafluoroethylene
polyurethane
polyvinyl acetate
polyvinyl alcohol
polyvinyl chloride
polyvinylidene fluoride
synthetic polyamide
synthetic polyisoprene

synthetic polyamide
UF: polyamide
BT: synthetic organic polymer
NT: aramid
nylon
nylon 11
nylon 4
nylon 46
nylon 6
nylon 66
polyphenylene terephthalamide

synthetic polyisoprene
BT: synthetic organic polymer
NT: synthetic rubber

synthetic resin
UF: resin
BT: thermosetting polymer
NT: epoxy resin

synthetic rubber
UF: rubber
BT: synthetic polyisoprene

synthetic-organic dye
UF: synthetic organic dye
BT: organic dye
synthetic dye

T shirt
USE: T-shirt

T-shirt
UF: T shirt
BT: shirt

tafetta fabric
BT: woven fabric

take down mechanism
USE: take-down mechanism

take up mechanism
USE: take-up mechanism

take-down mechanism
UF: take down mechanism

BT: machine component

take-up mechanism
UF: take up mechanism
BT: machine component

tanglelacing
BT: texturing

tape
BT: narrow fabric

tapestry
BT: textile art

tariff structure
BT: international trade

tarpaulin
BT: architectural textile

tasar silk
BT: silk

tear resistance
BT: resistant property

technical textile
BT: product by property or purpose
NT: phase-change material
shape memory material
superabsorbent material

technology
BT: Facet G: art, science and technology
NT: biotechnology
electronics
information technology
mechatronics
nanotechnology
robotics
scanning technology

temperature
BT: thermal property

temperature control
BT: control

temperature effect
BT: thermal phenomena

temple
BT: loom component

tenacity
BT: mechanical property

tensile property
BT: mechanical property

tension
BT: stress

tension control
BT: process control

tensioner
BT: machine component

tent
BT: architectural textile

terminology
BT: science in general

terpolymer
BT: polymer by monomer sequence

terry fabric
BT: pile fabric

testing
BT: methodology
NT: high-volume-instrument testing
loading (testing)
soil burial test

testing instrument
BT: instrument
NT: manikin

textile art
UF: art
BT: art, design and culture
NT: tapestry

textile clothing industry
USE: textile-clothing industry

textile conservation
BT: repair

textile history
BT: history

textile industry
BT: textile-clothing industry

textile machinery industry
BT: textile-clothing industry

textile-clothing industry
UF: textile clothing industry
BT: industry
NT: carpet industry
clothing industry
embroidery industry
knitting industry
nonwoven fabric industry
spinning industry
textile industry
textile machinery industry
weaving industry

texturing

BT: yarn manufacture
NT: deknitting
draw-texturing
false twist
tanglelacing

texturing machine
BT: machine
NT: crimping machine
draw-texturing machine

theoretical analysis
BT: analysis

theory
BT: science in general

thermal analysis
BT: analysis

thermal conductivity
BT: conductivity
thermal property

thermal expansion
BT: expansion

thermal insulation
BT: thermal phenomena

thermal phenomena
BT: physical phenomena
NT: cooling
heat recovery
heat transfer
temperature effect
thermal insulation
thermal property
thermodynamics

thermal property
BT: physical property
thermal phenomena
NT: absorption isotherm
adsorption isotherm
melting point
temperature
thermal conductivity
thermal stability
thermoplastic property
thermosetting property

thermal stability
BT: stability
thermal property

thermal treatment
BT: heat treatment

thermochromic dye
BT: dye by property or form

thermodynamics
BT: dynamic or static phenomena
mechanics
thermal phenomena

thermogravimetry
BT: measuring

thermoplastic fiber
UF: thermoplastic fibre
BT: fiber by property or purpose

thermoplastic fibre
USE: thermoplastic fiber

thermoplastic polymer
BT: polymer by physical property

thermoplastic property
BT: thermal property

thermosetting polymer
BT: polymer by physical property
NT: synthetic resin

thermosetting property
BT: thermal property

thickening agent
BT: additive

thickness
BT: measure or dimension
NT: count

thinning
BT: finishing

threading
BT: materials handling

three-dimensional textile
BT: constructed textile

tie
BT: garment accessory

tie dyeing
BT: dyeing

tire
UF: tyre
BT: vehicle component

tire cord
UF: tyre cord
BT: automotive textile
cord
vehicle component

tissue engineering
BT: engineering
medical process

titanium
BT: chemical element

titanium chloride
BT: inorganic compound

titanium dioxide
BT: oxide

titanium nitride
BT: inorganic compound

titanium oxide
BT: oxide

toe closing
BT: garment manufacture
making-up
sewing

toe closing machine
BT: machine

top dyeing
BT: dyeing

top making
BT: yarn manufacture

torque
BT: stress

toughness
BT: durability

tourniquet
BT: medical textile

tow
BT: fiber by form

tow conversion
BT: spinning

towel
BT: furnishings

toxicity
BT: health and disease

toy
BT: product by item type

trade
BT: trade and commerce
NT: international trade
protectionism

trade agreement
BT: international trade

trade and commerce
BT: society and economics
NT: branding
consumption
cost
counterfeiting
market
marketing
sales
supply chain
trade
trade organization

trade organization
BT: trade and commerce
NT: trade union

trade union
BT: trade organization

traditional textile
BT: product by property or purpose

train
BT: transportation vehicle

training
BT: society and economics

transfer printing
BT: printing
NT: heat transfer printing

transmission electron microscopy
BT: electron microscopy

transparency
BT: optical property

transport
BT: materials handling

transport process
BT: dynamic or static phenomena
NT: breathability
convection
diffusion
heat transfer
mass transfer
moisture transfer

transportation vehicle
BT: product by item type
NT: aircraft
train
vehicle component
vessel

trash
BT: waste

traveler
BT: machine component

traverse mechanism

BT: machine component

triacetate
BT: cellulose acetate

triacetate fabric
BT: regenerated cellulose fabric

triacetate fiber
UF: triacetate fibre
BT: regenerated cellulose fiber

triacetate fibre
USE: triacetate fiber

triacetate yarn
BT: regenerated cellulose yarn

tribology
BT: physics

trichromatic dye
BT: dye by property or form

trichromatic dyeing
BT: dyeing

trim
BT: product component
NT: bead
eyelet
sequin

trimming
BT: making-up

trousers
BT: outerwear
NT: jeans

tube
BT: machine component

tubular knitting machine
BT: knitting machine

tubular material
BT: constructed textile

tuft
BT: fabric component

tufting
BT: fabric manufacture

tufting machine
BT: machine

turbine
BT: equipment by item type

turnover
BT: sales

tussah silk
BT: silk

twist
BT: yarn property
NT: s-twist
z-twist

twisting
BT: yarn manufacture
NT: plying

twisting machine
BT: machine

tyre
USE: tire

tyre cord
USE: tire cord

ultrasonics
BT: physics

ultraviolet radiation
BT: radiation

ultraviolet resistance
BT: electromagnetic resistance

umbrella
BT: accessories

underwear
BT: garment by item type
NT: brassiere
corset
hosiery
lingerie

uniform
BT: outerwear

uniformity
BT: physical property

union dyeing
BT: dyeing

unwinding
BT: winding

upholstery
BT: furnishings
NT: seat cover

urban design
BT: design

use
BT: Facet B: operations
NT: asbestos replacement

end use
high temperature application
household use
industrial use
medical use
nuclear application
offshore application

vacuum
BT: pressure

vapor
UF: vapour
BT: gas

vapour
USE: vapor

vat dye
BT: dye by application or reaction

vat dyeing
BT: dyeing

vegetable dye
BT: natural dye

vegetable fiber
UF: vegetable fibre
BT: cellulosic fiber
NT: abaca fiber
agavaceae fiber
bamboo fiber
banana fiber
bast fiber
coir fiber
cotton fiber
peat fiber
pineapple leaf fiber
sisal fiber
Spanish broom fiber

vegetable fiber fabric
UF: vegetable fibre fabric
BT: cellulosic fiber fabric
NT: bast fiber fabric
coir fabric
cotton fabric
sisal fabric

vegetable fiber industry
BT: natural fiber industry
NT: cotton industry
flax industry
jute industry
linen industry

vegetable fiber yarn
UF: vegetable fibre yarn
BT: cellulosic fiber yarn
NT: bast fiber yarn
coir yarn
cotton yarn
sisal yarn

vegetable fibre
USE: vegetable fiber

vegetable fibre fabric
USE: vegetable fiber fabric

vegetable fibre yarn
USE: vegetable fiber yarn

vehicle component
BT: product component
transportation vehicle
NT: air bag
airfoil
brake (vehicle)
car seat
seat belt
tire
tire cord

velour fabric
BT: pile fabric

velvet fabric
BT: pile fabric

ventilation
BT: airflow
Facet B: operations

vessel
BT: transportation vehicle

vibration
BT: dynamic or static phenomena

vibration damping
BT: control

vicuna fiber
UF: vicuna fibre
BT: keratin fiber

vicuna fibre
USE: vicuna fiber

vinyl polymer
BT: polymer by chemical composition
NT: polyacrylonitrile
polymethyl methacrylate
polystyrene
polyvinyl acetate
polyvinyl alcohol
polyvinyl chloride
polyvinylidene fluoride

virus resistance
BT: biological resistance

viscoelasticity
BT: elasticity

viscose rayon
BT: regenerated cellulose

viscose rayon fabric
BT: rayon fiber fabric

viscose rayon fiber
UF: viscose rayon fibre
BT: rayon fiber

viscose rayon fibre
USE: viscose rayon fiber

viscose rayon wool blend
USE: viscose-rayon-wool blend

viscose rayon yarn
BT: rayon fiber yarn

viscose-rayon blend
BT: manufactured fiber blend
NT: acrylic-viscose-rayon blend
cotton-viscose-rayon blend
flax-viscose-rayon blend
linen-viscose-rayon blend
nylon-viscose-rayon blend
polyester-viscose-rayon blend
polypropylene-viscose-rayon blend
silk-viscose-rayon blend
viscose-rayon-wool blend

viscose-rayon-wool blend
UF: viscose rayon wool blend
BT: viscose-rayon blend
wool blend

viscosity
BT: mechanical property

visibility
BT: optical property

volatile organic compound
BT: organic compound

voluminizing
BT: finishing

vortex spinning
BT: spinning

wadding
BT: padding (material)

wader
BT: footwear

wage
BT: labor and employment

waistband
BT: garment component

wall covering
BT: furnishings

warp
BT: fabric component

warp knitting
BT: knitting

warp knitting machine
BT: knitting machine

warp preparation
BT: preparation
yarn manufacture
NT: sizing
warping

warp yarn
BT: yarn by property or purpose

warping
BT: warp preparation
NT: beaming

warping machine
BT: winding machine

washfastness
BT: dye property

washing
BT: wet processing
NT: afterwashing
conditioning
rinsing
scouring

washing machine
BT: wet processing machine
NT: scouring machine

waste
BT: general materials by property or purpose
NT: effluent
fly
lint
nep
trash
wastewater

waste containment
BT: waste management

waste disposal
BT: waste management

waste management
BT: environmental management
NT: waste containment
waste disposal

wastewater
BT: waste
water

water
BT: liquid
NT: wastewater
water vapor

water conservation
BT: environmental management

water consumption
BT: consumption

water jet
BT: jet

water jet loom
USE: water-jet loom

water pollution
BT: pollution

water resistance
BT: hydrophobic property

water treatment
BT: environmental treatment
NT: chlorination
dechlorination
decolorizing
desalination

water vapor
UF: water vapour
BT: water
NT: steam

water vapour
USE: water vapor

water-jet loom
UF: water jet loom
BT: shuttleless loom

waterproofing
BT: proofing

wave shed loom
USE: wave-shed loom

wave-shed loom
UF: wave shed loom
BT: loom

wax
BT: general materials by form

wear
BT: damage
NT: abrasion
pilling

wear resistance
BT: resistant property
NT: abrasion resistance

weather resistance
BT: resistant property

weavability
BT: yarn property

weave
BT: fabric construction

weaving
BT: fabric manufacture
NT: beating-up
drawing-in
greige weaving
hand weaving
leno
multiphase weaving
picking
reeding
weft insertion

weaving industry
BT: textile-clothing industry

web
BT: nonwoven fabric

web forming
BT: nonwoven fabric manufacture

web forming machine
BT: nonwoven fabric machine

webbing
BT: woven fabric

weft
BT: fabric component
NT: pick

weft insertion
BT: weaving

weft knitting
BT: knitting

weight
BT: measure or dimension
NT: add-on
molecular weight

welding
BT: bonding

welfare provision
BT: health and welfare

wet finishing

BT: finishing
NT: creping
decatizing
stonewashing

wet laying
BT: nonwoven fabric manufacture

wet processing
BT: general manufacturing process
NT: bleaching
boiling
degumming
desizing
deweighting
dyeing
impregnating
washing
wetting

wet processing machine
BT: machine
NT: bleaching machine
crabbing machine
dyeing machine
stentering machine
washing machine

wetcleaning
BT: cleaning

wetcleaning machine
BT: cleaning machine

wetting
BT: chemical process
wet processing

whisker
BT: fiber by form
NT: ceramic whisker

whitening
BT: bleaching

whole garment knitting
USE: whole-garment knitting

whole-garment knitting
UF: whole garment knitting
BT: garment manufacture
knitting

wicking
BT: chemical process

wig
BT: garment accessory
headwear

winch dyeing
BT: dyeing

winding
BT: yarn manufacture
NT: back-winding
lap forming
reeling
rewinding
unwinding

winding machine
BT: machine
NT: warping machine

wipe
BT: product by item type

womens employment
BT: employment

womenswear
BT: garment by user

wool
BT: animal material
NT: cashmere
fleece
lambswool
merino wool
yak wool

wool blend
BT: animal fiber blend
NT: acrylic-wool blend
cotton-wool blend
flax-wool blend
linen-wool blend
nylon-wool blend
polyester-wool blend
polypropylene-wool blend
silk-wool blend
viscose-rayon-wool blend

wool fabric
BT: keratin fiber fabric

wool fiber
UF: wool fibre
BT: keratin fiber

wool fibre
USE: wool fiber

wool grease
BT: animal material
NT: lanolin

wool industry
BT: animal fiber industry

wool yarn
BT: keratin fiber yarn

working conditions
BT: labor and employment

workwear
BT: garment by property or purpose

worsted fabric
BT: fabric by form

worsted yarn
BT: yarn by form

woven fabric
BT: fabric by form
NT: canvas
chenille fabric
denim fabric
double cloth
galloon
gauze
muslin
ribbon
tafetta fabric
webbing

wrap spinning
BT: spinning

X ray analysis
USE: X-ray analysis

X-ray analysis
UF: X ray analysis
Xray analysis
BT: image analysis

Xray analysis
USE: X-ray analysis

yak wool
BT: wool

yarn
BT: materials
NT: yarn by form
yarn by origin or composition
yarn by property or purpose

yarn brake
BT: braking system

yarn breakage
BT: breaking

yarn by form
BT: yarn
NT: BCF yarn
fancy yarn
knitted yarn
microfiber yarn
partially oriented yarn
plied yarn
worsted yarn

yarn by origin or composition
BT: yarn
NT: blended yarn
manufactured fiber yarn
natural fiber yarn
rubber yarn

yarn by property or purpose
BT: yarn
NT: carpet yarn
warp yarn

yarn construction
BT: structural property
yarn property

yarn diameter
BT: diameter
yarn property

yarn manufacture
BT: manufacture
NT: cabling
coiling
commingling
intermingling
spinning
splicing
texturing
top making
twisting
warp preparation
winding
yarn preparation

yarn modification
BT: modification

yarn preparation
BT: preparation
yarn manufacture
NT: carding
combing
drafting
roving

yarn property
BT: physicochemical property by material
NT: count
denier
knittability
linear density
spinnability
twist
weavability
yarn construction
yarn diameter

yellowing
BT: color change

yield
BT: agricultural production
NT: clip (yield)

Young modulus
BT: elastic modulus

zein fiber
UF: zein fibre
BT: natural fiber

zein fibre
USE: zein fiber

zeta potential
BT: mathematical expression

zinc oxide
BT: oxide

zipper
BT: fastener

THESAURUS TREE STRUCTURE

Facet A: industry and organization

. industry
. . agricultural industry
. . . agricultural production
. . . . crop production
. . . . livestock production
. . . . yield
. . . . . clip (yield)
. . . animal husbandry
. . . cultivation
. . . farming
. . . . organic farming
. . . harvesting
. . . . cotton picking
. . . pest and disease management
. . . . disease management
. . . . pest control
. . . selective breeding
. . . sericulture
. . composites industry
. . construction industry
. . drycleaning industry
. . dye manufacturing industry
. . dyeing industry
. . finishing industry
. . industrial enterprise
. . . buyer
. . . manufacturer
. . . supplier
. . laundering industry
. . manufactured fiber industry
. . . synthetic fiber industry
. . manufacturing
. . natural fiber industry
. . . animal fiber industry
. . . . cashmere industry
. . . . silk industry
. . . . wool industry
. . . vegetable fiber industry
. . . . cotton industry
. . . . flax industry
. . . . jute industry
. . . . linen industry
. . printing industry
. . service sector
. . . customer service
. . . financial service
. . . retailing
. . textile-clothing industry
. . . carpet industry
. . . clothing industry
. . . embroidery industry
. . . knitting industry
. . . nonwoven fabric industry
. . . spinning industry
. . . textile industry
. . . textile machinery industry
. . . weaving industry
. management
. . business strategy
. . . acquisition
. . . business merger
. . . business plan
. . . cost reduction
. . . joint venture
. . costing
. . environmental management
. . . emission reduction
. . . energy conservation
. . . environmental protection
. . . pollution control
. . . . noise control
. . . waste management
. . . . waste containment
. . . . waste disposal
. . . water conservation
. . forecasting
. . legislation
. . . intellectual property
. . . . copyright
. . . . patent
. . . . royalty
. . . standard
. . . . benchmarking
. . logistics
. . optimization
. . organizational structure
. . personnel management
. . planning
. . . business plan
. . . decision making
. . production management
. . . color management
. . . . color matching
. . . down time
. . . plant layout
. . . production control
. . . quality
. . . . quality assurance
. . . . quality control
. . . quick response
. . . right first time approach
. . professional practise
. . regulation
. . safety
. . . consumer safety
. society and economics
. . economic conditions
. . . boom
. . . globalization
. . economic strategy
. . education
. . ethnicity
. . financial provision
. . health and welfare
. . . health care
. . . occupational health
. . . welfare provision
. . investment
. . labor and employment
. . . buyer
. . . employment

. . . . womens employment
. . . labor
. . . . child labor
. . . . labor relations
. . . supplier
. . . wage
. . . working conditions
. . production
. . . agricultural production
. . . . crop production
. . . . livestock production
. . . . yield
. . . . . clip (yield)
. . . computer aided production
. . . productivity
. . profitability
. . social history
. . . industrial revolution
. . social impact
. . trade and commerce
. . . branding
. . . . eco-labeling
. . . consumption
. . . . energy consumption
. . . . . power consumption
. . . . water consumption
. . . cost
. . . counterfeiting
. . . market
. . . . competitiveness
. . . . demand
. . . . e-commerce
. . . . niche market
. . . . price
. . . . supply
. . . marketing
. . . . advertising
. . . . merchandizing
. . . sales
. . . . turnover
. . . supply chain
. . . . buying
. . . . distribution
. . . . outsourcing
. . . . sourcing
. . . trade
. . . . international trade
. . . . . dumping
. . . . . exports
. . . . . free trade
. . . . . imports
. . . . . tariff structure
. . . . . trade agreement
. . . . protectionism
. . . trade organization
. . . . trade union
. . training

Facet B: operations
. air conditioning
. control
. . color management
. . . color matching
. . computer control
. . electronic control
. . erosion control
. . moisture management
. . odor control
. . perspiration control
. . pest control
. . pollution control
. . . noise control
. . process control
. . . automation
. . . dosing
. . . metering
. . . tension control
. . production control
. . quality control
. . temperature control
. . vibration damping
. cooling
. . quenching
. drainage
. environmental treatment
. . aerobic treatment
. . anaerobic treatment
. . biological treatment
. . effluent treatment
. . ozone treatment
. . water treatment
. . . chlorination
. . . dechlorination
. . . decolorizing
. . . desalination
. heating
. inspection
. insulation
. irradiation
. lighting
. machine function
. . bobbin changing
. . can changing
. . clamping
. . doffing
. . let-off motion
. . needle penetration
. . needle selection
. . sewing speed
. maintenance
. manufacture
. . dye manufacture
. . fabric manufacture
. . . braiding
. . . knitting
. . . . casting off
. . . . circular knitting
. . . . crocheting
. . . . flat knitting
. . . . hand knitting
. . . . jacquard knitting
. . . . linking
. . . . narrowing
. . . . raschel knitting
. . . . warp knitting
. . . . weft knitting

. . . . whole-garment knitting
. . . lace making
. . . laminating
. . . layering
. . . nonwoven fabric manufacture
. . . . crosslapping
. . . . dry laying
. . . . . air laying
. . . . felting
. . . . hydroentangling
. . . . needling
. . . . spunbonding
. . . . stitch-bonding
. . . . web forming
. . . . wet laying
. . . quilting
. . . tufting
. . . weaving
. . . . beating-up
. . . . drawing-in
. . . . greige weaving
. . . . hand weaving
. . . . leno
. . . . multiphase weaving
. . . . picking
. . . . reeding
. . . . weft insertion
. . fiber manufacture
. . . blending
. . . drawing
. . . . autoleveling
. . . . draw-texturing
. . . . spin-drawing
. . . electrospinning
. . . extrusion
. . . . flash spinning
. . . . melt blowing
. . . . melt spinning
. . . fiber preparation
. . . . decorticating
. . . . defoliation
. . . . delignifying
. . . . ginning
. . . . hackling
. . . . lint removal
. . . . opening
. . . . . scutching
. . . . retting
. . . fibrillation
. . garment manufacture
. . . millinery
. . . toe closing
. . . whole-garment knitting
. . general manufacturing process
. . . aftertreatment
. . . . afterclearing
. . . . afterwashing
. . . bonding
. . . . fusing
. . . . spunbonding
. . . . stitch-bonding
. . . . welding
. . . comminuting
. . . compacting
. . . continuous process
. . . . continuous dyeing
. . . cutting
. . . . pattern cutting
. . . . slitting
. . . encapsulation
. . . fabrication
. . . finishing
. . . . abrading
. . . . acid treatment
. . . . . carbonizing
. . . . afterclearing
. . . . aging
. . . . . flash aging
. . . . alkali treatment
. . . . application (finishing)
. . . . . spray application
. . . . biofinishing
. . . . . biopolishing
. . . . . enzymatic treatment
. . . . . . cellulase treatment
. . . . bulking
. . . . cationizing
. . . . chemical vapor deposition
. . . . cleaning
. . . . . drycleaning
. . . . . stain removal
. . . . . wetcleaning
. . . . coarsening
. . . . coating
. . . . . back-coating
. . . . . edge coating
. . . . . plating
. . . . . . electroless plating
. . . . . . electroplating
. . . . creasing
. . . . curing
. . . . delustering
. . . . deodorizing
. . . . dry finishing
. . . . . crimping
. . . . . defuzzing
. . . . . emerizing
. . . . . grinding
. . . . . milling
. . . . . raising
. . . . . . napping
. . . . . shearing
. . . . . singeing
. . . . . stentering
. . . . electrolysis
. . . . etching
. . . . . chemical etching
. . . . . sputter etching
. . . . flocking
. . . . grafting
. . . . heat treatment
. . . . . annealing
. . . . . drying
. . . . . heat bonding
. . . . . heat setting
. . . . . . boarding

. . . . . . calendering
. . . . embossing
. . . . . . crabbing
. . . . . sintering
. . . . . steaming
. . . . . thermal treatment
. . . . laundering
. . . . . home laundering
. . . . . industrial laundering
. . . . loading (finishing)
. . . . mercerizing
. . . . metallization
. . . . padding (finishing)
. . . . plasma treatment
. . . . press finishing
. . . . pressing
. . . . . ironing
. . . . pretreatment
. . . . printing
. . . . . devore
. . . . . digital printing
. . . . . direct printing
. . . . . discharge printing
. . . . . ink-jet printing
. . . . . pigment printing
. . . . . reactive printing
. . . . . resist printing
. . . . . rotary printing
. . . . . screen printing
. . . . . transfer printing
. . . . . . heat transfer printing
. . . . proofing
. . . . . waterproofing
. . . . shrinking
. . . . stripping
. . . . thinning
. . . . voluminizing
. . . . wet finishing
. . . . . creping
. . . . . decatizing
. . . . . stonewashing
. . . knotting
. . . lubricating
. . . making-up
. . . . labeling
. . . . . eco-labeling
. . . . laying-up
. . . . lining
. . . . made to measure
. . . . marking
. . . . pattern cutting
. . . . piecing
. . . . spreading
. . . . toe closing
. . . . trimming
. . . materials selection
. . . mixing
. . . modification
. . . . chemical modification
. . . . fabric modification
. . . . fiber modification
. . . . genetic modification
. . . . surface modification
. . . . . patterning
. . . . yarn modification
. . . molding
. . . . injection molding
. . . . pultrusion
. . . . resin transfer molding
. . . one step process
. . . . one step dyeing
. . . open-width process
. . . . open-width dyeing
. . . preparation
. . . . fiber preparation
. . . . . decorticating
. . . . . defoliation
. . . . . delignifying
. . . . . ginning
. . . . . hackling
. . . . . lint removal
. . . . . opening
. . . . . . scutching
. . . . . retting
. . . . warp preparation
. . . . . sizing
. . . . . warping
. . . . . . beaming
. . . . yarn preparation
. . . . . carding
. . . . . . autoleveling
. . . . . combing
. . . . . drafting
. . . . . . gilling
. . . . . roving
. . . sewing
. . . . embroidering
. . . . gathering
. . . . hand sewing
. . . . overlock sewing
. . . . toe closing
. . . wet processing
. . . . bleaching
. . . . . brightening
. . . . . whitening
. . . . boiling
. . . . . kiering
. . . . degumming
. . . . . boiling-off
. . . . desizing
. . . . deweighting
. . . . dyeing
. . . . . backstaining
. . . . . batch dyeing
. . . . . batik
. . . . . beam dyeing
. . . . . chutki
. . . . . continuous dyeing
. . . . . dye fixing
. . . . . ikat dyeing
. . . . . jet dyeing
. . . . . jig dyeing
. . . . . kettle dyeing
. . . . . leveling
. . . . . mass dyeing
. . . . . melt dyeing

. . . . . mordant dyeing
. . . . . one step dyeing
. . . . . open-width dyeing
. . . . . pad dyeing
. . . . . pad-batch dyeing
. . . . . pad-steam dyeing
. . . . . piece dyeing
. . . . . resist dyeing
. . . . . rope dyeing
. . . . . space dyeing
. . . . . spin dyeing
. . . . . spray dyeing
. . . . . supercritical carbon dioxide dyeing
. . . . . tie dyeing
. . . . . top dyeing
. . . . . trichromatic dyeing
. . . . . union dyeing
. . . . . vat dyeing
. . . . . winch dyeing
. . . . impregnating
. . . . washing
. . . . . afterwashing
. . . . . conditioning
. . . . . rinsing
. . . . . scouring
. . . . wetting
. . glass manufacture
. . papermaking
. . rope making
. . yarn manufacture
. . . cabling
. . . coiling
. . . commingling
. . . intermingling
. . . spinning
. . . . air-jet spinning
. . . . compact spinning
. . . . core spinning
. . . . direct spinning
. . . . friction spinning
. . . . open-end spinning
. . . . ring spinning
. . . . rotor spinning
. . . . spin-drawing
. . . . tow conversion
. . . . vortex spinning
. . . . wrap spinning
. . . splicing
. . . texturing
. . . . deknitting
. . . . draw-texturing
. . . . false twist
. . . . tanglelacing
. . . top making
. . . twisting
. . . . plying
. . . warp preparation
. . . . sizing
. . . . warping
. . . . . beaming
. . . winding
. . . . back-winding
. . . . lap forming
. . . . reeling
. . . . rewinding
. . . . unwinding
. . . yarn preparation
. . . . carding
. . . . . autoleveling
. . . . combing
. . . . drafting
. . . . . gilling
. . . . roving
. materials handling
. . agitation
. . bale plucking
. . batching
. . folding
. . storage
. . threading
. . transport
. medical process
. . bone repair
. . drug delivery
. . tissue engineering
. monitoring
. . on-line monitoring
. recycling
. repair
. . restoration
. . textile conservation
. separation
. . decorticating
. . dialysis
. . filtration
. . . dust filtration
. . . fluid filtration
. . . . blood filtration
. . . . gas filtration
. . . . . air filtration
. . flocculation
. . layer separation
. . opening
. . . scutching
. . phase separation
. . removal
. . . afterclearing
. . . dechlorination
. . . decolorizing
. . . defoliation
. . . defuzzing
. . . degumming
. . . . boiling-off
. . . delignifying
. . . demineralization
. . . desalination
. . . desizing
. . . distillation
. . . dye removal
. . . extraction method
. . . lint removal
. . . oil removal
. . . pollutant removal
. . . purification
. . . stain removal
. . . stripping

. sterilization
. use
. . asbestos replacement
. . end use
. . high temperature application
. . household use
. . industrial use
. . medical use
. . nuclear application
. . offshore application
. ventilation

Facet C: materials, products and resources
. chemical
. . chemical by form
. . . monomer
. . . nanotube
. . . polymer
. . . . polymer by chemical composition
. . . . . fluoropolymer
. . . . . . perfluoropolymer
. . . . . inorganic polymer
. . . . . . polysiloxane
. . . . . . silicone
. . . . . organic polymer
. . . . . . biopolymer
. . . . . . natural organic polymer
. . . . natural polyamide
. . . . polypeptide
. . . . protein
. . . . albumin
. . . . collagen
. . . . DNA
. . . . enzyme
. . . . protease
. . . . fibroin
. . . . keratin
. . . . zein
. . . . natural polyisoprene
. . . . natural rubber
. . . . polyamine
. . . . polysaccharide
. . . . alginate
. . . . alginic acid
. . . . cellulose
. . . . chitin
. . . . chitosan
. . . . starch
. . . . . . semisynthetic organic polymer
. . . . polylactic acid
. . . . polylactide
. . . . regenerated cellulose
. . . . cellulose acetate
. . . . triacetate
. . . . cellulose xanthate
. . . . viscose rayon
. . . . . . synthetic organic polymer
. . . . acrylic polymer
. . . . polyacrylamide
. . . . polyacrylate
. . . . polyacrylic acid
. . . . polyacrylonitrile
. . . . polymethyl methacrylate
. . . . modacrylic
. . . . polyaniline
. . . . polybenzimidazole
. . . . polybenzoxazole
. . . . polyphenylene benzobisoxazole
. . . . polyphenylene benzobisthiazole
. . . . polybutadiene
. . . . polydimethylsiloxane
. . . . polyester
. . . . polybutylene terephthalate
. . . . polycaprolactone
. . . . polycarbonate
. . . . polyethylene naphthalate
. . . . polyethylene terephthalate
. . . . polypropylene terephthalate
. . . . polytrimethylene terephthalate
. . . . polyether
. . . . polyarylether ketone
. . . . polyether ether ketone
. . . . polyether imide
. . . . polyether ketone
. . . . polyether sulfone
. . . . polyethylene glycol
. . . . polyphenylene oxide
. . . . polyethylene imine
. . . . polyimide
. . . . polyether imide
. . . . polyketone
. . . . polyarylether ketone
. . . . polyether ether ketone
. . . . polyether ketone
. . . . polyolefin
. . . . polybutylene
. . . . polyethylene
. . . . polypropylene
. . . . polyphenol
. . . . polyphenylene
. . . . polypyrrole
. . . . polystyrene
. . . . polysuccinimide
. . . . polysulfide
. . . . polyarylene sulfide
. . . . polyphenylene sulfide
. . . . polysulfone
. . . . polyether sulfone
. . . . polytetrafluoroethylene
. . . . polyurethane
. . . . elastane
. . . . spandex
. . . . polyvinyl acetate
. . . . polyvinyl alcohol
. . . . polyvinyl chloride
. . . . polyvinylidene fluoride
. . . . synthetic polyamide
. . . . aramid
. . . . nylon
. . . . nylon 11
. . . . nylon 4
. . . . nylon 46
. . . . nylon 6
. . . . nylon 66
. . . . polyphenylene terephthalamide

. . . . synthetic polyisoprene
. . . . synthetic rubber
. . . . . vinyl polymer
. . . . . . polyacrylonitrile
. . . . . . polymethyl methacrylate
. . . . . . polystyrene
. . . . . . polyvinyl acetate
. . . . . . polyvinyl alcohol
. . . . . . polyvinyl chloride
. . . . . . polyvinylidene fluoride
. . . . polymer by monomer sequence
. . . . . copolymer
. . . . . . copolymer by structure
. . . . block copolymer
. . . . graft copolymer
. . . . . homopolymer
. . . . . oligomer
. . . . . terpolymer
. . . . polymer by physical property
. . . . . elastomer
. . . . . thermoplastic polymer
. . . . . thermosetting polymer
. . . . . . synthetic resin
. . . . epoxy resin
. . . . polymer by polymerization reaction
. . . . . addition polymer
. . . . . condensation polymer
. . . . polymer by stereochemical configuration
. . . . . atactic polymer
. . . . . isotactic polymer
. . . . . syndiotactic polymer
. . chemical by property or purpose
. . . acid
. . . additive
. . . . absorber
. . . . activator
. . . . adsorber
. . . . . activated carbon
. . . . afterclearing agent
. . . . antifibrillating agent
. . . . antifoaming agent
. . . . antioxidant
. . . . binding agent
. . . . bonding agent
. . . . . adhesive agent
. . . . buffering agent
. . . . catalyst
. . . . cationizing agent
. . . . complexing agent
. . . . cross linking agent
. . . . dispersant
. . . . dyeing assistant
. . . . . accelerating agent
. . . . . carrier
. . . . . coupling agent
. . . . . fixing agent
. . . . . leveling agent
. . . . . mordant
. . . . finishing agent
. . . . . abrasive agent
. . . . . bleaching agent
. . . . . bleaching assistant
. . . . . brightening agent
. . . . . coating agent
. . . . . conditioning agent
. . . . . delustering agent
. . . . . oxidizing agent
. . . . . sizing agent
. . . . . softening agent
. . . . inhibitor
. . . . plasticizing agent
. . . . printing assistant
. . . . . carrier
. . . . reducing agent
. . . . stabilizing agent
. . . . surface-active agent
. . . . swelling agent
. . . . thickening agent
. . . alkali
. . . detergent
. . . . conditioning agent
. . . dye
. . . . dye by application or reaction
. . . . . acid dye
. . . . . anionic dye
. . . . . azo dye
. . . . . basic dye
. . . . . cationic dye
. . . . . chrome dye
. . . . . direct dye
. . . . . disperse dye
. . . . . reactive dye
. . . . . vat dye
. . . . dye by origin or composition
. . . . . indigo dye
. . . . . metal-complex dye
. . . . . natural dye
. . . . . . vegetable dye
. . . . . organic dye
. . . . . . synthetic-organic dye
. . . . . sulfur dye
. . . . . synthetic dye
. . . . . . anthraquinone dye
. . . . . . azo dye
. . . . . . synthetic-organic dye
. . . . dye by property or form
. . . . . leuco dye
. . . . . photochromic dye
. . . . . thermochromic dye
. . . . . trichromatic dye
. . . fertilizer
. . . finishes
. . . . coatings
. . . lubricant
. . . . spin finish
. . . pesticide
. . . . fungicide
. . . . herbicide
. . . pigment
. . . precursor
. . . refrigerant
. . . solvent
. . chemical element or compound
. . . chemical compound
. . . . fluorochemical
. . . . inorganic compound

. . . . . ammonia
. . . . . ammonium carbonate
. . . . . boric acid
. . . . . copper sulfate
. . . . . hydrogen peroxide
. . . . . oxide
. . . . . . carbon dioxide
. . . . . . titanium dioxide
. . . . . . titanium oxide
. . . . . . zinc oxide
. . . . . titanium chloride
. . . . . titanium nitride
. . . . organic compound
. . . . . acetate
. . . . . acrylonitrile
. . . . . alcohol
. . . . . alkoxylate
. . . . . amine
. . . . . aromatic amine
. . . . . caprolactam
. . . . . formaldehyde
. . . . . hydrocarbon
. . . . . . perchloroethylene
. . . . . lipid
. . . . . organic acid
. . . . . . acrylic acid
. . . . . . amino acid
. . . . . . polycarboxylic acid
. . . . . volatile organic compound
. . . chemical element
. . . . boron
. . . . carbon
. . . . nickel
. . . . oxygen
. . . . silicon
. . . . silver
. . . . titanium
. materials
. . blend
. . . manufactured fiber blend
. . . . acrylic blend
. . . . . acrylic-cotton blend
. . . . . acrylic-polyester blend
. . . . . acrylic-viscose-rayon blend
. . . . . acrylic-wool blend
. . . . aramid blend
. . . . . aramid-nylon blend
. . . . . aramid-polyester blend
. . . . nylon blend
. . . . . aramid-nylon blend
. . . . . nylon-polyester blend
. . . . . nylon-polyethylene blend
. . . . . nylon-polypropylene blend
. . . . . nylon-silk blend
. . . . . nylon-viscose-rayon blend
. . . . . nylon-wool blend
. . . . polyester blend
. . . . . acrylic-polyester blend
. . . . . aramid-polyester blend
. . . . . cotton-polyester blend
. . . . . flax-polyester blend
. . . . . linen-polyester blend
. . . . . nylon-polyester blend
. . . . . polyester-polyethylene blend
. . . . . polyester-polypropylene blend
. . . . . polyester-silk blend
. . . . . polyester-viscose-rayon blend
. . . . . polyester-wool blend
. . . . polyolefin blend
. . . . . cotton-polypropylene blend
. . . . . nylon-polyethylene blend
. . . . . nylon-polypropylene blend
. . . . . polyester-polyethylene blend
. . . . . polyester-polypropylene blend
. . . . . polyethylene-polypropylene blend
. . . . . polypropylene-viscose-rayon blend
. . . . . polypropylene-wool blend
. . . . viscose-rayon blend
. . . . . acrylic-viscose-rayon blend
. . . . . cotton-viscose-rayon blend
. . . . . flax-viscose-rayon blend
. . . . . linen-viscose-rayon blend
. . . . . nylon-viscose-rayon blend
. . . . . polyester-viscose-rayon blend
. . . . . polypropylene-viscose-rayon blend
. . . . . silk-viscose-rayon blend
. . . . . viscose-rayon-wool blend
. . . natural fiber blend
. . . . animal fiber blend
. . . . . silk blend
. . . . . . cotton-silk blend
. . . . . . nylon-silk blend
. . . . . . polyester-silk blend
. . . . . . silk-viscose-rayon blend
. . . . . . silk-wool blend
. . . . . wool blend
. . . . . . acrylic-wool blend
. . . . . . cotton-wool blend
. . . . . . flax-wool blend
. . . . . . linen-wool blend
. . . . . . nylon-wool blend
. . . . . . polyester-wool blend
. . . . . . polypropylene-wool blend
. . . . . . silk-wool blend
. . . . . . viscose-rayon-wool blend
. . . . cellulosic fiber blend
. . . . . cotton blend
. . . . . . acrylic-cotton blend
. . . . . . cotton-flax blend
. . . . . . cotton-linen blend
. . . . . . cotton-polyester blend
. . . . . . cotton-polypropylene blend
. . . . . . cotton-silk blend
. . . . . . cotton-viscose-rayon blend
. . . . . . cotton-wool blend
. . . . . flax blend
. . . . . . cotton-flax blend
. . . . . . flax-hemp blend
. . . . . . flax-jute blend
. . . . . . flax-polyester blend
. . . . . . flax-viscose-rayon blend
. . . . . . flax-wool blend
. . . . . hemp blend
. . . . . . flax-hemp blend
. . . . . jute blend
. . . . . . flax-jute blend

. . . . . linen blend
. . . . . . cotton-linen blend
. . . . . . linen-polyester blend
. . . . . . linen-viscose-rayon blend
. . . . . . linen-wool blend
. . . polymer blend
. . fabric
. . . fabric by form
. . . . braid
. . . . knitted fabric
. . . . . crocheted fabric
. . . . lace fabric
. . . . . bobbin lace
. . . . . lace
. . . . laminate
. . . . mesh
. . . . . gauze
. . . . . net
. . . . . scrim
. . . . microfiber fabric
. . . . multiaxial fabric
. . . . nonwoven fabric
. . . . . batting
. . . . . bonded fiber fabric
. . . . . felt
. . . . . . press felt
. . . . . needled fabric
. . . . . web
. . . . pile fabric
. . . . . corduroy fabric
. . . . . terry fabric
. . . . . velour fabric
. . . . . velvet fabric
. . . . relief fabric
. . . . worsted fabric
. . . . woven fabric
. . . . . canvas
. . . . . . sailcloth
. . . . . chenille fabric
. . . . . denim fabric
. . . . . double cloth
. . . . . galloon
. . . . . gauze
. . . . . muslin
. . . . . ribbon
. . . . . tafetta fabric
. . . . . webbing
. . . fabric by origin or composition
. . . . blended fabric
. . . . manufactured fiber fabric
. . . . . acrylic fabric
. . . . . alginate fabric
. . . . . carbon fiber fabric
. . . . . . graphite fiber fabric
. . . . . ceramic fiber fabric
. . . . . chitin fabric
. . . . . chitosan fabric
. . . . . elastane fabric
. . . . . . spandex fabric
. . . . . elastomeric fabric
. . . . . glass fiber fabric
. . . . . metal fiber fabric
. . . . . mineral fiber fabric
. . . . . . asbestos fabric
. . . . . . basalt fabric
. . . . . modacrylic fabric
. . . . . polyamide fabric
. . . . . . aramid fabric
. . . . . . nylon fabric
. . . . . polyester fabric
. . . . . . polyethylene terephthalate fabric
. . . . . polyether ether ketone fabric
. . . . . polyimide fabric
. . . . . polyolefin fabric
. . . . . . polyethylene fabric
. . . . . . polypropylene fabric
. . . . . polytetrafluoroethylene fabric
. . . . . polyurethane fabric
. . . . . polyvinyl alcohol fabric
. . . . . polyvinyl chloride fabric
. . . . . regenerated cellulose fabric
. . . . . . acetate fabric
. . . . . . rayon fiber fabric
. . . . lyocell fabric
. . . . modal fabric
. . . . viscose rayon fabric
. . . . . . triacetate fabric
. . . . . synthetic fiber fabric
. . . . . . shingosen fabric
. . . . natural fiber fabric
. . . . . animal fiber fabric
. . . . . . keratin fiber fabric
. . . . alpaca fabric
. . . . angora fabric
. . . . cashmere fabric
. . . . mohair fabric
. . . . wool fabric
. . . . . . silk fabric
. . . . . cellulosic fiber fabric
. . . . . . vegetable fiber fabric
. . . . bast fiber fabric
. . . . flax fiber fabric
. . . . linen fabric
. . . . hemp fabric
. . . . jute fabric
. . . . kenaf fabric
. . . . ramie fabric
. . . . coir fabric
. . . . cotton fabric
. . . . sisal fabric
. . . . rubber fabric
. . . fabric by property or purpose
. . . . apparel fabric
. . . . backing fabric
. . . . barrier fabric
. . . . chintz
. . . . coated fabric
. . . . crepe fabric
. . . . dyed fabric
. . . . . batik
. . . . filter fabric
. . . . furnishing fabric
. . . . interfacing
. . . . interlining
. . . . moire fabric
. . . . multilayer fabric

. . . . narrow fabric
. . . . . tape
. . . . papermakers fabric
. . . . . press felt
. . . . printed fabric
. . . . protective fabric
. . . . reinforced fabric
. . . . sailcloth
. . . . smart fabric
. . . . spacer fabric
. . . fabric component
. . . . openings
. . . . . perforation
. . . . . shed
. . . . . slit
. . . . pile
. . . . . nap
. . . . selvedge
. . . . tuft
. . . . warp
. . . . weft
. . . . . pick
. . fiber
. . . fiber by form
. . . . bicomponent fiber
. . . . fibrid
. . . . fibril
. . . . filament
. . . . . hollow filament
. . . . . microfilament
. . . . . monofilament
. . . . . multifilament
. . . . . plexifilament
. . . . hollow fiber
. . . . microfiber
. . . . nanofiber
. . . . noil
. . . . sliver
. . . . staple fiber
. . . . tow
. . . . whisker
. . . . . ceramic whisker
. . . fiber by origin or composition
. . . . blended fiber
. . . . manufactured fiber
. . . . . acrylic fiber
. . . . . alginate fiber
. . . . . basalt fiber
. . . . . boron fiber
. . . . . . boron-nitride fiber
. . . . . carbon fiber
. . . . . . graphite fiber
. . . . . ceramic fiber
. . . . . . silicon carbide fiber
. . . . . chitin fiber
. . . . . chitosan fiber
. . . . . elastane fiber
. . . . . . spandex fiber
. . . . . elastomeric fiber
. . . . . glass fiber
. . . . . melamine resin fiber
. . . . . metal fiber
. . . . . . aluminum fiber
. . . . . mineral fiber
. . . . . . alumina fiber
. . . . . . aluminosilicate fiber
. . . . . . asbestos fiber
. . . . . modacrylic fiber
. . . . . novolac fiber
. . . . . polyamide fiber
. . . . . . aramid fiber
. . . . . . nylon fiber
. . . . . polybenzazole fiber
. . . . . polyester fiber
. . . . . . polybutylene terephthalate fiber
. . . . . . polyethylene terephthalate fiber
. . . . . polyether ether ketone fiber
. . . . . polyimide fiber
. . . . . polylactic acid fiber
. . . . . polyolefin fiber
. . . . . . polyethylene fiber
. . . . . . polypropylene fiber
. . . . . polyphenylene sulfide fiber
. . . . . polytetrafluoroethylene fiber
. . . . . polyurethane fiber
. . . . . polyvinyl alcohol fiber
. . . . . polyvinyl chloride fiber
. . . . . regenerated cellulose fiber
. . . . . . acetate fiber
. . . . . . rayon fiber
. . . . cupro fiber
. . . . lyocell fiber
. . . . modal fiber
. . . . polynosic fiber
. . . . viscose rayon fiber
. . . . . . triacetate fiber
. . . . . synthetic fiber
. . . . natural fiber
. . . . . animal fiber
. . . . . . keratin fiber
. . . . alpaca fiber
. . . . angora fiber
. . . . cashmere fiber
. . . . mohair fiber
. . . . vicuna fiber
. . . . wool fiber
. . . . . . silk fiber
. . . . . cellulosic fiber
. . . . . . vegetable fiber
. . . . abaca fiber
. . . . agavaceae fiber
. . . . furcraea fiber
. . . . bamboo fiber
. . . . banana fiber
. . . . bast fiber
. . . . flax fiber
. . . . linen
. . . . hemp fiber
. . . . hibiscus fiber
. . . . jute fiber
. . . . kenaf fiber
. . . . ramie fiber
. . . . coir fiber
. . . . cotton fiber
. . . . peat fiber
. . . . pineapple leaf fiber

. . . . sisal fiber
. . . . Spanish broom fiber
. . . . . zein fiber
. . . . organic fiber
. . . . rubber fiber
. . . fiber by property or purpose
. . . . binder fiber
. . . . carpet fiber
. . . . high-modulus fiber
. . . . piezoelectric fiber
. . . . thermoplastic fiber
. . general materials
. . . general materials by form
. . . . colloid
. . . . . aerosol
. . . . . emulsion
. . . . film
. . . . fluid
. . . . . gas
. . . . . . air
. . . . . . vapor
. . . . . liquid
. . . . . . liquid crystal
. . . . . . moisture
. . . . . . oil
. . . . . . water
. . . . wastewater
. . . . water vapor
. . . . steam
. . . . foam
. . . . gel
. . . . . hydrogel
. . . . liposome
. . . . liquor
. . . . membrane
. . . . particle
. . . . powder
. . . . wax
. . . general materials by origin or composition
. . . . animal material
. . . . . feather
. . . . . fiber medulla
. . . . . hair
. . . . . . camel hair
. . . . . . dog hair
. . . . . . goat hair
. . . . . . horsehair
. . . . . . human hair
. . . . . . mohair
. . . . . sericin
. . . . . silk
. . . . . . eri silk
. . . . . . muga silk
. . . . . . mulberry silk
. . . . . . raw silk
. . . . . . spider silk
. . . . . . tasar silk
. . . . . . tussah silk
. . . . . skin
. . . . . . leather
. . . . . . skin (human)
. . . . . . snakeskin
. . . . . wool
. . . . . . cashmere
. . . . . . fleece
. . . . . . lambswool
. . . . . . merino wool
. . . . . . yak wool
. . . . . wool grease
. . . . . . lanolin
. . . . ceramics
. . . . composite
. . . . . fiber reinforced composite
. . . . . . preform
. . . . . . prepreg
. . . . . matrix
. . . . . . ceramic matrix
. . . . . nanocomposite
. . . . concrete
. . . . cotton wool
. . . . geological material
. . . . . asbestos
. . . . . basalt
. . . . glass
. . . . metal
. . . . . alloy
. . . . mineral wool
. . . . plant material
. . . . . coir
. . . . . cork
. . . . . latex
. . . . . natural resin
. . . . . natural rubber
. . . . . plant extract
. . . . plastic
. . . . . acrylic
. . . . raw material
. . . . . raw silk
. . . general materials by property or purpose
. . . . biomimetic material
. . . . contaminant
. . . . . dust
. . . . . . mote
. . . . . foreign body
. . . . . impurity
. . . . . . honeydew
. . . . dye mixture
. . . . electret
. . . . fuel
. . . . handling pack
. . . . . bale
. . . . . fiber bundle
. . . . . hank
. . . . . lap
. . . . . package
. . . . ink
. . . . . printing ink
. . . . insulating material
. . . . packaging
. . . . padding (material)
. . . . . fiberfill
. . . . . stuffing
. . . . . wadding
. . . . polymer melt
. . . . polymer solution
. . . . printing paste

. . . . reinforcement
. . . . . soil reinforcement
. . . . sample
. . . . spinning solution
. . . . substrate
. . . . waste
. . . . . effluent
. . . . . fly
. . . . . lint
. . . . . nep
. . . . . trash
. . . . . wastewater
. . yarn
. . . yarn by form
. . . . BCF yarn
. . . . fancy yarn
. . . . . boucle yarn
. . . . . chenille yarn
. . . . knitted yarn
. . . . microfiber yarn
. . . . partially oriented yarn
. . . . plied yarn
. . . . worsted yarn
. . . yarn by origin or composition
. . . . blended yarn
. . . . manufactured fiber yarn
. . . . . acrylic yarn
. . . . . alginate yarn
. . . . . basalt yarn
. . . . . carbon fiber yarn
. . . . . . graphite fiber yarn
. . . . . ceramic fiber yarn
. . . . . chitin yarn
. . . . . chitosan yarn
. . . . . elastane yarn
. . . . . . spandex yarn
. . . . . elastomeric yarn
. . . . . glass fiber yarn
. . . . . metal fiber yarn
. . . . . mineral fiber yarn
. . . . . . asbestos yarn
. . . . . modacrylic yarn
. . . . . polyamide yarn
. . . . . . aramid yarn
. . . . . . nylon yarn
. . . . . polyester yarn
. . . . . . polyethylene terephthalate yarn
. . . . . polyether ether ketone yarn
. . . . . polyimide yarn
. . . . . polyolefin yarn
. . . . . . polyethylene yarn
. . . . . . polypropylene yarn
. . . . . polytetrafluoroethylene yarn
. . . . . polyurethane yarn
. . . . . polyvinyl alcohol yarn
. . . . . polyvinyl chloride yarn
. . . . . regenerated cellulose yarn
. . . . . . acetate yarn
. . . . . . rayon fiber yarn
. . . . lyocell yarn
. . . . modal yarn
. . . . viscose rayon yarn
. . . . . . triacetate yarn
. . . . . synthetic fiber yarn
. . . . natural fiber yarn
. . . . . animal fiber yarn
. . . . . . keratin fiber yarn
. . . . alpaca yarn
. . . . angora yarn
. . . . cashmere yarn
. . . . mohair yarn
. . . . wool yarn
. . . . . . silk yarn
. . . . . cellulosic fiber yarn
. . . . . . vegetable fiber yarn
. . . . bast fiber yarn
. . . . flax fiber yarn
. . . . linen yarn
. . . . hemp yarn
. . . . jute yarn
. . . . kenaf yarn
. . . . ramie yarn
. . . . coir yarn
. . . . cotton yarn
. . . . sisal yarn
. . . . rubber yarn
. . . yarn by property or purpose
. . . . carpet yarn
. . . . warp yarn
. product
. . product by item type
. . . accessories
. . . . bag
. . . . garment accessory
. . . . . belt
. . . . . . abdominal support belt
. . . . . glove
. . . . . mask
. . . . . tie
. . . . . wig
. . . . luggage
. . . . umbrella
. . . aerial
. . . banner
. . . battery separator
. . . brush
. . . filter
. . . . cigarette filter
. . . . dust filter
. . . . fluid filter
. . . . . blood filter
. . . . . gas filter
. . . . . . air filter
. . . flag
. . . fuel cell
. . . furnishings
. . . . bedding
. . . . . blanket
. . . . . mattress
. . . . . quilt
. . . . . sleeping bag
. . . . blinds
. . . . curtain
. . . . . net curtain
. . . . cushion
. . . . floor covering

. . . . . carpet
. . . . . . axminster carpet
. . . . . . carpet backing
. . . . . . carpet tile
. . . . . . saxony carpet
. . . . . mat
. . . . . . kilim
. . . . . rug
. . . . towel
. . . . upholstery
. . . . . seat cover
. . . . wall covering
. . . furniture
. . . . seat
. . . . . car seat
. . . garment
. . . . garment by item type
. . . . . footwear
. . . . . . gaiter
. . . . . . sock
. . . . . . wader
. . . . . garment accessory
. . . . . . belt
. . . . abdominal support belt
. . . . . . glove
. . . . . . mask
. . . . . . tie
. . . . . . wig
. . . . . headwear
. . . . . . mask
. . . . . . wig
. . . . . outerwear
. . . . . . bib
. . . . . . dress
. . . . . . jacket
. . . . . . sari
. . . . . . shirt
. . . . T-shirt
. . . . . . skirt
. . . . . . suit
. . . . . . sweater
. . . . . . trousers
. . . . jeans
. . . . . . uniform
. . . . . underwear
. . . . . . brassiere
. . . . . . corset
. . . . . . hosiery
. . . . sock
. . . . . . lingerie
. . . . garment by property or purpose
. . . . . casualwear
. . . . . knitwear
. . . . . maternitywear
. . . . . protective clothing
. . . . . . body armor
. . . . . reversible garment
. . . . . sleepwear
. . . . . sportswear
. . . . . . activewear
. . . . . . skiwear
. . . . . support garment
. . . . . . abdominal support belt
. . . . . . brassiere
. . . . . . corset
. . . . . swimwear
. . . . . . diving suit
. . . . . workwear
. . . . garment by user
. . . . . babieswear
. . . . . childrenswear
. . . . . menswear
. . . . . womenswear
. . . . garment component
. . . . . collar
. . . . . cuff
. . . . . gusset
. . . . . insole
. . . . . interfacing
. . . . . interlining
. . . . . shoulder pad
. . . . . sleeve
. . . . . waistband
. . . garment hanger
. . . kite
. . . liner
. . . . clay liner
. . . . diskette liner
. . . mannequin
. . . mop
. . . pipe
. . . rope
. . . . cable
. . . . cord
. . . . . tire cord
. . . . lariat
. . . sewing thread
. . . toy
. . . transportation vehicle
. . . . aircraft
. . . . train
. . . . vehicle component
. . . . . air bag
. . . . . airfoil
. . . . . brake (vehicle)
. . . . . car seat
. . . . . seat belt
. . . . . tire
. . . . . tire cord
. . . . vessel
. . . wipe
. . product by property or purpose
. . . aerospace textile
. . . . hot-air balloon
. . . . parachute
. . . agricultural textile
. . . architectural textile
. . . . awning
. . . . canopy
. . . . emergency shelter
. . . . roofing
. . . . tarpaulin
. . . . tent
. . . artificial textile
. . . . artificial fur
. . . . artificial hair

. . . . artificial leather
. . . . artificial turf
. . . automotive textile
. . . . air bag
. . . . seat belt
. . . . tire cord
. . . beauty product
. . . constructed textile
. . . . three-dimensional textile
. . . . tubular material
. . . disposable product
. . . geosynthetics
. . . . clay liner
. . . . geocell
. . . . geogrid
. . . . geomembrane
. . . . geonet
. . . . geotextile
. . . historical textile
. . . horticultural textile
. . . household textile
. . . hygiene product
. . . . dental floss
. . . . diaper
. . . . incontinence pad
. . . . sanitary product
. . . . . sanitary napkin
. . . industrial textile
. . . marine textile
. . . medical textile
. . . . bandage
. . . . blood filter
. . . . brace (medical textile)
. . . . dressing
. . . . implant
. . . . orthosis
. . . . prosthesis
. . . . splint
. . . . suture
. . . . tourniquet
. . . military textile
. . . . camouflage
. . . organic product
. . . sports goods
. . . . bicycle frame
. . . . fishing line
. . . sports surface
. . . . artificial turf
. . . surgical product
. . . technical textile
. . . . phase-change material
. . . . shape memory material
. . . . superabsorbent material
. . . traditional textile
. . product component
. . . backing
. . . . backing fabric
. . . . carpet backing
. . . binding
. . . fastener
. . . . buttonhole
. . . . zipper
. . . garment component
. . . . collar
. . . . cuff
. . . . gusset
. . . . insole
. . . . interfacing
. . . . interlining
. . . . shoulder pad
. . . . sleeve
. . . . waistband
. . . hem
. . . knot
. . . label
. . . openings
. . . . perforation
. . . . shed
. . . . slit
. . . pleat
. . . pocket
. . . . air pocket
. . . seal
. . . . gasket
. . . seam
. . . stitch
. . . trim
. . . . bead
. . . . eyelet
. . . . sequin
. . . vehicle component
. . . . air bag
. . . . airfoil
. . . . brake (vehicle)
. . . . car seat
. . . . seat belt
. . . . tire
. . . . tire cord
. resource
. . renewable resource
. . sustainable resource

Facet D: phenomena

. behavioral or psychological phenomena
. . color perception
. . comfort
. . decision making
. . esthetic property
. . gender identity
. biological phenomena
. . biological characteristic
. . . body shape
. . . body size
. . . breed
. . . maturity
. . biological property
. . . bioactive property
. . . biocompatibility
. . . biodegradability
. . fouling
. . health and disease
. . . biohazard
. . . carcinogenicity
. . . disease
. . . . allergy
. . . toxicity

. chemical phenomena
. . chemical process
. . . coagulating
. . . controlled release
. . . corrosion
. . . coupling
. . . cross linking
. . . decomposition
. . . degradation
. . . . enzymatic degradation
. . . . photochemical degradation
. . . desalination
. . . desorption
. . . dispersion
. . . dissolution
. . . distillation
. . . dye migration
. . . dye penetration
. . . dye uptake
. . . exhaustion
. . . . dye exhaustion
. . . ion exchange
. . . kinetics
. . . nucleation
. . . odor emitting
. . . phase separation
. . . purification
. . . soil release
. . . sorption
. . . . absorption
. . . . adsorption
. . . wetting
. . . wicking
. . chemical property
. . . absorbency
. . . adsorbency
. . . chemical composition
. . . . formula
. . . concentration
. . . crystalline orientation
. . . . degree of orientation
. . . crystallinity
. . . flammability
. . . hydrophilic property
. . . molecular structure
. . . molecular weight
. . . oleophilic property
. . . pH
. . . polymer property
. . . . degree of polymerization
. . . solubility
. . chemical reaction
. . . acetylation
. . . amination
. . . aminolysis
. . . chlorination
. . . deacetylation
. . . dechlorination
. . . electrolysis
. . . hydrolysis
. . . oxidation
. . . photolysis
. . . polymerization
. . . . copolymerization
. . . pyrolysis
. . . synthesis
. environmental phenomena
. . contamination
. . environmental impact
. . indoor air quality
. . pollution
. . . air pollution
. . . water pollution
. . sustainability
. mathematical phenomena
. . mathematical analysis
. . . finite element analysis
. . . Fourier transform analysis
. . . fractal analysis
. . . fuzzy logic
. . mathematical expression
. . . absorption isotherm
. . . add-on
. . . adsorption isotherm
. . . birefringence
. . . blend ratio
. . . constant
. . . . elastic modulus
. . . . . Young modulus
. . . . initial modulus
. . . draw ratio
. . . friction coefficient
. . . micronaire value
. . . moisture regain
. . . pull-out
. . . zeta potential
. . measure or dimension
. . . bulk
. . . crease recovery
. . . cross section
. . . degree of orientation
. . . degree of polymerization
. . . density
. . . . denier
. . . . linear density
. . . diameter
. . . . fiber diameter
. . . . yarn diameter
. . . fiber curvature
. . . flow rate
. . . glass transition temperature
. . . length
. . . . fiber length
. . . . gauge length
. . . shape
. . . . balloon (yarn)
. . . . body shape
. . . . cross-sectional shape
. . . size
. . . . body size
. . . . particle size
. . . . pore size
. . . thickness
. . . . count
. . . weight
. . . . add-on

. . . . molecular weight
. physical phenomena
. . adhesion
. . . fiber-matrix adhesion
. . damage
. . . aging
. . . . flash aging
. . . fracture
. . . . breaking
. . . . . end breakage
. . . . . yarn breakage
. . . . cracking
. . . wear
. . . . abrasion
. . . . pilling
. . disintegration
. . . debonding
. . . delamination
. . . fraying
. . distortion
. . . bending
. . . buckling
. . . cockling
. . . necking
. . . puckering
. . . relaxing
. . dynamic or static phenomena
. . . condensation
. . . elongation
. . . . breaking elongation
. . . energy
. . . . radiation
. . . . . infrared radiation
. . . . . microwave radiation
. . . . . radio frequency
. . . . . ultraviolet radiation
. . . expansion
. . . . hygral expansion
. . . . thermal expansion
. . . flotation
. . . flow
. . . . fiber flow
. . . . flow rate
. . . . fluid flow
. . . . . gas flow
. . . . . . airflow
. . . . ventilation
. . . . . jet
. . . . . . air jet
. . . . . . water jet
. . . fluid dynamics
. . . . aerodynamics
. . . . computational fluid dynamics
. . . . hydrodynamics
. . . force
. . . . centrifugal force
. . . . friction
. . . . impact
. . . . pressure
. . . . . air pressure
. . . . . vacuum
. . . . stress
. . . . . compression
. . . . . fatigue
. . . . . tension
. . . . . torque
. . . inflation
. . . kinetics
. . . shrinkage
. . . speed
. . . . sewing speed
. . . stability
. . . . dimensional stability
. . . . thermal stability
. . . thermodynamics
. . . transport process
. . . . breathability
. . . . convection
. . . . diffusion
. . . . . reverse osmosis
. . . . heat transfer
. . . . mass transfer
. . . . moisture transfer
. . . . . moisture release
. . . vibration
. . electrical phenomena
. . . electrical property
. . . . dielectric property
. . . . electrical conductivity
. . . electricity
. . . . static electricity
. . optical phenomena
. . . color change
. . . . fading
. . . . metamerism
. . . . yellowing
. . . frosting
. . . optical property
. . . . fluorescence
. . . . holographic property
. . . . iridescence
. . . . luminescence
. . . . luster
. . . . reflective property
. . . . transparency
. . . . visibility
. . . shade variation
. . physical property
. . . acoustic property
. . . adhesive property
. . . . stickiness
. . . airtight
. . . anisotropy
. . . appearance
. . . . color
. . . . esthetic property
. . . . evenness
. . . . fabric appearance
. . . . garment appearance
. . . . hairiness
. . . . pattern (configuration)
. . . . . applique
. . . . . embroidery
. . . . . . beadwork
. . . . . . hand embroidery
. . . . . inlay

. . . . . patchwork
. . . . streakiness
. . . breathability
. . . buoyancy
. . . conductivity
. . . . electrical conductivity
. . . . thermal conductivity
. . . defect
. . . . bar
. . . . barre
. . . . streakiness
. . . electrical property
. . . . dielectric property
. . . . electrical conductivity
. . . handle
. . . . hairiness
. . . humidity
. . . inflatable
. . . loft
. . . machinability
. . . mechanical property
. . . . bending property
. . . . breaking point
. . . . compressive property
. . . . crease recovery
. . . . deformation
. . . . . elongation
. . . . . . breaking elongation
. . . . . strain
. . . . . . creep
. . . . drape
. . . . durability
. . . . . toughness
. . . . elastic property
. . . . . elastic modulus
. . . . . . Young modulus
. . . . . elastic recovery
. . . . . elasticity
. . . . . . viscoelasticity
. . . . elastoplasticity
. . . . extensibility
. . . . flexibility
. . . . shear property
. . . . stiffness
. . . . strength
. . . . . breaking strength
. . . . stress-strain property
. . . . stretch
. . . . tenacity
. . . . tensile property
. . . . viscosity
. . . moisture content
. . . optical property
. . . . fluorescence
. . . . holographic property
. . . . iridescence
. . . . luminescence
. . . . luster
. . . . reflective property
. . . . transparency
. . . . visibility
. . . permeability
. . . porosity
. . . structural property
. . . . cord construction
. . . . fabric construction
. . . . . bias
. . . . . honeycomb construction
. . . . . pitch
. . . . . seamless
. . . . . weave
. . . . fine structure
. . . . garment construction
. . . . . seamless
. . . . loop
. . . . microstructure
. . . . . microstructural property
. . . . sheath-core structure
. . . . yarn construction
. . . thermal property
. . . . absorption isotherm
. . . . adsorption isotherm
. . . . melting point
. . . . temperature
. . . . thermal conductivity
. . . . thermal stability
. . . . thermoplastic property
. . . . thermosetting property
. . . uniformity
. . physicomechanical property
. . thermal phenomena
. . . cooling
. . . . quenching
. . . heat recovery
. . . heat transfer
. . . temperature effect
. . . thermal insulation
. . . thermal property
. . . . absorption isotherm
. . . . adsorption isotherm
. . . . melting point
. . . . temperature
. . . . thermal conductivity
. . . . thermal stability
. . . . thermoplastic property
. . . . thermosetting property
. . . thermodynamics
. physicochemical property
. . physicochemical property by characteristic
. . . dyeability
. . . easy care property
. . . glass transition temperature
. . . high performance
. . . launderability
. . . printability
. . . reproducibility
. . . resistant property
. . . . antifelting property
. . . . antiyellowing property
. . . . ballistic resistance
. . . . barrier property
. . . . biological resistance
. . . . . antifungal property
. . . . . antimicrobial property
. . . . . insect resistance
. . . . . . moth resistance

. . . . . virus resistance
. . . . bullet resistance
. . . . chemical resistance
. . . . . chlorine resistance
. . . . corrosion resistance
. . . . crease resistance
. . . . cut resistance
. . . . electrical resistance
. . . . . antistatic property
. . . . electromagnetic resistance
. . . . . infrared resistance
. . . . . ultraviolet resistance
. . . . flame resistance
. . . . heat resistance
. . . . hydrophobic property
. . . . . water resistance
. . . . impact resistance
. . . . oil resistance
. . . . pill resistance
. . . . puncture resistance
. . . . shrink resistance
. . . . slip resistance
. . . . soil resistance
. . . . stab resistance
. . . . stain resistance
. . . . tear resistance
. . . . wear resistance
. . . . . abrasion resistance
. . . . weather resistance
. . . surface property
. . physicochemical property by material
. . . composite property
. . . . fiber-matrix adhesion
. . . . fiber-matrix interface
. . . . interfacial property
. . . . interlaminar property
. . . dye property
. . . . colorfastness
. . . . crockfastness
. . . . dye levelness
. . . . dye recipe
. . . . lightfastness
. . . . washfastness
. . . fabric property
. . . . fabric appearance
. . . . fabric construction
. . . . . bias
. . . . . honeycomb construction
. . . . . pitch
. . . . . seamless
. . . . . weave
. . . . moire
. . . . sett
. . . fiber property
. . . . birefringence
. . . . crimp
. . . . fiber arrangement
. . . . . fiber orientation
. . . . . . degree of orientation
. . . . fiber curvature
. . . . fiber diameter
. . . . fiber distribution
. . . . fiber fineness
. . . . fiber length
. . . . filament property
. . . . linear density
. . . . micronaire value
. . . garment property
. . . . fit
. . . . garment appearance
. . . . garment construction
. . . . . seamless
. . . yarn property
. . . . count
. . . . denier
. . . . knittability
. . . . linear density
. . . . spinnability
. . . . twist
. . . . . s-twist
. . . . . z-twist
. . . . weavability
. . . . yarn construction
. . . . yarn diameter

Facet E: equipment, facilities and structures

. equipment
. . equipment by item type
. . . applicator
. . . . squeegee
. . . autoclave
. . . bath
. . . . dye bath
. . . boiler
. . . computer
. . . . hardware
. . . . neural network
. . . . software
. . . . . computer program
. . . condenser
. . . fixed bed
. . . fluidized bed
. . . instrument
. . . . detector
. . . . sensor
. . . . testing instrument
. . . . . manikin
. . . laser
. . . machine
. . . . bale plucking machine
. . . . baling machine
. . . . batching machine
. . . . blending machine
. . . . bonding machine
. . . . . spunbonding machine
. . . . . stitch-bonding machine
. . . . braiding machine
. . . . carding machine
. . . . combing machine
. . . . . hackling machine
. . . . compacting machine
. . . . cotton picking machine
. . . . cutting machine
. . . . decorticating machine
. . . . dosing machine
. . . . drafting machine

. . . . . gilling machine
. . . . drawing machine
. . . . drawing-in machine
. . . . finishing machine
. . . . . abrading machine
. . . . . calendering machine
. . . . . cleaning machine
. . . . . . drycleaning machine
. . . . . . wetcleaning machine
. . . . . coating machine
. . . . . decatizing machine
. . . . . drying machine
. . . . . . mangle
. . . . . dyeing-finishing machine
. . . . . emerizing machine
. . . . . flocking machine
. . . . . heat setting machine
. . . . . . boarding machine
. . . . . laundering machine
. . . . . mercerizing machine
. . . . . pressing machine
. . . . . . ironing machine
. . . . . printing machine
. . . . . raising machine
. . . . . . napping machine
. . . . . shearing machine
. . . . . shrinking machine
. . . . . singeing machine
. . . . . spreading machine
. . . . . steaming machine
. . . . folding machine
. . . . ginning machine
. . . . inspection machine
. . . . knitting machine
. . . . . cabling machine
. . . . . crochet machine
. . . . . linking machine
. . . . . tubular knitting machine
. . . . . warp knitting machine
. . . . labeling machine
. . . . lace machine
. . . . laminating machine
. . . . loom
. . . . . carpet loom
. . . . . dobby loom
. . . . . jacquard loom
. . . . . loom component
. . . . . . dobby
. . . . . . gripper
. . . . . . jacquard
. . . . . . rapier
. . . . . . reed
. . . . . . shuttle
. . . . . . temple
. . . . . shuttle loom
. . . . . . handloom
. . . . . shuttleless loom
. . . . . . air-jet loom
. . . . . . rapier loom
. . . . . . water-jet loom
. . . . . wave-shed loom
. . . . machine component
. . . . . alignment mechanism
. . . . . antitwist device
. . . . . antiwrap device
. . . . . apron
. . . . . balance tube
. . . . . balloon limiter
. . . . . beam
. . . . . bearings
. . . . . blade
. . . . . bobbin
. . . . . bobbin holder
. . . . . braking system
. . . . . . yarn brake
. . . . . brush
. . . . . calender bowl
. . . . . cam
. . . . . chute
. . . . . comb
. . . . . cone
. . . . . creel
. . . . . cutter
. . . . . cylinder
. . . . . . doffer
. . . . . . licker-in
. . . . . die
. . . . . drawframe
. . . . . drive
. . . . . drop wire
. . . . . drum
. . . . . embroidery frame
. . . . . filter
. . . . . . cigarette filter
. . . . . . dust filter
. . . . . . fluid filter
. . . . blood filter
. . . . gas filter
. . . . air filter
. . . . . fourdrinier wire
. . . . . gasket
. . . . . gear
. . . . . godet
. . . . . guide bar
. . . . . harness
. . . . . heald
. . . . . heald frame
. . . . . hook
. . . . . housing
. . . . . jack
. . . . . jet
. . . . . . air jet
. . . . . . water jet
. . . . . knife
. . . . . lifting device
. . . . . loom component
. . . . . . dobby
. . . . . . gripper
. . . . . . jacquard
. . . . . . rapier
. . . . . . reed
. . . . . . shuttle
. . . . . . temple
. . . . . machine clothing
. . . . . . card clothing
. . . . . . press felt

. . . . . magnet
. . . . . mandrel
. . . . . motor
. . . . . mounting
. . . . . navel
. . . . . needle
. . . . . . knitting needle
. . . . . needle bar
. . . . . needle bed
. . . . . nip
. . . . . nipper
. . . . . nozzle
. . . . . pirn
. . . . . presser foot
. . . . . pulse jet
. . . . . ring
. . . . . roller
. . . . . rotor
. . . . . sinker
. . . . . sliver can
. . . . . spindle
. . . . . spool
. . . . . stop motion
. . . . . stuffer box
. . . . . suction device
. . . . . take-down mechanism
. . . . . take-up mechanism
. . . . . tensioner
. . . . . traveler
. . . . . traverse mechanism
. . . . . tube
. . . . nonwoven fabric machine
. . . . . crosslapping machine
. . . . . melt blowing machine
. . . . . needling machine
. . . . . spunbonding machine
. . . . . stitch-bonding machine
. . . . . web forming machine
. . . . opening machine
. . . . . bale opening machine
. . . . . scutching machine
. . . . papermaking machine
. . . . quilting machine
. . . . roving machine
. . . . sewing machine
. . . . . embroidery machine
. . . . sizing machine
. . . . spinneret
. . . . spinning machine
. . . . texturing machine
. . . . . crimping machine
. . . . . draw-texturing machine
. . . . toe closing machine
. . . . tufting machine
. . . . twisting machine
. . . . wet processing machine
. . . . . bleaching machine
. . . . . crabbing machine
. . . . . dyeing machine
. . . . . . pad
. . . . . stentering machine
. . . . . washing machine
. . . . . . scouring machine
. . . . winding machine
. . . . . warping machine
. . . pump
. . . reactor
. . . stencil
. . . turbine
. . equipment by property or purpose
. . . color matching system
. . . control system
. . . cooling device
. . . data management system
. . . electromagnetic device
. . . materials handling system
. . . . conveyor belt
. . . . guiding system
. . . . . feed
. . . . . guide bar
. . . . . sliver guide
. . . measuring system
. . . . dosing system
. . . monitoring system
. . . pattern preparation system
. . . piezoelectric device
. . . pneumatic device
. facility or structure
. . bridge
. . cleanroom
. . mill
. . museum

Facet F: organisms and biological structures

. biological structure or form
. . cocoon
. . seed
. organism
. . animal
. . . alpaca
. . . bollworm
. . . maggot
. . . rabbit
. . . . angora
. . . sheep
. . . . merino sheep
. . . silkworm
. . . . Bombyx mori
. . microorganism
. . . bacterium
. . plant
. . . cotton
. . . flax
. . . hemp
. . . indigo
. . . jute
. . . kapok
. . . kenaf
. . . mulberry
. . . ramie
. . . sisal

Facet G: art, science and technology

. applied science
. . agriculture
. . . agronomy

. . engineering
. . . civil engineering
. . . process engineering
. . . tissue engineering
. . ergonomics
. art, design and culture
. . costume
. . design
. . . computer aided design
. . . customization
. . . interior design
. . . pattern (plan)
. . . repeat
. . . urban design
. . fashion
. . handicraft
. . textile art
. . . tapestry
. life science
. . ecology
. . psychology
. physical science
. . chemistry
. . . electrochemistry
. . . polymer science
. . physics
. . . cryogenics
. . . mechanics
. . . . fluid mechanics
. . . . . fluid dynamics
. . . . . . aerodynamics
. . . . . . computational fluid dynamics
. . . . . . hydrodynamics
. . . . . hydraulics
. . . . kinetics
. . . . rheology
. . . . thermodynamics
. . . tribology
. . . ultrasonics
. science in general
. . morphology
. . terminology
. . theory
. social science
. . cultural anthropology
. . ethics
. . history
. . . social history
. . . . industrial revolution
. . . textile history
. technology
. . biotechnology
. . . genetic modification
. . electronics
. . information technology
. . . fiber optics
. . mechatronics
. . nanotechnology
. . robotics
. . scanning technology

Facet H: information and communication
. communication
. . business communication
. . color communication
. . communication medium
. . . communication event
. . . . conference
. . . . . conference (AATCC)
. . . . . conference (ATIRA)
. . . . . conference (Beltwide Cotton)
. . . . . conference (BTRA)
. . . . . conference (CIRCOT)
. . . . . conference (COTTECH)
. . . . . conference (DWI)
. . . . . conference (IAF)
. . . . . conference (IFVTCC)
. . . . . conference (International Man-made Fibre)
. . . . . conference (International Silk Association)
. . . . . conference (International Wool Research)
. . . . . conference (ITMF)
. . . . . conference (Material World)
. . . . . conference (Medical Textiles)
. . . . . conference (NITRA)
. . . . . conference (SAMPE)
. . . . . conference (SDC)
. . . . . conference (SITRA)
. . . . . conference (TIFCON)
. . . . . conference (TIWC)
. . . . exhibition
. . . . . exhibition (ATME-I)
. . . . . exhibition (BITMEX)
. . . . . exhibition (Bobbin Show)
. . . . . exhibition (CAD Expo)
. . . . . exhibition (CITME)
. . . . . exhibition (Domotex)
. . . . . exhibition (Expo Detergo)
. . . . . exhibition (Expofil)
. . . . . exhibition (Heimtextil)
. . . . . exhibition (IDEA)
. . . . . exhibition (IFAI)
. . . . . exhibition (IHE)
. . . . . exhibition (IKME)
. . . . . exhibition (IMB)
. . . . . exhibition (INDA)
. . . . . exhibition (INDA-TEC)
. . . . . exhibition (INDEX)
. . . . . exhibition (Interfloor)
. . . . . exhibition (ITMA)
. . . . . exhibition (ITME-I)
. . . . . exhibition (JEC)
. . . . . exhibition (JIAM)
. . . . . exhibition (SEAMS)
. . . . . exhibition (Techtextil)
. . . . . exhibition (Texitech)
. . . media
. . . . internet
. . . publication
. . . . annual report
. . . . buyers guide
. . . . exhibition review
. . . . professional association profile
. . global communication
. information
. . company information
. . database

. . directory information
. . economic information
. . . economic data
. . information processing
. . . classification
. . . classing
. . . data processing
. . . . electronic data interchange
. . information technology
. . . fiber optics

Facet I: research and methodology
. methodology
. . age determination
. . analysis
. . . chemical analysis
. . . . amino acid analysis
. . . . chromatography
. . . failure analysis
. . . image analysis
. . . . magnetic resonance imaging
. . . . X-ray analysis
. . . mathematical analysis
. . . . finite element analysis
. . . . Fourier transform analysis
. . . . fractal analysis
. . . . fuzzy logic
. . . microscopy
. . . . electron microscopy
. . . . . scanning electron microscopy
. . . . . transmission electron microscopy
. . . qualitative analysis
. . . quantitative analysis
. . . spectroscopy
. . . . electron spin resonance
. . . . Raman spectroscopy
. . . structural analysis
. . . . microstructural analysis
. . . theoretical analysis
. . . thermal analysis
. . assessment
. . . objective assessment
. . . performance assessment
. . . risk assessment
. . calculation
. . calibration
. . color matching
. . digital method
. . identification method
. . measuring
. . . anthropometry
. . . . body scanning
. . . calorimetry
. . . colorimetry
. . . spectrophotometry
. . . thermogravimetry
. . prediction
. . . color prediction
. . . life prediction
. . sampling
. . simulation
. . . centrifuge modeling
. . . computer simulation
. . . model
. . . . mathematical model
. . testing
. . . high-volume-instrument testing
. . . loading (testing)
. . . . biaxial loading
. . . soil burial test
. research (academic)
. research and development
. . consumer research
. . product development
. . . customization

USE OF SUBJECT INDEXING PER YEAR

	1999	2000	2001	2002	2003	2004
abaca fiber	1	0	0	1	0	0
abdominal support belt	3	1	1	0	3	3
abrading machine	0	1	0	1	0	1
abrading	2	4	6	1	4	2
abrasion resistance	29	40	32	33	41	22
abrasion	11	20	14	13	16	10
abrasive agent	3	3	2	0	0	0
absorbency	133	123	104	134	117	114
absorber	28	27	13	9	10	8
absorption isotherm	1	0	0	0	1	0
absorption	42	27	52	47	33	28
accelerating agent	4	1	3	3	5	0
accessories	31	22	12	9	13	5
acetate fabric	2	0	0	0	0	1
acetate fiber	8	3	0	1	0	0
acetate yarn	2	0	0	0	0	0
acetate	0	1	9	1	1	0
acetylation	2	3	4	3	2	1
acid dye	28	34	31	34	44	28
acid treatment	7	20	19	12	10	9
acid	0	0	0	0	6	9
acoustic property	25	20	22	30	17	12
acoustics	0	0	1	0	0	0
acquisition	1	34	38	54	28	30
acrylic-cotton blend	0	1	1	0	5	0
acrylic-polyester blend	0	0	2	0	1	0
acrylic-viscose-rayon blend	0	0	1	0	0	0
acrylic-wool blend	0	2	4	0	1	0
acrylic acid	1	3	12	15	11	6
acrylic fabric	4	3	6	6	5	5
acrylic fiber	30	29	29	22	25	17
acrylic yarn	6	11	5	5	6	5
acrylic	6	4	11	16	21	11
acrylonitrile	5	4	11	19	12	9
activated carbon	0	0	5	10	12	12
activator	6	4	4	4	4	6
activewear	10	3	18	18	13	15
actuator	1	0	0	0	0	0
add-on	2	4	5	4	5	4
addition polymer	0	0	0	0	0	1
additive	41	67	69	68	44	56
adhesion	52	41	53	63	53	44

	1999	2000	2001	2002	2003	2004
adhesive agent	64	41	55	36	55	40
adhesive property	10	50	31	28	30	25
adornment	0	0	0	0	1	0
adsorbency	12	24	25	21	7	16
adsorber	0	0	0	0	1	2
adsorption isotherm	0	0	0	1	0	0
adsorption	0	1	3	19	37	29
advertising	0	0	0	1	0	8
aerial	0	0	0	2	0	1
aerobic treatment	0	2	1	3	1	0
aerodynamics	6	8	12	12	10	4
aerogel	0	0	0	0	0	1
aerosol	6	3	4	1	2	1
aerospace textile	42	35	30	32	34	42
afterclearing agent	0	0	1	0	0	0
afterclearing	1	0	0	0	1	2
aftertreatment	12	14	16	13	15	10
afterwashing	5	3	5	2	1	2
agavaceae fiber	0	1	0	0	0	1
age determination	0	0	0	0	2	0
aging	8	15	19	34	34	28
agitation (laundering)	0	1	0	0	0	0
agitation	0	0	6	10	6	5
agricultural auction	0	0	0	0	0	1
agricultural textile	11	13	11	6	12	8
agriculture	49	121	26	12	12	4
agronomic information	0	0	0	0	2	0
air-jet loom	29	60	29	18	27	27
air-jet spinning	14	14	13	12	11	13
air bag	47	42	27	34	26	35
air conditioning	17	19	25	22	14	19
air entanglement	0	0	1	0	0	0
air filter	3	15	7	3	1	6
air filtration	5	13	7	6	3	6
air jet	25	34	34	19	27	24
air laying	27	24	32	37	23	19
air pocket	1	4	2	3	0	3
air pollution	2	9	4	10	8	8
air pressure	3	11	8	7	4	7
air	11	19	9	5	7	3
aircraft	20	17	13	31	21	27
airflow	65	88	61	44	46	28
airfoil	2	1	2	1	0	1

	1999	2000	2001	2002	2003	2004
airtight	1	0	1	0	1	1
alcohol	1	1	4	6	6	5
alginate fabric	2	2	0	0	0	0
alginate fiber	4	3	2	0	0	4
alginate	1	4	6	5	3	12
alignment mechanism	0	9	0	6	10	8
alkali treatment	16	15	19	18	28	15
alkali	5	3	14	16	11	12
allergy	9	7	9	11	8	6
alloy	1	9	14	16	15	8
alpaca fabric	0	0	0	0	0	2
alpaca fiber	4	1	5	5	2	4
alpaca yarn	0	1	0	0	0	0
alpaca	1	2	8	19	17	16
alumina fiber	0	3	6	2	1	2
aluminosilicate fiber	0	0	1	0	0	0
aluminum fiber	1	0	2	1	2	1
aluminum oxide	0	0	0	0	0	1
amination	2	1	0	0	0	0
amine	0	0	0	0	2	5
amino acid analysis	1	2	3	4	0	0
amino acid	0	0	0	4	7	6
aminolysis	0	1	0	1	1	1
ammonia	8	6	8	6	7	6
ammonium carbonate	0	0	0	0	0	1
ammonium chloride	0	0	0	0	1	0
ammonium	0	0	0	0	0	1
anaerobic treatment	1	2	1	3	3	0
analysis	192	56	15	17	4	7
angora fabric	1	0	0	0	0	0
angora fiber	0	0	3	0	0	0
angora yarn	0	0	1	0	0	0
angora	1	0	0	2	3	1
animal fiber fabric	2	1	2	0	0	0
animal fiber yarn	0	1	3	2	0	0
animal fiber	9	7	7	14	10	10
animal husbandry	0	0	0	0	0	4
anionic dye	9	12	8	6	3	6
anisotropy	17	17	28	32	13	9
annealing	4	8	14	22	14	7
annual report	0	32	21	44	38	18
anthraquinone dye	5	6	6	10	9	3
anthropometry	12	27	10	7	9	4

	1999	2000	2001	2002	2003	2004
antibiotics	0	0	0	0	0	1
antifelting property	4	7	6	4	5	1
antifibrillating agent	0	1	4	1	3	0
antifoaming agent	1	1	2	0	6	2
antifungal property	10	10	10	10	9	8
antimicrobial property	90	90	110	125	118	95
antioxidant	1	2	1	2	7	9
antistatic property	19	21	21	14	23	20
antiyellowing property	0	6	5	3	7	2
apparel fabric	65	76	110	135	160	135
appearance	5	4	6	10	19	15
application (finishing)	32	28	11	16	28	24
applicator	11	6	1	3	1	6
applique	3	8	10	7	5	4
apron	3	4	3	3	4	3
aquisition	0	6	0	0	0	0
aramid fabric	5	16	2	4	4	4
aramid fiber	39	39	23	36	22	30
aramid yarn	4	4	3	2	3	3
aramid	5	4	4	2	4	2
architectural textile	77	102	75	75	74	95
aromatic amine	0	8	7	4	1	2
artificial fur	1	0	2	3	2	2
artificial hair	1	2	2	2	1	2
artificial leather	13	21	18	13	13	31
artificial textile	0	0	0	0	0	1
artificial turf	0	9	11	9	7	16
asbestos fiber	0	2	0	0	0	0
asbestos replacement	3	1	3	0	0	0
asbestos	0	1	0	0	1	0
assessment	0	0	0	0	0	2
atomic force microscopy	0	0	0	0	0	1
autoclave	6	2	3	2	5	3
autoleveling	1	6	2	0	5	0
automation	133	189	181	166	138	93
automotive textile	178	220	147	150	163	112
awning	12	18	21	20	16	11
axminster carpet	1	7	7	1	6	4
azo dye	33	39	40	45	37	46
babieswear	5	7	3	6	1	3
back-coating	7	8	1	5	6	3
backing fabric	9	10	7	7	10	19
backing	12	18	10	21	15	12

	1999	2000	2001	2002	2003	2004
backpack	0	0	0	1	0	0
backstaining	0	2	2	3	1	1
backwashing	0	0	0	1	0	0
bacterium	3	10	14	23	14	2
bag	15	20	11	21	18	14
balance of trade	0	0	0	0	1	1
balance tube	0	1	2	6	7	2
bale opening machine	5	2	3	3	3	0
bale plucking machine	2	0	2	0	0	0
bale plucking	3	1	0	0	0	0
bale	12	2	4	4	3	1
baling machine	3	2	2	0	0	0
ballistic resistance	1	27	21	15	20	16
balloon (yarn)	7	4	13	5	0	4
balloon limiter	3	2	2	1	2	0
bamboo fiber	2	1	1	0	4	1
banana fiber	3	8	1	1	3	3
bandage	28	28	34	30	41	35
banner	10	15	14	9	13	8
bar	1	1	2	0	1	4
barre	0	1	0	1	0	1
barrier fabric	30	24	17	13	19	24
barrier property	24	29	26	25	28	30
basalt fabric	0	1	0	1	0	1
basalt fiber	4	4	5	1	4	4
basalt yarn	0	0	0	1	0	0
basalt	0	0	1	0	1	0
basic dye	1	5	3	6	7	10
bast fiber fabric	0	1	1	0	0	0
bast fiber yarn	0	1	0	0	0	0
bast fiber	8	16	5	6	3	8
batch dyeing	4	3	13	1	4	5
batching machine	2	2	1	1	0	0
batching	4	9	5	9	9	5
bath	2	9	2	4	1	6
batik	0	0	1	3	6	3
battery separator	4	4	2	2	3	4
batting	17	9	14	12	13	16
BCF yarn	0	10	11	9	14	17
bead	0	1	0	2	5	1
beadwork	0	0	0	0	7	2
beam dyeing	2	1	1	2	1	1
beam	10	8	10	2	14	7

	1999	2000	2001	2002	2003	2004
beaming	3	3	2	6	4	3
bearings	14	12	5	8	8	11
beating-up	2	1	7	5	2	5
beauty product	0	0	0	7	9	7
bedding	25	17	20	18	20	21
belt	36	36	29	16	5	4
benchmarking	0	2	7	8	4	5
bending property	14	16	23	12	13	8
bending	8	2	9	12	15	19
bias	0	1	5	5	7	8
biaxial loading	12	3	6	7	3	1
bib	0	4	1	0	2	0
bicomponent fiber	61	37	31	26	46	58
bicycle frame	0	2	2	5	1	1
binder fiber	0	0	0	0	0	10
binder	0	0	1	0	0	0
binding agent	30	42	15	29	27	21
binding	0	0	0	0	4	0
bioactive fiber	0	0	0	0	0	1
bioactive property	0	0	0	1	0	2
biochemistry	0	0	0	1	0	0
biocide	0	0	0	0	0	3
biocompatibility	25	15	22	11	20	13
biodegradability	51	48	78	125	108	52
biofinishing	6	4	8	9	5	2
biofouling	0	0	1	0	0	0
biohazard	0	0	1	10	12	5
biological characteristic	0	0	0	0	0	1
biological resistance	15	0	0	3	6	5
biological treatment	14	23	18	19	18	12
biology	0	0	0	0	0	1
biomaterial	0	0	0	0	2	0
biomechanics	0	0	0	1	1	0
biomimetic material	0	0	0	0	2	3
biopolishing	4	8	3	4	4	1
biopolymer	0	0	0	0	10	13
biotechnology	21	29	37	72	40	24
birefringence	2	9	9	17	10	3
bisu waste	1	0	0	0	0	0
blade	11	14	7	19	23	15
blanket	14	5	16	13	15	8
bleaching agent	16	22	15	39	39	13
bleaching assistant	1	1	0	1	0	0

	1999	2000	2001	2002	2003	2004
bleaching machine	3	2	3	4	1	2
bleaching	65	70	73	80	71	44
blend ratio	8	15	22	35	27	7
blended fabric	51	28	20	12	22	13
blended fiber	58	33	20	31	34	22
blended yarn	35	42	25	21	17	16
blending machine	1	7	4	2	2	3
blending	23	43	143	145	26	16
blinds	5	3	8	6	4	2
block copolymer	1	9	46	85	90	30
blood filter	3	3	0	7	0	2
blood filtration	3	5	2	1	0	0
blouse	0	0	0	1	0	0
boarding machine	2	0	1	5	1	2
boarding	3	0	1	4	0	1
boat	0	0	1	0	0	0
bobbin changing	6	2	3	0	0	0
bobbin holder	5	5	1	1	0	0
bobbin lace	0	1	1	3	2	1
bobbin	29	37	22	18	27	19
body armor	19	17	11	2	7	4
body scanning	0	15	11	12	14	7
body shape	28	29	14	18	14	5
body size	0	28	24	24	21	8
boiler	6	1	4	2	3	3
boiling-off	0	0	0	1	1	2
boiling	1	2	5	5	6	4
bollworm	2	40	4	0	0	1
Bombyx mori	12	7	12	18	25	14
bonded fiber fabric	78	27	13	9	5	3
bonding agent	40	5	4	10	3	4
bonding machine	3	2	1	2	2	2
bonding	42	43	43	42	54	40
bone repair	7	8	2	0	9	6
boom	0	1	0	2	3	6
boric acid	0	0	0	0	0	1
boron-nitride fiber	0	0	0	0	1	0
boron fiber	2	1	2	2	2	2
boron	1	2	2	3	3	0
boucle yarn	1	4	6	3	0	0
bouyancy	0	0	0	5	3	2
brace (medical textile)	0	4	4	2	1	1
brace	0	2	0	0	0	0

	1999	2000	2001	2002	2003	2004
braid	18	24	18	20	20	25
braiding machine	17	3	6	12	7	7
braiding	24	14	15	19	16	18
braille	0	0	0	0	1	0
brake (vehicle)	7	4	7	3	3	7
braking system	0	0	2	6	6	6
branding	19	28	32	52	51	64
brassiere	26	18	21	20	8	10
breaking elongation	17	23	32	18	27	8
breaking point	2	1	4	4	2	1
breaking strength	18	34	39	27	24	9
breaking	6	4	4	10	6	7
breathability	56	41	42	32	46	43
breed	25	20	15	15	22	4
breeding	0	0	0	0	0	2
bridge	6	13	13	10	17	12
brightening agent	4	6	8	8	2	14
brightening	1	3	2	9	6	1
brush	2	4	4	8	4	3
buckling	17	11	20	26	22	25
buffering agent	0	2	1	2	0	2
building material	0	0	0	0	0	1
bulk	18	19	26	21	15	11
bulking	1	9	5	5	1	0
bullet resistance	12	8	3	7	5	1
buoyancy	0	0	0	0	0	1
business communication	0	0	0	1	3	15
business merger	0	3	21	13	11	6
business plan	0	50	17	51	32	22
business strategy	0	8	27	119	159	231
button	0	0	0	0	1	1
buttonhole	3	10	12	4	7	8
buyer	2	2	4	3	6	3
buyers guide	17	24	22	34	19	17
buying	1	12	13	20	27	13
cable	14	5	19	9	11	11
cabling machine	4	0	2	2	3	1
cabling	3	1	6	2	1	1
calculation	0	0	0	1	18	18
calender bowl	3	3	2	1	0	0
calendering machine	9	9	16	3	2	4
calendering	19	15	22	11	12	5
calibration	6	11	5	12	9	6

	1999	2000	2001	2002	2003	2004
calorimetry	6	12	19	52	39	17
cam	13	22	18	5	19	4
camel hair	1	0	3	1	2	0
camelid	0	0	0	0	0	2
camouflage	8	8	6	7	12	16
can changing	2	3	6	3	1	0
canopy	7	10	7	9	2	7
canvas	5	6	8	12	4	5
caprolactam	7	8	21	18	7	9
car seat	3	9	9	9	11	4
carbon dioxide	15	12	19	28	22	17
carbon fiber fabric	15	11	6	9	10	5
carbon fiber yarn	2	3	2	4	5	2
carbon fiber	158	133	147	166	167	192
carbon	19	11	19	21	36	42
carbonizing	10	13	10	10	9	4
carcinogenicity	3	5	6	3	6	0
card clothing	15	12	7	7	4	3
card	13	38	16	9	20	2
carding machine	56	65	29	21	34	15
carding	65	64	37	31	23	13
carpet backing	17	23	31	21	28	14
carpet fiber	12	25	18	21	18	9
carpet industry	35	69	37	35	38	20
carpet loom	8	6	9	0	6	5
carpet tile	5	2	6	3	6	2
carpet yarn	18	23	19	12	12	9
carpet	154	127	104	78	64	78
carrier	14	24	8	9	5	3
cashew nut	0	0	0	0	1	0
cashmere fabric	3	3	1	1	2	2
cashmere fiber	4	4	3	4	3	6
cashmere industry	0	4	13	6	5	4
cashmere yarn	0	1	2	0	0	1
cashmere	10	11	9	12	16	4
casting off	1	1	0	0	1	0
casting on	0	0	0	0	1	0
casualwear	6	4	3	8	8	6
catalyst	9	20	59	129	102	35
cationic dye	16	16	6	4	4	7
cationizing agent	0	0	0	0	2	2
cationizing	0	0	2	0	8	7
cellulase treatment	21	18	13	10	13	9

	1999	2000	2001	2002	2003	2004
cellulose acetate	0	0	0	1	7	9
cellulose triacetate	0	0	0	0	0	1
cellulose	49	39	87	98	95	70
cellulosic fiber fabric	78	64	33	22	33	25
cellulosic fiber yarn	8	5	5	0	0	2
cellulosic fiber	108	72	64	81	88	65
centrifugal force	7	4	4	5	8	7
centrifuge modeling	0	3	2	0	1	0
ceramic fiber fabric	2	3	0	0	1	0
ceramic fiber yarn	0	1	0	0	0	0
ceramic fiber	25	19	21	9	13	10
ceramic matrix	8	17	21	10	10	12
ceramic whisker	0	1	0	0	1	0
ceramics	8	8	20	20	20	15
chelation	0	0	1	0	0	0
chemical analysis	14	18	4	18	12	7
chemical composition	218	115	109	158	154	128
chemical etching	0	0	0	0	0	1
chemical modification	105	90	69	83	67	47
chemical property	0	6	8	9	10	10
chemical reaction	0	0	0	0	3	5
chemical resistance	23	35	17	19	28	15
chemical vapor deposition	0	3	4	4	3	2
chemical	60	31	56	59	45	26
chemistry	0	0	0	0	0	3
chenille fabric	1	1	2	4	5	3
chenille yarn	3	3	3	4	7	8
child labor	0	0	0	2	3	1
childrenswear	21	19	11	18	17	12
chintz	0	0	0	0	3	0
chitin fiber	1	0	0	2	2	0
chitin	5	2	6	12	8	3
chitosan fabric	1	0	0	0	0	0
chitosan fiber	1	1	1	1	2	1
chitosan	19	11	28	41	57	33
chlorination	0	0	0	0	1	1
chlorine resistance	1	0	5	2	6	2
chromatography	7	15	29	47	23	9
chrome dye	5	1	3	0	1	1
chromium	0	0	0	0	0	1
chute	0	0	0	1	0	0
cigarette filter	0	3	3	4	0	3
circuit board	0	0	0	1	0	0

WORLD TEXTILES Subject Indexing

	1999	2000	2001	2002	2003	2004
circular knitting	89	99	114	80	75	61
citric acid	0	0	0	0	0	1
civil engineering	76	55	84	70	61	66
clamping	25	21	5	13	9	11
classification	16	48	60	70	45	20
classing	3	2	1	0	4	1
clay liner	4	4	12	3	4	3
cleaning machine	8	7	4	11	4	13
cleaning	80	102	89	69	70	96
cleanroom	4	10	5	6	5	7
climate control	0	0	0	0	0	2
clip (yield)	0	0	0	0	0	2
clip	2	5	1	6	6	1
clothing industry	109	117	90	113	121	95
coagulating	20	11	11	11	10	8
coalescence	0	0	0	0	1	0
coarsening	1	2	0	0	1	0
coated fabric	72	71	53	52	56	62
coating agent	102	62	22	11	27	21
coating ink	0	0	0	0	1	0
coating machine	15	17	8	15	16	11
coating	146	136	144	123	145	120
cocoon	26	19	15	16	26	16
coiling	4	8	1	6	2	2
coir fiber	3	1	2	2	3	1
coir yarn	0	1	0	0	2	1
coir	0	2	2	2	0	1
collagen	12	3	4	7	11	2
collar	10	14	7	0	2	7
colloid	0	0	2	1	3	3
color change	12	17	11	14	21	9
color communication	0	0	0	0	10	14
color management	20	14	34	22	24	17
color matching system	11	16	8	6	2	0
color matching	9	15	9	15	12	7
color measurement	0	0	0	0	1	0
color perception	4	11	8	7	4	4
color prediction	2	10	4	3	1	0
color	69	113	120	129	198	100
colorfastness	50	90	113	105	91	48
colorimetry	53	56	43	40	36	33
comb	3	6	2	3	2	2
combing machine	14	14	4	11	7	8

	1999	2000	2001	2002	2003	2004
combing	8	21	13	24	13	9
combustion	0	0	0	0	0	1
comfort	121	126	149	138	119	72
commingling	3	4	4	4	3	2
comminuting	0	0	0	1	0	0
communication	0	0	0	0	5	7
compact spinning	0	0	16	24	42	16
compacting machine	2	4	4	4	0	0
compacting	6	13	12	14	9	11
company information	949	907	741	933	680	716
comparative study	0	0	0	0	0	1
competitiveness	61	120	108	120	127	145
complexing agent	2	5	3	4	0	3
composite property	40	89	80	49	70	63
composite	234	104	221	248	232	263
composites industry	0	0	0	0	19	30
composition	1	0	0	0	0	0
compression	0	0	0	2	35	37
compressive property	42	61	61	58	31	25
compressor	0	0	0	0	1	0
computational fluid dynamics	0	0	1	2	3	7
computational model	0	0	0	0	0	1
computer aided design	95	124	94	126	96	88
computer aided production	92	77	52	47	32	29
computer control	0	6	17	39	24	14
computer program	77	80	89	63	62	30
computer simulation	33	105	117	86	98	51
computer	51	23	25	16	11	6
concentration	6	32	37	46	13	9
concrete	22	18	36	31	19	35
condensation	0	0	0	0	0	3
condenser	10	20	23	8	7	4
conditioning agent	2	1	4	8	3	1
conditioning	10	10	8	4	8	5
conductivity	26	21	41	77	39	28
cone	3	6	5	8	0	1
conference (Aachen)	0	0	0	51	1	0
conference (AATCC)	5	4	0	0	1	0
conference (ATIRA)	0	2	0	0	0	0
conference (Beltwide Cotton)	563	478	0	1	0	0
conference (BTRA)	0	1	0	0	0	0
conference (COTTECH)	3	0	0	0	0	0
conference (DWI)	1	0	0	0	0	0

	1999	2000	2001	2002	2003	2004
conference (International Man-made Fibre)	3	7	8	1	7	3
conference (ITMA)	0	0	0	0	1	0
conference (ITMF)	4	17	1	0	3	1
conference (NITRA)	17	1	0	0	0	0
conference (SAMPE)	1	2	1	0	0	0
conference (SITRA)	0	1	0	0	0	0
conference (TIFCON)	3	1	1	1	0	0
conference (TIWC)	2	0	1	0	0	2
conference exhibition	0	0	0	0	1	0
conference report	0	0	0	0	1	26
conference	171	203	353	152	157	125
conservation	0	1	0	0	0	1
constructed fabric	1	0	0	0	0	0
constructed textile	7	7	6	4	5	6
construction industry	0	0	0	0	1	5
consumer research	48	77	61	71	56	43
consumer safety	8	9	7	12	17	2
consumption	27	44	51	82	65	56
contaminant	0	0	0	0	0	2
contamination	27	30	36	33	30	23
continuous dyeing	12	8	5	5	9	7
continuous process	35	31	21	26	26	10
control system	226	160	156	113	129	113
control	43	48	52	65	38	6
controlled release	0	2	17	19	16	8
convection	0	4	1	1	1	5
conveyor belt	19	35	27	32	36	25
cooling device	17	2	1	0	0	0
cooling	21	25	23	28	19	16
copolymer	17	43	139	160	160	72
copolymerization	9	8	56	100	96	27
copper sulfate	0	0	0	1	0	0
copper	0	0	0	0	2	1
copyright	2	7	6	13	15	10
cord construction	6	17	11	10	15	12
cord	16	17	23	17	13	10
corduroy fabric	1	1	1	3	2	1
core spinning	7	10	18	9	11	9
cork	0	0	0	0	1	1
corn fiber	0	0	0	0	1	0
corporate branding	0	0	0	0	1	0
corrosion resistance	21	25	9	21	11	11

	1999	2000	2001	2002	2003	2004
corrosion	0	1	0	6	7	1
corset	1	1	1	2	2	1
cosmetic textile	0	1	3	2	0	0
cost reduction	0	0	0	1	20	15
cost	81	202	196	187	172	86
costing	20	16	10	10	8	5
costume	0	0	0	2	21	21
cotton-flax blend	3	1	1	1	2	1
cotton-linen blend	3	1	0	1	3	1
cotton-polyester blend	18	37	20	28	32	21
cotton-polypropylene blend	0	1	2	3	3	0
cotton-silk blend	0	0	0	0	2	3
cotton-viscose-rayon blend	0	3	2	3	4	2
cotton-wool blend	3	1	2	0	3	2
cotton blend	0	0	0	0	0	1
cotton fabric	112	162	154	156	149	113
cotton fiber	98	91	68	65	84	39
cotton hemp blend	1	0	0	0	0	0
cotton industry	89	487	33	40	31	22
cotton nylon blend	1	0	0	0	0	0
cotton wool	3	1	1	0	0	1
cotton yarn	46	55	46	53	38	35
cotton	632	111	98	92	71	53
count	9	7	5	12	13	11
counterfeiting	3	10	4	11	14	13
coupling agent	2	9	11	14	9	5
coupling	5	9	12	13	12	2
crabbing machine	1	0	0	0	1	0
crabbing	2	1	2	1	0	0
crack propagation	0	0	0	0	1	0
cracking	13	22	38	36	26	24
crease recovery	9	16	16	21	12	8
crease resistance	28	28	46	37	20	23
creasing	9	12	13	6	13	8
credit provision	0	0	0	0	0	1
creel	7	9	14	9	15	7
creep	27	19	27	25	29	16
crepe fabric	3	1	0	3	1	1
crepe	2	4	6	2	1	0
creping	3	2	0	1	1	4
crime prevention	0	0	0	0	0	1
crimp	22	20	21	27	19	27
crimping machine	2	4	0	5	0	3

	1999	2000	2001	2002	2003	2004
crimping	11	8	7	13	7	11
crochet machine	8	9	5	4	3	4
crocheted fabric	3	2	2	0	4	3
crocheting	3	2	6	1	2	5
crockfastness	10	6	4	6	5	0
crop production	0	0	0	0	6	9
cross-sectional shape	25	16	24	31	31	14
cross linking agent	22	13	16	23	12	9
cross linking	41	41	96	129	139	77
cross section	23	17	31	34	38	25
crosslapping machine	1	2	2	0	2	0
crosslapping	2	1	3	0	4	1
crosslining	1	0	0	0	0	0
cryogenics	0	0	0	0	1	4
crystalline orientation	5	4	9	15	18	12
crystallinity	33	47	112	217	179	58
crystallization	0	0	0	1	0	3
cuff	2	8	2	5	2	4
cultivation	382	157	40	37	35	11
cultural anthropology	0	0	6	22	17	14
culture	0	0	0	4	0	0
cupro fiber	2	1	0	0	0	0
curing	24	28	36	61	53	40
curtain	15	8	14	2	8	9
cushion	37	33	29	11	15	12
customer service	0	0	1	0	0	9
customization	0	2	17	17	11	16
cut resistance	12	22	13	13	20	14
cutter	17	23	19	11	8	10
cutting machine	13	16	20	11	15	12
cutting	55	52	53	52	48	50
cylinder	4	15	12	9	15	9
damage	63	47	68	57	54	44
damping	0	0	0	0	0	2
data management system	23	30	15	24	21	11
data processing	23	25	23	17	14	10
database	4	17	14	16	14	6
dating	0	0	0	1	0	0
deacetylation	0	0	0	0	2	0
debonding	7	9	7	9	8	1
decatizing machine	3	9	0	1	2	0
decatizing	3	10	4	1	1	2
dechlorination	0	1	0	0	0	0

	1999	2000	2001	2002	2003	2004
decision making	0	0	0	2	0	5
decolorizing	11	16	29	31	20	19
decomposition	0	0	0	16	25	10
decoration	0	0	0	0	1	0
decorticating machine	1	0	0	0	0	0
decorticating	0	0	2	1	1	1
defect	43	53	89	64	67	36
defoliation	8	5	0	1	0	0
deformation	38	60	95	82	106	44
defuzzing	0	1	0	1	1	1
degradation	18	37	58	96	74	31
degreasing	0	0	0	0	3	0
degree of orientation	0	1	5	2	5	1
degree of polymerization	4	1	6	11	7	3
degumming	11	6	8	6	2	8
deknitting	0	0	2	0	1	0
delamination	22	16	41	27	28	29
delignifying	1	1	2	0	1	0
delustering agent	0	2	0	1	1	0
delustering	1	2	0	0	3	1
demand	0	77	67	95	92	65
demineralization	0	0	1	2	1	0
demographic survey	0	0	0	0	0	1
denier	10	12	21	23	21	30
denim fabric	32	29	28	38	33	42
density	17	38	41	38	41	26
dental floss	6	4	4	7	0	0
deodorizing	9	18	11	9	7	4
depolymerization	0	0	0	0	3	1
desalination	0	0	3	0	2	2
design	312	282	223	229	317	220
desizing	20	14	13	13	12	7
desorption	0	2	5	7	4	1
detector	37	42	63	66	49	36
detergent	30	40	69	103	93	85
detwisting	0	0	0	0	1	0
devore	0	1	0	1	0	0
deweighting	0	0	1	0	2	1
dialysis	1	2	0	2	1	1
diameter	0	0	0	0	0	1
diaper	30	28	20	35	47	33
die	11	10	7	9	9	5
dielectric property	6	4	15	28	27	14

	1999	2000	2001	2002	2003	2004
diet	0	0	0	0	1	0
diffusion	2	7	28	30	26	8
digital method	11	18	26	42	64	42
digital printing	32	27	57	32	37	32
dimensional stability	10	57	40	39	74	21
direct dye	22	20	18	18	29	8
direct printing	0	1	0	3	0	2
direct spinning	0	1	0	0	0	0
directory information	32	35	22	29	20	14
disaster relief	0	0	0	0	1	0
discharge printing	10	5	3	6	5	6
disease management	1	18	5	2	0	2
disease resistance	0	0	1	3	0	0
disease	4	2	3	10	14	4
dispersant	7	10	7	10	9	7
disperse dye	63	46	58	52	58	48
dispersion	12	12	9	10	22	29
disposable product	53	48	36	40	46	53
dissolution	0	0	0	1	3	12
distillation	0	1	0	1	1	1
distortion	3	1	2	4	4	3
distribution	36	43	27	61	31	28
diving suit	7	3	2	1	0	2
DNA	1	1	3	8	9	7
dobby loom	4	4	2	2	1	3
dobby	6	1	9	2	2	0
doffer	9	3	4	3	1	1
doffing	10	3	9	4	3	3
dog hair	0	0	0	1	0	0
dosing machine	2	3	0	9	3	3
dosing system	7	14	5	2	1	0
dosing	6	10	11	9	11	17
double cloth	0	4	1	1	3	2
down time	0	11	5	9	9	1
drafting machine	12	9	9	8	7	1
drafting	33	30	35	28	30	23
drag	0	0	0	0	0	1
drainage	17	23	19	23	16	12
drape	21	43	34	17	15	23
draw-texturing machine	4	2	2	3	1	1
draw-texturing	7	5	9	5	2	6
draw ratio	6	3	6	9	11	4
drawframe	15	20	16	12	9	5

	1999	2000	2001	2002	2003	2004
drawing-in machine	5	2	3	3	0	0
drawing-in	3	3	6	3	0	1
drawing machine	6	12	7	6	4	6
drawing	46	44	57	53	48	29
dress code	0	0	0	0	1	0
dress	14	26	22	42	12	13
dressing	50	50	43	44	54	40
drive	78	86	73	54	57	41
drop wire	1	2	2	1	0	0
drug delivery	0	0	0	12	14	3
drum	17	23	35	34	51	47
dry finishing	4	4	2	3	6	3
dry laying	6	2	1	0	1	2
drycleaning industry	1	6	2	0	0	0
drycleaning machine	10	19	7	9	7	3
drycleaning	25	35	31	23	21	14
dryer	0	0	0	1	0	0
drying machine	20	43	42	34	51	71
drying	39	54	51	49	61	33
ductility	0	0	0	1	0	0
dumping	2	3	1	7	10	1
durability	27	26	32	34	42	42
dust filter	3	7	3	2	6	3
dust filtration	5	10	0	3	3	2
dust	40	21	13	8	9	7
dye bath	20	11	26	14	25	11
dye exhaustion	7	17	25	17	28	8
dye fixing	30	27	21	28	32	15
dye house	0	1	0	0	0	0
dye levelness	10	10	7	11	15	2
dye manufacture	0	0	0	0	1	21
dye manufacturing industry	2	3	1	6	4	3
dye manufacturing	21	17	22	18	20	9
dye migration	7	4	4	19	13	5
dye mixture	23	41	31	12	17	9
dye penetration	10	5	5	7	4	2
dye properties	1	0	0	0	0	0
dye property	59	84	80	39	48	57
dye recipe	9	10	22	15	10	2
dye removal	4	9	12	11	11	14
dye uptake	22	28	25	31	28	14
dye	83	73	132	97	107	82
dyeability	43	63	65	50	65	40

	1999	2000	2001	2002	2003	2004
dyed fabric	5	5	0	0	1	4
dyed textile	0	21	0	0	0	0
dyeing-finishing machine	4	8	2	13	5	3
dyeing assistant	34	28	23	15	6	6
dyeing industry	76	29	26	21	11	7
dyeing machine	81	48	38	54	73	44
dyeing	298	319	276	306	299	277
e-commerce	0	6	34	30	16	18
earthquake resistant	0	0	0	0	1	0
easy care property	32	24	26	25	21	15
eco-labeling	10	6	9	7	5	5
ecology	61	67	51	52	26	13
economic conditions	0	0	0	0	0	7
economic data	161	116	118	130	115	80
economic development	0	0	0	0	0	1
economic information	272	422	267	345	361	252
economic strategy	3	100	137	145	116	103
edge coating	2	1	0	2	0	0
education	45	52	49	67	64	58
effluent treatment	61	74	82	65	65	46
effluent	17	22	15	11	8	3
elastane fabric	6	7	5	11	4	0
elastane fiber	15	13	23	16	11	4
elastane yarn	12	13	18	31	11	13
elastane	7	6	7	13	12	17
elasterell-p	0	0	0	0	1	0
elastic modulus	14	24	51	31	23	12
elastic property	0	0	0	0	0	3
elastic recovery	5	8	9	22	10	19
elasticity	86	102	114	89	71	62
elastomer	26	37	54	89	82	48
elastomeric fabric	14	28	25	29	28	26
elastomeric fiber	13	13	8	20	17	17
elastomeric yarn	16	11	11	21	20	15
elastoplasticity	0	0	0	2	1	1
electrect	0	0	1	1	0	0
electret	8	5	7	9	2	8
electrical circuit	0	0	0	0	0	1
electrical conductivity	26	37	39	35	41	23
electrical property	11	24	36	17	21	23
electrical resistance	0	2	5	14	15	4
electricity	7	26	24	21	24	17
electrochemistry	0	1	1	17	32	23

	1999	2000	2001	2002	2003	2004
electroless plating	0	0	0	0	1	3
electrolysis	0	4	2	2	5	2
electromagnetic device	19	18	8	6	5	6
electromagnetic radiation	0	0	0	0	0	1
electromagnetic resistance	5	7	5	15	13	9
electron microscopy	6	6	8	9	1	5
electron spin resonance	0	0	0	0	1	0
electronic control	27	37	22	35	31	25
electronic data interchange	2	11	4	5	3	4
electronics	31	70	54	69	75	40
electroplating	0	1	1	1	0	1
electrospinning	0	0	3	8	24	36
electrostatics	0	0	0	0	0	1
elongation	25	20	29	31	27	11
embossing	14	13	10	12	6	5
embroidering	21	13	43	42	29	47
embroidery frame	11	8	5	8	9	13
embroidery industry	4	8	16	18	14	15
embroidery machine	42	30	34	37	27	37
embroidery	59	41	126	102	99	108
emergency shelter	0	0	0	0	0	2
emerizing machine	3	3	1	1	0	0
emerizing	6	1	2	0	2	0
emission reduction	0	1	8	13	19	15
employment	36	44	52	74	74	37
emulsion	9	13	25	36	26	15
encapsulation	1	14	11	14	20	17
end breakage	1	2	0	6	3	1
end use	0	59	59	86	111	60
energy conservation	56	77	43	51	37	23
energy consumption	2	15	24	24	23	8
energy	7	26	19	29	25	23
engineering	0	4	0	0	9	23
engraving	0	0	0	0	0	1
environmental impact	1	0	7	33	77	39
environmental management	0	0	0	2	0	5
environmental protection	162	161	204	150	85	58
enzymatic degradation	7	9	9	7	7	3
enzymatic treatment	64	78	73	65	58	34
enzyme	24	28	39	49	42	15
epoxy resin	20	18	40	29	60	30
equipment	0	0	0	0	0	1
ergonomics	10	16	15	9	10	7

WORLD TEXTILES Subject Indexing

	1999	2000	2001	2002	2003	2004
eri silk	2	2	2	2	17	5
erosion control	8	8	8	10	6	3
esthetic property	0	0	2	17	15	17
etching	2	5	4	4	3	0
ethics	0	0	1	6	10	6
ethnicity	0	0	0	0	0	1
ethnology	0	0	1	1	1	0
evenness	12	18	13	15	10	7
exhaustion	11	13	10	12	7	3
exhibition (ATME-I)	0	15	25	1	0	5
exhibition (Bobbin Show)	3	1	0	0	0	0
exhibition (CAD Expo)	5	2	0	0	0	0
exhibition (CITME)	0	4	3	7	2	3
exhibition (Domotex)	7	17	2	2	9	5
exhibition (Expofil)	5	8	6	3	14	12
exhibition (Heimtextil)	5	9	3	3	7	14
exhibition (IDEA)	0	0	9	0	0	7
exhibition (IFAI)	0	0	0	0	0	3
exhibition (IHE)	0	4	0	0	1	0
exhibition (IKME)	0	0	0	0	0	8
exhibition (IMB)	0	41	4	1	13	7
exhibition (INDA)	1	3	1	1	0	0
exhibition (INDEX)	10	1	0	14	0	0
exhibition (ITMA)	226	57	19	11	101	120
exhibition (ITME-I)	0	3	19	0	2	6
exhibition (JEC)	6	4	3	3	5	4
exhibition (JIAM)	6	0	0	1	0	0
exhibition (Techtextil)	23	15	31	11	35	8
exhibition review	0	0	2	1	116	124
exhibition	81	138	190	215	168	125
expansion	0	0	0	0	0	1
exports	110	158	149	234	173	95
extensibility	8	22	12	22	19	10
extraction method	0	0	6	11	17	15
extruder	0	0	0	0	0	1
extrusion	111	81	76	104	113	99
eyelet	2	4	5	2	1	0
fabric appearance	43	47	51	41	40	21
fabric construction	286	449	285	219	302	161
fabric manufacture	109	100	59	50	90	152
fabric modification	16	158	84	57	87	78
fabric performance	0	1	0	0	0	0
fabric property	292	369	268	145	158	106

	1999	2000	2001	2002	2003	2004
fabric	196	67	69	37	48	8
fabrication	0	0	0	27	70	76
fading	9	12	22	11	14	13
failure analysis	0	0	0	0	1	20
false twist	34	22	39	27	22	31
fancy yarn	17	18	10	11	19	14
farming	0	0	4	36	18	7
fashion	143	160	183	226	249	217
fastener	68	52	43	46	47	37
fatigue resistance	0	0	0	0	0	1
fatigue	31	18	33	34	21	19
feasibility study	0	0	0	0	0	1
feather	0	3	2	1	3	5
feed	94	123	98	84	58	67
felt	24	25	24	15	18	13
felting	0	5	5	9	4	7
fertilizer	44	26	11	7	3	4
fiber-matrix adhesion	6	21	14	11	14	8
fiber-matrix interface	24	21	17	13	14	9
fiber arrangement	11	13	6	13	17	13
fiber bundle	25	26	32	51	32	22
fiber curvature	1	6	4	14	6	2
fiber diameter	47	44	53	64	42	27
fiber distribution	7	18	11	12	11	11
fiber fineness	24	3	16	31	24	23
fiber flow	2	9	1	3	1	7
fiber length	49	46	40	35	46	22
fiber manufacture	141	280	231	201	281	356
fiber medulla	0	0	1	1	0	1
fiber modification	57	101	116	56	63	66
fiber optics	0	1	5	14	19	6
fiber orientation	18	35	52	44	44	45
fiber preparation	1	12	43	37	44	37
fiber property	281	394	317	210	188	138
fiber reinforced composite	583	528	514	540	605	696
fiber	134	62	60	46	44	16
fiberfill	5	7	3	7	6	4
fibre properties	1	0	0	0	0	0
fibrid	0	3	0	5	0	0
fibril	13	15	11	15	13	6
fibrillation	27	24	25	19	20	19
fibroin	0	5	5	8	6	10
filament property	14	32	12	3	10	10

WORLD TEXTILES Subject Indexing

	1999	2000	2001	2002	2003	2004
filament	107	52	70	63	42	52
film	53	41	109	228	221	88
filter fabric	54	96	54	78	34	59
filter	52	37	62	27	30	21
filtration	59	83	89	68	48	52
financial provision	0	2	1	12	12	17
financial services	0	0	0	1	4	2
fine structure	60	44	27	37	35	18
finishes	56	52	69	66	51	47
finishing agent	64	67	80	77	99	109
finishing industry	45	38	15	11	15	22
finishing machine	52	47	34	55	54	44
finishing	243	214	237	233	218	188
finite element analysis	0	0	0	3	33	52
fishing line	4	2	2	0	4	1
fishnet	0	0	0	0	1	0
fit	13	20	19	37	25	21
fixed bed	0	0	0	0	0	1
fixing agent	3	9	3	9	7	6
flag	4	4	9	4	6	4
flame resistance	99	105	108	128	143	104
flammability	12	31	26	40	31	19
flash aging	0	1	0	0	1	0
flash spinning	1	0	0	0	3	3
flat knitting	68	65	50	33	55	42
flax-jute blend	0	0	0	0	1	0
flax-polyester blend	0	1	0	0	1	0
flax-polypropylene blend	0	0	0	0	1	0
flax fiber fabric	0	1	0	1	3	1
flax fiber	18	22	23	22	33	20
flax industry	2	2	1	1	0	1
flax	9	12	10	14	9	7
fleece	21	21	26	27	18	12
flexibility	33	34	36	33	28	27
flocculation	3	1	8	3	3	0
flocking machine	2	9	3	1	3	2
flocking	21	70	22	32	23	18
floor covering	48	68	35	34	38	36
flotation	9	3	6	0	3	2
flow rate	0	1	19	11	13	6
flow	0	0	0	0	0	1
fluid dynamics	0	0	0	0	1	1
fluid filter	6	9	5	2	0	2

	1999	2000	2001	2002	2003	2004
fluid filtration	3	4	0	0	0	1
fluid flow	13	32	24	24	15	19
fluid mechanics	0	0	0	0	0	1
fluid	0	0	0	0	0	7
fluidized bed	0	0	0	0	0	1
fluorescence	15	17	40	46	39	19
fluorination	0	0	0	0	1	0
fluorocarbon	0	0	1	0	0	0
fluorochemical	19	32	16	11	13	9
fluoropolymer	0	0	0	1	16	24
fly	8	2	1	0	0	0
foam	37	42	13	29	34	16
folding machine	6	1	1	1	0	4
folding	25	16	13	12	8	15
footwear	94	99	85	65	51	57
force	0	0	0	0	1	9
forecasting	61	74	25	58	60	12
foreign body	2	1	9	3	11	4
formaldehyde-free agent	0	0	0	0	2	1
formaldehyde	12	20	14	13	19	4
formula	20	77	67	49	48	44
fouling	0	0	0	3	0	3
fourdrinier wire	2	0	1	0	0	0
Fourier transform analysis	7	7	22	15	14	4
fractal analysis	5	4	2	3	3	1
fracture mechanics	0	0	0	0	0	1
fracture	17	23	48	44	45	19
fragrance	0	0	1	0	0	0
framing	0	0	0	0	0	1
franchising	0	0	0	0	0	4
fraying	1	2	0	3	0	2
free trade	0	0	0	0	14	16
friction coefficient	15	16	15	15	17	8
friction spinning	10	16	11	8	12	7
friction	32	37	38	52	51	37
frosting	0	0	0	0	1	0
fuel cell	0	0	0	0	0	3
fuel	0	0	0	0	3	2
fungicide	0	4	1	0	0	0
furcrea fiber	0	0	1	0	0	0
Furcrea	0	1	0	0	0	0
furnishing fabric	46	60	69	59	49	45
furnishings	0	0	1	6	7	12

	1999	2000	2001	2002	2003	2004
furniture	15	12	2	20	36	15
fusing	10	13	15	16	15	6
fuzzy logic	7	16	15	6	8	7
gaiter	0	0	1	3	0	1
galloon	0	3	0	1	0	1
gamma radiation	0	0	0	0	1	0
gamma ray	0	1	0	0	0	0
garment appearance	0	0	0	0	5	9
garment component	0	0	0	0	1	10
garment construction	17	117	96	122	105	71
garment hanger	28	42	32	15	8	4
garment manufacture	217	169	105	83	114	117
garment property	67	104	31	25	20	18
garment	361	240	156	114	81	48
gas filter	8	4	4	6	3	2
gas filtration	7	6	6	6	2	0
gas flow	3	7	6	7	3	3
gas	9	16	20	28	25	12
gasket	1	1	1	2	2	1
gathering	4	3	1	0	1	1
GATT	0	0	2	0	0	0
gauge length	2	8	6	6	12	5
gauze	3	4	6	4	7	4
gear	7	4	1	3	7	5
gel	0	0	0	0	0	8
Gematex	0	1	0	0	0	0
gender identity	0	0	0	0	0	3
genetic modification	117	104	22	40	13	9
geocell	3	0	5	0	1	0
geogrid	1	6	18	3	8	6
geomembrane	6	10	25	8	15	7
geometric model	0	0	0	0	0	1
geometry	0	0	0	0	1	0
geonet	1	3	1	0	0	0
geosynthetics	45	58	57	54	51	58
geotextile	79	91	110	49	56	54
gilling machine	0	2	0	0	0	0
gilling	3	2	2	0	0	1
ginning machine	3	7	0	0	4	2
ginning	25	35	6	5	4	0
glass fiber fabric	31	24	28	38	8	23
glass fiber yarn	11	10	3	7	6	2
glass fiber	169	124	138	136	159	146

	1999	2000	2001	2002	2003	2004
glass transition temperature	8	25	43	70	42	11
glass	6	4	6	6	15	10
glitter	0	0	1	0	0	0
global communication	0	0	0	0	1	4
globalization	41	43	32	49	41	39
glove	74	91	16	6	25	16
goat hair	4	6	1	2	6	3
goat	0	0	0	0	0	1
godet	2	4	3	6	1	8
graft copolymer	6	14	22	56	41	14
grafting	18	18	44	88	74	33
graphite fiber fabric	0	0	0	0	0	1
graphite fiber yarn	0	0	0	1	2	1
graphite fiber	7	9	19	7	9	13
greige weaving	0	3	3	3	0	0
grinding	0	2	4	5	6	5
gripper	6	22	25	9	5	15
guide bar	32	41	24	8	4	3
guideline	0	0	0	0	1	0
guiding system	0	5	19	47	47	43
gusset	5	9	4	0	0	1
hackling machine	0	0	0	0	2	0
hackling	1	1	1	3	2	4
hair	0	0	1	0	2	1
hairiness	29	28	29	21	28	9
hand embroidery	0	0	7	1	12	3
hand knitting	1	1	2	2	2	3
hand sewing	1	1	3	1	0	1
hand spinning	0	0	0	0	0	2
hand weaving	0	8	4	6	11	5
handicraft	0	0	3	10	21	22
handle	65	65	77	65	64	42
handling pack	0	0	0	0	0	1
handloom	2	1	4	1	7	5
hank	1	2	2	1	1	0
hardward	0	1	0	0	0	0
hardware	1	6	4	3	4	5
harness	6	14	10	3	5	4
harvesting	12	18	5	11	6	0
headwear	55	63	15	23	19	12
heald frame	8	4	3	3	3	1
heald	12	7	4	3	5	4
health and disease	0	0	0	0	0	1

	1999	2000	2001	2002	2003	2004
health care	0	0	0	0	1	3
heat bonding	43	31	17	16	14	10
heat recovery	1	6	3	4	7	1
heat resistance	43	45	48	54	48	31
heat setting machine	6	8	9	8	4	5
heat setting	30	26	29	25	31	19
heat transfer printing	12	7	3	10	6	8
heat transfer	12	19	33	29	27	19
heat treatment	62	39	25	29	32	29
heating	51	63	70	72	66	43
heavy metal	0	0	0	0	0	1
heddle	0	5	6	6	1	2
helmet	4	2	0	0	0	0
hem	9	12	8	1	5	2
hemostatic property	0	0	1	0	0	2
hemp fabric	2	0	2	0	2	0
hemp fiber	15	10	7	7	15	19
hemp yarn	1	1	2	2	0	1
hemp	5	10	8	3	4	3
herbicide	10	17	4	3	1	1
hessian	0	0	0	0	1	0
hibiscus fiber	0	0	1	0	0	1
Hibiscus	0	1	0	0	0	0
high-modulus fiber	8	12	8	5	4	8
high-volume-instrument testing	11	3	0	1	0	2
high-volume-instrument	0	0	0	0	1	0
high performance	1	68	101	96	99	81
high speed	0	0	0	0	1	0
high temperature application	0	0	0	0	9	15
historical textile	4	29	38	39	56	32
history	0	0	0	1	0	1
hollow fiber	40	27	34	61	31	33
hollow filament	1	6	0	0	0	0
holographic property	1	5	1	1	0	0
home laundering	4	15	7	12	10	20
honeycomb construction	3	8	14	5	7	14
honeydew	0	0	1	5	2	0
hook	14	23	12	13	15	3
horsehair	0	0	1	0	2	0
horticultural textile	0	4	12	3	5	2
horticulture	0	0	0	0	0	1
hosiery	62	66	72	78	57	48

	1999	2000	2001	2002	2003	2004
hot-air balloon	1	1	0	1	1	0
household textile	57	59	50	31	33	44
household use	0	0	2	2	0	0
housing	12	14	6	11	28	23
human hair	6	10	4	2	3	2
human rights	0	0	0	0	1	1
humic acid	0	0	0	0	0	1
humidity	8	21	29	21	19	22
hydraulics	4	13	9	8	6	6
hydrodynamics	0	0	3	13	9	9
hydroentangling	45	57	52	41	51	47
hydroextraction	0	0	0	0	1	0
hydrogel	0	0	0	2	4	16
hydrogen peroxide	0	0	0	0	7	15
hydrogenation	0	0	0	0	1	1
hydrolysis	16	23	27	33	30	19
hydrophilic property	42	37	43	34	38	30
hydrophobic property	24	35	40	20	28	15
hygiene product	162	128	85	87	86	94
hygral expansion	2	5	3	3	6	1
identification method	0	0	0	0	0	26
ikat dyeing	0	1	1	0	1	3
image analysis	38	48	50	36	34	23
impact resistance	18	23	14	18	22	16
impact	21	23	23	27	32	21
implant	0	5	21	11	26	22
imports	50	80	65	76	70	36
impregnating	45	50	58	61	59	52
impurity	13	15	9	12	11	2
incontinence pad	9	9	6	9	14	7
indigo dye	10	14	9	8	9	9
indigo	8	3	8	7	4	4
indoor air quality	3	3	0	7	5	6
industrial application	0	0	0	0	2	0
industrial enterprise	0	0	0	0	0	4
industrial laundering	4	2	3	4	4	0
industrial policy	0	0	0	0	0	2
industrial revolution	0	0	0	0	2	1
industrial textile	182	235	264	201	161	115
industrial use	0	25	20	50	44	47
inflatable	0	2	0	0	17	32
inflation	15	22	17	23	11	5
information technology	65	65	79	82	58	48

	1999	2000	2001	2002	2003	2004
infrared radiation	12	6	14	6	9	7
infrared resistance	1	2	1	1	2	1
inhibitor	0	2	5	5	13	2
initial modulus	0	6	7	4	6	2
injection molding	7	15	25	31	21	21
ink-jet printing	50	55	82	78	64	64
ink	12	22	9	6	8	2
inlay	3	2	3	4	2	0
inorganic compound	0	0	0	0	1	1
inorganic fiber	0	0	0	0	0	5
inorganic polymer	0	0	0	0	1	0
insect resistance	23	8	7	3	3	3
insole	0	0	12	14	15	10
inspection machine	23	30	15	23	5	11
inspection	14	24	14	23	16	11
instability	0	0	0	0	1	0
insulating material	56	47	34	46	33	49
insulation	30	11	21	18	14	19
insurance system	0	0	0	0	1	2
intellectual property	0	0	0	8	17	9
interfacial property	11	34	50	43	40	26
interfacing	5	9	6	1	0	1
interference	0	0	0	0	1	0
interior design	1	8	24	38	23	27
interlace	2	0	0	0	0	0
interlaminar property	8	5	15	8	8	6
interlining	15	16	13	14	3	4
intermetallic alloy	0	0	0	0	0	1
intermingling	14	10	9	7	8	7
international trade	44	104	114	122	193	146
internet	34	79	83	54	45	29
investment	60	121	95	134	110	78
invoice	0	0	0	1	0	0
ion exchange	8	4	23	9	12	16
iridescence	0	1	2	1	0	1
iron oxide	0	0	0	0	0	1
iron	0	0	0	0	0	1
ironing machine	13	25	5	12	28	28
ironing	12	24	14	14	22	16
irradiation	5	16	21	40	60	23
jack	3	1	0	0	1	2
jacket	13	7	11	6	12	8
jacquard knitting	8	10	6	5	7	3

	1999	2000	2001	2002	2003	2004
jacquard loom	21	21	3	13	8	9
jacquard	20	16	19	16	22	14
jeans	10	15	6	11	8	19
jet dyeing	7	4	13	5	17	8
jet printing	1	0	0	0	0	0
jet	0	0	0	0	0	5
jig dyeing	3	5	2	1	2	0
joint venture	44	66	40	35	33	31
jute fabric	3	14	4	5	6	4
jute fiber	14	15	11	12	12	12
jute industry	2	0	3	1	1	2
jute yarn	3	7	3	4	5	6
jute	6	12	8	8	5	4
kapok	0	2	2	1	2	0
kenaf fabric	1	1	0	0	0	0
kenaf fiber	6	7	1	3	6	4
kenaf yarn	0	1	0	0	1	0
kenaf	4	5	1	1	1	0
keratin fiber	17	25	12	6	7	2
keratin	1	5	5	6	2	10
kerosene	0	0	0	0	0	1
Keular fiber	1	0	0	0	0	0
kiering	1	3	3	3	1	1
kilim	1	0	4	1	0	0
kinetics	1	9	114	128	150	57
kite	0	0	0	0	0	2
knife	9	10	10	4	2	6
knittability	1	2	12	1	1	3
knitted fabric	258	306	319	283	318	214
knitted yarn	0	1	1	2	12	8
knitting industry	23	20	19	30	37	31
knitting machine	210	214	246	180	187	146
knitting needle	33	19	26	33	27	11
knitting	61	64	97	59	69	48
knitwear	39	48	61	81	72	53
knot	0	5	5	1	3	7
knotting	8	4	10	5	6	7
label	12	28	13	19	17	26
labeling machine	4	1	1	3	0	4
labeling	25	23	10	22	17	30
labor relations	0	0	0	0	2	2
labor	30	38	53	35	68	32
laboratory	0	0	0	0	0	4

	1999	2000	2001	2002	2003	2004
lac dye	0	0	0	0	0	1
lace fabric	8	14	5	3	6	9
lace machine	4	4	2	0	3	9
lace making	10	12	9	13	23	30
lace	9	10	18	10	18	17
lambswool	0	1	2	1	0	0
laminate	193	130	181	131	156	117
laminating machine	5	8	13	7	11	8
laminating	19	33	41	32	32	12
lanolin	0	1	0	1	0	0
lap forming	4	7	6	1	0	1
lap	5	1	1	5	5	3
lariat	1	2	0	1	1	0
laser	11	15	21	17	23	14
lateral deflection	0	1	0	0	0	0
latex	19	10	16	45	24	15
launderability	21	42	47	34	31	18
laundering industry	2	2	0	0	2	0
laundering machine	7	24	53	46	129	162
laundering	59	69	101	160	114	103
layer separation	0	1	1	0	1	0
layering	13	34	38	52	76	62
laying-up	3	10	3	6	8	6
leakage	0	0	0	0	0	1
leather	0	0	0	2	9	15
legislation	15	28	34	34	38	31
leno technique	0	0	0	2	0	0
leno	4	9	9	4	17	6
let-off motion	6	4	11	5	1	1
leuco dye	0	2	0	1	1	1
leveling agent	4	4	4	4	6	2
leveling	2	2	9	3	2	4
liability	0	0	0	0	0	1
licensing	0	0	0	0	0	2
licker-in	1	1	0	1	3	0
life assessment	0	0	0	0	1	0
life prediction	0	1	7	17	10	12
lifting device	1	5	2	6	3	5
light scattering	1	0	0	0	0	0
lightfastness	34	23	26	36	35	26
lighting	0	0	0	0	1	2
Lindauer Dornier	0	0	0	1	0	0
linear density	29	36	23	35	26	7

	1999	2000	2001	2002	2003	2004
linen-polyester blend	1	1	0	0	2	1
linen-viscose-rayon blend	1	1	0	0	0	0
linen-wool blend	0	2	0	1	0	0
linen fabric	10	10	17	12	6	11
linen industry	7	3	1	4	5	3
linen yarn	8	2	0	5	6	0
linen	7	7	9	10	10	4
liner	6	22	14	14	13	9
lingerie	14	12	21	20	16	15
lining	25	30	34	22	25	19
linking machine	2	2	2	1	5	2
linking	3	0	2	2	6	5
lint removal	7	10	0	1	0	0
lint	13	13	10	5	2	2
lipid	0	0	0	0	2	1
liposome	5	2	3	2	1	2
liquid crystal	0	0	0	0	1	4
liquid	13	25	31	25	19	11
liquor	5	6	10	2	5	5
livestock production	0	10	22	52	34	13
loading (finishing)	0	0	0	0	0	2
loading (testing)	0	0	0	1	0	6
loading	34	39	67	56	55	33
loft	0	1	2	4	6	5
logistics	0	0	0	0	10	33
logo	0	0	0	0	1	0
loom component	77	121	78	53	41	72
loom function	1	1	0	0	0	0
loom	127	115	136	81	88	75
loop	16	31	24	9	24	17
lubricant	19	13	32	29	26	18
lubricating	16	10	6	16	12	18
lubrication	0	0	0	0	1	0
luggage	6	3	0	0	1	3
luminescence	2	9	11	17	21	7
luster	11	5	12	6	7	7
lyocell fabric	11	5	6	3	2	4
lyocell fiber	44	28	38	27	40	28
lyocell yarn	7	4	5	2	1	0
lyocell	10	6	10	4	8	4
machinability	0	0	0	0	0	3
machine clothing	17	27	17	10	19	24
machine component	330	342	227	145	173	186

WORLD TEXTILES Subject Indexing

	1999	2000	2001	2002	2003	2004
machine function	109	303	132	126	172	172
machine maintenance	0	0	0	0	1	0
machine	67	72	54	40	25	13
made to measure	0	3	4	5	3	5
maggot	0	0	0	1	0	0
magnet	15	13	17	10	16	10
magnetic property	0	0	0	1	0	0
magnetic resonance imaging	0	1	12	22	12	3
mail order	0	0	0	0	0	1
maintenance	24	18	38	42	30	28
making-up	57	153	101	66	56	19
management model	0	0	0	0	0	3
management	84	84	86	103	69	47
mandrel	0	0	5	6	6	4
mangle	3	2	3	0	1	2
manikin	0	0	0	0	3	2
mannequin	0	0	0	0	0	3
manufactured fiber fabric	8	8	13	1	2	3
manufactured fiber industry	34	42	42	43	32	18
manufactured fiber yarn	3	6	4	3	0	4
manufactured fiber	52	53	63	35	33	33
manufacturer	72	102	89	121	160	136
manufacturing	179	228	148	137	125	115
marine textile	30	18	31	25	26	27
market price	0	0	0	0	1	0
market	335	409	378	514	386	434
marketing	158	182	155	207	153	162
marking	5	3	5	9	4	4
mask	0	1	0	3	11	5
mass customization	0	0	0	1	0	0
mass dyeing	3	1	0	1	1	0
mass transfer	0	2	3	5	6	3
mat	44	43	33	19	28	40
material selection	0	0	0	0	2	3
materials handling system	33	40	27	13	11	15
materials handling	39	31	32	48	35	31
materials selection	0	0	0	0	6	7
maternitywear	0	0	0	0	1	2
mathematical analysis	94	163	166	190	169	99
mathematical expression	0	0	0	0	0	2
mathematical model	67	170	237	165	193	90
mathematical phenomena	0	0	0	0	0	5
matrix	34	39	45	28	21	22

	1999	2000	2001	2002	2003	2004
mattress	7	6	15	8	14	11
maturity	13	11	8	5	5	1
measuring system	130	162	77	74	66	35
measuring	70	37	42	78	87	42
mechanical property	198	237	293	329	391	242
mechanics	0	0	0	0	0	1
mechatronics	0	0	0	0	1	2
media	0	0	0	0	2	7
medical process	0	0	0	0	0	7
medical textile	308	299	266	231	247	209
medical use	0	0	0	3	47	27
medulla	0	0	1	0	2	0
melamine resin fiber	6	4	9	3	6	3
melt blowing machine	5	3	3	0	0	2
melt blowing	41	32	28	28	29	23
melt dyeing	0	0	0	0	1	0
melt spinning	74	80	81	77	61	54
melting point	32	19	33	43	38	15
membrane	57	76	111	133	62	77
menswear	29	26	24	27	29	38
mercerizing machine	4	4	3	2	4	1
mercerizing	11	19	13	11	12	8
merchandizing	0	0	1	13	8	11
merger	0	19	0	0	0	0
merino sheep	7	8	9	24	6	1
merino wool	21	14	13	24	21	10
mesh	0	0	0	16	21	28
metal-complex dye	10	9	15	9	8	10
metal fiber fabric	10	8	1	7	6	7
metal fiber yarn	1	4	4	4	11	1
metal fiber	19	27	35	19	18	17
metal matrix	0	0	0	0	0	3
metal oxide	0	0	0	0	0	1
metal	32	37	51	59	36	29
metallic property	0	0	0	0	0	2
metallization	6	2	8	3	8	4
metamerism	2	3	1	1	2	0
metering	3	3	3	3	8	10
methane	0	0	0	1	1	0
Michell Group	0	0	0	1	0	0
microcapsule	0	0	0	0	1	5
microclimate	0	0	0	0	1	0
microencapsulation	0	0	0	0	0	1

	1999	2000	2001	2002	2003	2004
microfiber fabric	5	15	7	2	8	15
microfiber yarn	0	3	3	0	0	0
microfiber	36	17	39	29	46	30
microfibril	0	0	0	0	0	1
microfilament	1	5	5	4	5	7
micron	0	0	1	0	0	0
micronaire value	2	3	2	6	7	1
microorganism	0	0	0	0	2	1
microscopy	9	13	13	22	21	9
microstructural analysis	12	19	42	57	31	14
microstructural property	0	0	0	0	18	3
microstructure	0	0	0	0	5	7
microwave radiation	0	0	0	0	1	2
microwave	0	0	1	0	1	0
milinery	0	1	0	0	0	0
military textile	1	19	18	20	35	24
mill	42	43	41	27	28	17
millinery	0	4	5	3	2	0
milling	6	5	3	4	3	4
mineral compound	0	0	0	0	0	1
mineral fiber fabric	4	1	0	0	0	0
mineral fiber yarn	1	1	0	0	0	0
mineral fiber	21	12	10	9	13	10
mineral wool	8	4	4	2	3	1
mixing	0	0	0	2	5	9
modacrylic fabric	0	1	0	0	0	0
modacrylic fiber	2	1	3	1	1	2
modacrylic yarn	0	0	0	0	0	1
modacrylic	0	0	1	0	1	0
modal fabric	0	1	0	0	0	1
modal fiber	1	2	0	4	5	1
modal yarn	0	1	0	0	1	0
modal	0	0	0	1	1	1
model	72	56	45	33	25	33
modeling	0	0	0	3	0	0
modernization	0	1	3	1	0	0
modification	0	0	0	0	0	4
mohair fabric	3	0	0	0	0	1
mohair fiber	3	4	1	2	2	0
mohair yarn	0	3	0	1	0	0
mohair	4	5	6	8	7	2
moire fabric	0	0	0	1	0	0
moire	0	0	1	1	0	0

	1999	2000	2001	2002	2003	2004
moisture content	10	23	21	29	21	8
moisture management	0	0	0	4	23	21
moisture regain	8	4	6	9	8	1
moisture release	0	1	2	3	2	2
moisture transfer	25	28	34	38	23	22
moisture	14	9	14	13	10	10
molding	67	69	56	58	85	69
molecular structure	22	63	101	157	106	47
molecular weight	12	39	80	110	85	25
monitoring system	36	77	50	45	28	31
monitoring	43	23	20	35	41	29
monofilament	24	27	23	32	30	30
monomer	2	9	38	15	22	6
mop	0	1	1	0	1	0
mordant dyeing	3	12	7	10	13	12
mordant	9	10	19	25	21	8
morphology	19	38	93	169	200	88
mote	1	2	1	0	1	0
moth resistance	2	2	1	1	3	1
motif	0	0	0	0	0	2
motivation	0	0	0	0	0	1
motor	21	35	33	32	21	18
mounting	16	19	6	5	4	8
muga silk	1	4	2	5	3	2
mulberry silk	9	7	5	9	5	3
mulberry	15	14	15	18	14	10
mullite	0	0	0	0	1	0
multiaxial fabric	10	15	16	13	11	12
multicomponent fiber	0	0	0	0	1	0
multifilament	0	1	12	27	23	17
multilayer fabric	94	72	64	43	43	49
multiphase weaving	13	4	6	9	7	1
museum	8	4	17	17	28	9
muslin	1	0	0	2	0	0
nanocomposite	0	0	0	8	134	77
nanofiber	0	0	0	2	20	46
nanofiltration	0	0	0	0	0	1
nanoparticle	0	0	0	0	1	0
nanotechnology	0	3	49	60	62	54
nanotube	0	0	0	0	17	41
nap	0	0	1	1	2	0
napping machine	2	3	0	0	0	0
napping	1	2	3	3	4	3

	1999	2000	2001	2002	2003	2004
narrow fabric	31	22	23	24	18	14
narrowing	0	0	0	0	3	0
natural dye	30	41	54	64	67	57
natural fiber fabric	9	15	4	4	5	4
natural fiber industry	5	9	8	8	4	3
natural fiber yarn	4	6	2	2	1	0
natural fiber	58	36	53	50	56	46
natural resin	0	0	0	0	0	1
natural rubber	0	0	0	0	0	7
navel	0	0	1	2	3	0
necking	6	5	0	6	1	2
needle bar	6	18	9	6	3	4
needle bed	15	21	13	8	14	2
needle penetration	3	10	4	5	7	9
needle selection	11	15	6	6	16	6
needle	68	65	71	45	27	32
needled fabric	34	67	31	38	30	14
needling machine	15	14	10	17	15	15
needling	38	27	23	38	21	30
nep	12	7	3	6	5	1
net curtain	2	1	0	5	1	2
net	21	21	22	12	25	11
neural network	7	22	26	23	22	20
never-dried cotton	1	0	0	0	0	0
niche market	0	0	1	0	0	5
nickel	0	0	0	0	0	2
nip	7	6	12	7	5	2
nipper	4	3	2	1	0	0
nitric acid	0	0	0	0	0	1
noil	2	1	0	0	1	2
noise control	24	22	13	27	19	20
nonformaldehyde agent	0	0	0	0	1	0
nonwoven fabric industry	57	49	52	50	35	48
nonwoven fabric machine	31	40	54	48	36	46
nonwoven fabric manufacture	98	140	91	102	104	133
nonwoven fabric	480	453	448	427	430	379
North American Free Trade Agreement	0	0	0	0	3	0
novolac fiber	0	1	1	0	0	1
nozzle	33	42	39	38	33	26
nuclear application	0	0	0	0	0	1
nucleation	0	0	0	1	0	1
nutrition	0	0	0	1	0	0

	1999	2000	2001	2002	2003	2004
nylon-polyester blend	1	1	0	1	0	2
nylon-polyethylene blend	2	0	0	0	0	0
nylon-viscose-rayon blend	0	0	0	0	1	0
nylon-wool blend	3	4	1	2	1	1
nylon 11	0	0	0	1	1	0
nylon 4	0	0	0	1	0	0
nylon 46	2	0	0	0	0	0
nylon 6	16	31	55	68	52	42
nylon 66	12	21	27	26	32	15
nylon fabric	25	26	18	14	7	15
nylon fiber	53	40	26	28	19	24
nylon yarn	14	17	7	5	7	3
nylon	33	25	29	21	44	19
objective assessment	6	24	25	25	7	7
occupational health	21	47	18	19	10	4
odor control	28	16	22	26	14	23
odor emitting	0	0	5	6	13	12
odor	0	4	0	0	0	0
off-shore	0	1	0	0	0	0
offshore application	0	0	0	0	1	2
oil removal	0	0	4	2	6	0
oil resistance	13	14	12	10	24	13
oil	0	1	0	0	3	4
oleophilic property	0	1	4	0	2	0
oleophobic property	0	1	0	0	0	0
oligomer	4	2	16	27	32	8
on-line monitoring	17	45	21	26	21	17
one step dyeing	1	3	2	3	2	4
one step process	0	7	4	9	16	5
open-end spinning	66	47	37	32	19	28
open-width dyeing	3	2	1	0	8	5
open-width process	3	5	2	9	5	4
opening machine	2	7	5	9	5	12
opening	16	18	9	9	10	9
operation	0	0	0	0	1	0
optical phenomena	0	0	0	0	0	1
optical property	0	0	0	0	7	27
optimization	1	6	13	91	77	66
organic compound	0	0	0	0	0	5
organic dye	7	4	11	2	3	1
organic farming	0	0	1	1	1	1
organic fiber	0	1	0	3	2	4
organic product	0	1	5	4	3	3

	1999	2000	2001	2002	2003	2004
organic solvent	0	0	0	1	0	0
organism	0	0	0	0	0	1
organization structure	0	2	0	0	0	0
organizational restructuring	0	0	0	0	1	5
organizational structure	0	34	64	112	57	41
orientation	10	9	8	26	16	5
orthosis	7	5	4	5	6	6
outerwear	22	28	32	24	21	8
outsourcing	1	19	10	18	25	13
overlock sewing	3	9	2	0	3	1
oxidation biological treatment	0	0	1	0	0	0
oxidation	45	41	40	49	65	40
oxide	0	0	0	0	0	1
oxidizing agent	8	15	10	5	9	4
oxygen	0	0	0	0	0	2
ozone treatment	10	12	14	9	16	8
package	32	31	23	23	23	16
packaging	29	38	22	35	25	18
pad-batch dyeing	1	13	5	4	3	5
pad-steam dyeing	0	1	1	0	2	1
pad dyeing	2	4	1	1	1	6
pad	10	17	11	17	9	8
padder	0	0	0	0	0	1
padding (finishing)	0	0	3	3	5	4
padding (garment)	0	0	0	1	0	0
padding (material)	0	0	0	0	0	6
padding (product)	0	0	0	8	3	4
padding	30	33	13	2	0	1
paint	0	0	0	0	1	0
papermakers fabric	82	64	63	53	54	71
papermaking machine	34	52	33	9	11	20
papermaking	42	64	54	37	39	26
parachute	1	1	4	3	4	1
partially oriented yarn	0	0	6	9	14	4
particle size	7	21	12	13	31	9
particle	26	23	17	25	40	15
patch	0	0	0	0	2	0
patchwork	0	0	0	0	0	3
patent (independent inventor)	0	0	0	0	0	3
patent	0	0	0	10	22	12
pater preparation system	0	0	1	0	0	0
pattern (configuration)	0	0	0	0	0	26
pattern (plan)	0	0	0	0	0	3

	1999	2000	2001	2002	2003	2004
pattern cutting	21	20	36	18	14	8
pattern preparation system	24	24	18	2	10	5
patterning	73	57	70	62	50	51
patterns	37	56	73	67	82	23
perchloroethylene	2	0	2	4	1	2
perfluoropolymer	0	0	0	1	4	0
perforation	15	9	8	25	5	10
performance assessment	0	3	19	56	86	58
perfume	0	1	0	0	0	0
permeability	79	85	80	75	76	49
personnel management	17	27	21	26	8	8
perspiration control	12	20	16	7	6	3
pervaporation	0	0	0	0	1	1
pest control	236	187	13	15	7	5
pesticide	73	89	17	21	6	2
pH	16	28	25	26	21	8
phase-change material	11	16	19	14	12	26
phase separation	0	4	23	29	35	8
photochemical degradation	6	8	22	17	27	9
photochromic dye	0	1	3	2	1	2
photography	0	0	0	0	1	1
photolysis	0	0	0	1	2	0
physical property	67	84	70	66	90	39
physicochemical property	0	0	2	14	13	12
physicomechanical property	0	3	4	11	3	1
physics	0	0	0	0	0	1
physiochemical property	0	0	3	0	1	0
pick	1	7	4	3	1	1
picking	3	7	0	3	2	3
piece dyeing	9	5	1	5	5	0
piecing	13	7	1	3	3	5
piezoelectric device	12	6	12	26	21	9
piezoelectric fiber	4	0	5	1	0	0
pigment printing	11	11	8	8	7	3
pigment	30	44	44	48	41	38
pile fabric	0	0	0	0	0	3
pile	37	41	49	43	37	31
pill resistance	7	12	12	12	8	6
pilling	3	13	14	14	14	7
pineapple leaf fiber	3	2	0	0	3	1
PIPD fiber	0	0	0	1	0	0
pipe	13	4	8	13	13	15
pirn	1	0	0	2	0	4

	1999	2000	2001	2002	2003	2004
pitch	3	1	1	2	2	2
planning	12	16	17	29	41	20
plant extract	0	0	0	0	4	1
plant layout	0	17	20	14	16	6
plasma treatment	38	37	32	48	54	30
plastic	21	25	48	65	81	71
plasticity	0	0	0	0	0	1
plasticizing agent	5	7	14	4	11	1
plating	0	0	0	0	2	2
pleat	5	7	3	3	3	2
plexifilament	7	7	0	5	2	3
plied yarn	22	14	6	10	8	6
plying	11	14	12	12	28	10
pneumatic device	21	35	14	15	24	21
pocket	22	27	16	13	6	12
policy development	0	0	0	0	0	1
political statement	0	0	0	0	0	2
pollutant removal	0	0	0	1	4	9
pollution control	29	49	44	53	39	37
pollution	13	9	11	25	15	5
polyacrylamide	0	0	2	8	14	2
polyacrylate	0	0	3	0	0	0
polyacrylic acid fiber	2	0	0	0	0	0
polyacrylic acid	0	0	2	20	10	6
polyacrylonitrile	4	6	15	19	20	23
polyamide fabric	0	0	0	0	0	2
polyamide fiber	0	0	0	0	3	9
polyamide yarn	0	0	0	0	1	5
polyamide	41	70	100	122	117	77
polyamine	0	0	0	0	3	0
polyaniline	0	0	0	11	21	14
polyaryletherketone	0	0	0	0	2	0
polybenzazole fiber	0	2	2	2	2	3
polybenzimidazole	0	0	0	3	2	1
polybutadiene	0	0	0	0	3	3
polybutylene terephthalate fiber	0	2	4	0	3	0
polybutylene terephthalate	3	3	15	10	20	8
polycaprolactone	0	0	0	3	20	5
polycarbonate	0	0	0	4	29	17
polycarboxylic acid	6	16	9	10	10	5
polydimethylsiloxane	0	0	0	0	9	3
polyester-polyethylene blend	0	1	0	0	1	2

	1999	2000	2001	2002	2003	2004
polyester-polypropylene blend	0	1	0	1	2	0
polyester-silk blend	0	1	2	0	0	0
polyester-viscose-rayon blend	7	8	9	10	7	4
polyester-wool blend	4	14	11	5	15	9
polyester fabric	96	68	52	59	52	48
polyester fiber	108	109	102	90	103	56
polyester yarn	42	47	49	28	57	42
polyester	71	59	101	160	144	102
polyether ether ketone	0	0	0	0	7	6
polyether ether	0	0	0	0	0	1
polyether imide	0	0	0	0	0	1
polyether ketone	0	0	0	1	0	0
polyether sulfone	0	0	0	2	0	2
polyetheretherketone fiber	0	1	0	0	0	0
polyetheretherketone	1	0	3	6	5	0
polyetherimide	0	0	0	2	0	0
polyetherketone	0	0	0	0	1	0
polyethylene-polypropylene blend	4	2	4	1	0	0
polyethylene fabric	6	6	3	3	5	1
polyethylene fiber	18	21	11	15	15	19
polyethylene glycol	0	0	0	0	7	7
polyethylene naphthalate	0	1	5	8	7	5
polyethylene oxide	0	0	0	0	4	9
polyethylene terephthalate fabric	4	5	5	2	5	7
polyethylene terephthalate fiber	21	22	31	19	27	27
polyethylene terephthalate yarn	7	11	7	1	1	2
polyethylene terephthalate	13	22	54	86	90	44
polyethylene yarn	2	4	1	2	0	2
polyethylene	17	42	104	174	153	63
polyimide fabric	0	1	0	0	0	0
polyimide fiber	0	0	1	0	0	2
polyimide	2	5	25	57	49	22
polyisoprene	0	0	0	1	5	2
polyketone	0	0	0	0	1	1
polylactic acid fiber	8	18	9	15	17	6
polylactic acid	2	9	18	24	39	12
polylactide	0	0	0	0	3	4
polymer blend	0	0	0	83	206	75
polymer melt	11	15	40	46	41	15
polymer property	19	57	179	161	70	32
polymer science	0	2	198	502	372	1888
polymer solution	0	0	7	38	23	15

	1999	2000	2001	2002	2003	2004
polymer	140	225	529	385	480	466
polymeric compound	0	0	1	46	6	0
polymeric	0	0	0	0	0	1
polymerization	26	43	197	386	316	144
polymethyl methacrylate	0	1	1	41	34	17
polynosic fiber	0	1	0	0	0	0
polyolefin fabric	10	6	1	1	3	0
polyolefin fiber	33	19	21	12	9	10
polyolefin yarn	3	3	2	0	0	0
polyolefin	22	27	31	52	31	32
polyoxymethylene	0	0	0	0	0	1
polypeptide	0	0	0	1	3	1
polypheneylene terephthalamide fiber	0	1	0	0	0	0
polyphenylene benzobisoxazole	0	0	0	0	0	2
polyphenylene oxide	0	0	0	0	2	0
polyphenylene sulfide	0	0	0	1	2	1
polyphenylene sulphide fiber	0	2	4	1	0	1
polyphenylene terephthalamide	0	0	0	1	0	0
polyphenylenebenzobisoxazole	0	0	0	1	0	0
polyphenylenebenzobisthiazole	0	0	0	1	0	0
polypropylene-viscose-rayon blend	0	0	0	1	1	0
polypropylene fabric	21	19	11	15	15	8
polypropylene fiber	71	78	48	45	44	39
polypropylene terephthalate	0	0	2	0	2	1
polypropylene yarn	8	8	18	5	2	4
polypropylene	31	57	144	176	176	101
polypyridobisimidazole	0	0	0	0	0	1
polypyrrole	0	0	0	0	0	2
polysaccharide	0	0	0	5	15	12
polysiloxane	0	0	0	0	7	7
polystyrene	0	0	7	101	81	23
polysuccinimide	0	0	0	0	1	0
polysulfide	0	0	0	0	0	1
polysulfone	0	0	0	0	1	4
polytetrafluoroethylene fabric	3	5	0	1	0	0
polytetrafluoroethylene fiber	11	8	4	1	0	0
polytetrafluoroethylene yarn	0	2	0	1	0	0
polytetrafluoroethylene	13	14	14	23	14	20
polytrimethylene terephthalate	0	9	26	43	43	37
polyurethane fabric	3	1	2	0	0	1
polyurethane fiber	8	6	11	2	5	4
polyurethane yarn	2	1	1	0	0	1

	1999	2000	2001	2002	2003	2004
polyurethane	32	39	78	106	86	74
polyvinyl acetate	0	0	0	0	6	3
polyvinyl alcohol fabric	2	2	1	1	0	0
polyvinyl alcohol fiber	14	5	6	4	4	7
polyvinyl alcohol yarn	0	1	0	0	0	0
polyvinyl alcohol	3	7	33	30	31	21
polyvinyl chloride fabric	4	3	2	0	1	1
polyvinyl chloride fiber	3	2	1	0	1	0
polyvinyl chloride yarn	1	0	0	0	0	0
polyvinyl chloride	9	17	19	41	42	19
polyvinylidene fluoride	0	0	0	0	0	2
pompon	1	0	0	0	0	0
pore size	5	13	9	11	12	6
porosity	52	47	41	44	55	31
powder	0	5	12	18	25	10
power consumption	2	7	2	11	8	3
precursor	13	8	17	25	30	13
prediction	0	0	0	0	1	2
predictive model	0	0	0	0	0	8
preform	21	23	29	23	26	31
preparation	59	24	29	7	3	6
prepreg	12	20	22	7	19	11
preservative	0	0	0	1	1	1
press felt	0	19	14	8	20	29
press finishing	1	13	16	23	20	13
press furnishing	0	0	0	0	1	0
presser foot	4	10	6	7	9	5
pressing machine	7	11	5	5	6	8
pressing	23	17	17	19	21	21
pressure	33	40	55	69	48	41
pretreatment	51	56	55	66	67	39
price	51	89	73	130	109	63
printability	6	7	9	5	5	7
printed fabric	2	2	0	0	3	12
printed textile	0	8	0	0	0	0
printing assistant	4	4	0	1	2	0
printing industry	15	11	5	3	6	13
printing ink	12	12	36	47	44	39
printing machine	35	42	42	36	36	38
printing paste	17	10	16	13	17	10
printing	94	58	118	83	95	76
process control	17	49	48	51	65	68
process engineering	0	3	5	2	3	3

	1999	2000	2001	2002	2003	2004
process	0	0	0	0	1	0
product component	0	0	0	0	0	1
product development	3	127	167	231	182	110
product life cycle	0	0	0	1	4	7
product	0	0	0	0	0	3
production control	6	14	14	14	12	2
production management	53	63	35	57	50	46
production	240	246	277	352	306	164
productivity	98	87	95	99	111	42
professional association profile	0	0	0	0	10	14
professional practise	0	0	1	1	26	35
profitability	1	9	37	88	61	64
proofing	22	22	27	18	23	15
prosthesis	32	15	25	19	20	16
protease	7	13	10	7	6	2
protectionism	0	0	0	0	3	3
protective clothing	235	227	141	100	143	82
protective fabric	96	109	73	41	56	57
protective structure	0	0	0	0	2	0
protein fiber	0	0	0	0	0	1
protein	4	10	20	30	36	40
psychology	0	0	0	5	1	10
puckering	6	8	8	6	3	3
pull-out	6	6	14	7	7	5
pulp	0	0	0	0	0	6
pulse jet	1	0	0	0	0	0
pultrusion	6	5	11	13	8	14
pump	0	0	1	1	20	8
pumping system	0	0	0	0	1	0
punching	0	0	0	0	1	0
puncture resistance	11	14	12	5	1	7
purification	0	0	2	11	12	10
pyrolysis	2	7	8	12	14	6
qualitative analysis	11	24	16	21	19	32
quality assurance	24	62	48	49	49	28
quality control	113	172	128	146	107	85
quality	107	145	182	141	69	27
quantitative analysis	28	48	19	24	22	15
quenching	4	7	10	15	5	2
quick response	3	13	8	7	8	11
quilt	0	0	0	1	12	16
quilting machine	5	11	13	5	2	10
quilting	10	11	17	19	17	19

	1999	2000	2001	2002	2003	2004
quota system	0	0	0	0	3	2
radiation resistance	0	0	0	0	0	1
radiation	11	16	16	25	26	23
radio frequency	7	7	2	3	9	16
rail	0	0	0	0	0	1
raising machine	6	7	2	2	1	0
raising	8	6	4	6	5	4
Raman spectroscopy	6	7	9	11	14	6
ramie fabric	0	0	3	2	2	0
ramie fiber	3	5	3	4	0	4
ramie yarn	0	0	0	1	0	1
ramie	0	0	3	2	1	0
rapier loom	44	36	30	24	25	25
rapier	12	9	7	4	3	4
raschel knitting	15	24	28	19	21	11
raw material	26	41	30	23	48	19
raw silk	0	0	3	7	9	8
rayon fiber	0	0	0	0	2	3
reactive dye	76	88	89	105	93	78
reactive printing	8	8	6	1	4	2
reactor	0	0	0	0	8	7
ready-to-wear	0	0	0	0	1	0
recycling	97	91	120	137	111	101
reducing agent	12	14	24	21	9	12
reed	8	16	11	9	11	5
reeding	1	0	0	0	0	0
reeling	17	17	13	12	14	7
reflective property	16	26	23	18	16	14
refrigerant	0	1	1	0	2	1
regenerated cellulose fabric	3	2	2	1	2	0
regenerated cellulose fiber	22	12	17	5	17	10
regenerated cellulose yarn	0	1	0	0	1	0
regenerated cellulose	0	4	15	9	15	9
regenerated protein	0	0	0	0	0	1
regeneration	0	0	0	0	1	0
regulation	0	8	31	27	31	31
regulations	30	40	0	0	0	0
reinforced fabric	18	23	6	16	7	21
reinforcement	120	107	118	95	81	60
relaxing	8	15	23	44	23	3
relief fabric	0	2	2	0	3	0
remazol dye	0	0	0	2	0	0
removal	0	0	0	0	0	4

	1999	2000	2001	2002	2003	2004
renewable resource	0	10	14	15	28	8
repair	9	13	13	19	10	24
repeat	10	6	16	18	3	6
repellent property	1	3	0	0	0	0
reproducibility	0	0	0	0	3	4
research (academic)	26	13	19	15	28	26
research and development	172	306	210	259	186	152
resin transfer molding	20	12	15	10	16	27
resin	77	59	92	78	82	46
resist dyeing	3	2	5	7	6	3
resist printing	0	0	2	2	5	2
resistant property	0	0	0	0	0	1
resistivity	0	0	0	0	0	1
restoration	0	1	0	5	8	9
restructuring	0	0	0	0	2	2
retailing	89	96	59	87	86	173
retting	4	12	6	8	8	0
reusability	0	0	0	0	3	1
reverse osmosis	0	0	1	1	1	3
reversible garment	0	0	0	0	2	2
reversible	0	2	0	0	0	0
rewinding	5	3	7	4	1	0
rheology	22	19	50	57	35	32
ribbon	16	11	4	10	2	3
right first time approach	0	0	0	0	2	4
ring spinning	67	71	64	50	58	39
ring	5	11	6	6	15	6
rinsing	7	11	10	13	12	8
risk assessment	0	0	0	2	1	4
RMD	0	1	0	0	0	0
road traffic accident	0	0	0	0	0	1
road	0	0	0	1	0	0
robotics	0	0	0	3	11	9
roller	31	58	46	46	39	26
roofing	20	27	25	24	27	28
rope dyeing	6	3	7	1	7	1
rope making	13	11	6	9	8	14
rope	35	22	20	16	12	11
rotary printing	2	8	4	3	3	5
rotor spinning	62	49	57	62	63	32
rotor	35	25	24	19	11	18
roving machine	19	9	5	0	2	1
roving	28	18	19	16	12	16

	1999	2000	2001	2002	2003	2004
royalties	0	0	0	1	1	0
rubber fabric	0	1	2	0	0	1
rubber yarn	2	0	0	0	0	0
rubber	17	32	61	119	126	40
rug	13	13	19	14	7	4
s-twist yarn	0	1	0	0	0	0
s-twist	0	1	3	0	0	0
safety	67	108	82	74	55	41
sailcloth	1	6	9	1	5	4
sales	64	139	114	174	132	90
salt	0	0	0	1	0	0
sample	8	5	8	14	13	4
sampler	0	0	0	0	0	1
sampling	34	26	16	22	18	3
sanitary napkin	16	6	2	1	4	3
sanitary product	0	14	23	32	18	21
sari	1	3	4	2	0	2
saxony carpet	0	0	0	0	1	0
scanning electron microscopy	11	20	15	9	19	11
scanning technology	0	11	7	13	12	7
scouring machine	1	5	4	2	1	5
scouring	33	32	30	39	28	15
screen printing	16	19	15	25	21	29
scrim	7	6	10	6	8	14
sculpture	0	0	0	0	0	1
scutching machine	7	3	4	3	1	2
scutching	7	5	5	7	4	3
seal	18	28	19	19	26	23
seam	63	56	72	67	61	36
seamless	0	21	24	43	25	16
seat belt	0	0	0	0	3	2
seat cover	5	9	6	14	8	3
seat	5	9	4	5	13	4
seaweed	0	0	0	0	1	0
security review	0	0	0	0	0	2
seed	7	10	8	5	5	1
selective breeding	0	0	2	4	0	2
self healing property	0	0	1	0	0	0
selvedge	25	24	26	15	9	9
sensor	50	61	39	67	47	34
separation	0	23	31	29	30	35
sequin	1	0	0	0	0	0
sericin	0	0	0	0	1	4

WORLD TEXTILES Subject Indexing

	1999	2000	2001	2002	2003	2004
sericulture	58	57	68	55	82	67
service sector	1	0	0	0	0	1
sett	0	0	6	2	0	2
sewing machine	101	144	132	82	103	115
sewing speed	0	0	4	2	3	0
sewing thread	38	39	60	39	56	42
sewing	87	100	111	79	69	51
shade variation	4	16	20	21	13	3
shape memory material	0	0	7	18	14	12
shape	0	0	0	0	0	4
shear property	19	20	13	15	6	5
shearing (finishing)	0	0	0	0	0	1
shearing (livestock)	0	1	1	6	2	2
shearing machine	6	5	2	4	3	1
shearing	11	8	5	6	8	4
sheath-core structure	47	57	42	31	46	47
shed	26	16	22	36	12	17
shedding	0	0	0	0	0	1
sheep	6	12	13	23	26	5
sheet	0	0	0	0	1	0
shielding	0	0	0	0	1	0
shingosen fabric	6	3	0	1	2	1
shirt	17	19	21	20	23	16
shoulder pad	1	2	0	0	1	0
shrink resistance	17	21	17	13	26	10
shrinkage	34	41	39	41	57	23
shrinking machine	5	4	4	4	1	0
shrinking	3	5	10	9	5	4
shuttle loom	4	2	5	2	2	3
shuttle	4	10	8	3	3	6
shuttleless loom	50	32	29	28	15	16
silica fiber	0	0	0	0	2	0
silicon carbide fiber	39	29	28	22	15	14
silicon dioxide	0	0	0	0	0	1
silicon	0	1	19	10	8	6
silicone	0	0	6	11	9	16
silk-wool blend	1	0	0	0	0	0
silk fabric	35	22	29	40	30	29
silk fiber	22	21	25	17	19	29
silk industry	55	51	52	47	52	60
silk yarn	8	5	6	7	6	5
silk	25	26	28	31	26	30
silkworm	34	16	20	27	53	23

	1999	2000	2001	2002	2003	2004
silver	13	5	6	14	11	21
singeing machine	1	3	2	1	0	0
singeing	2	1	4	1	5	1
sinker	8	6	16	6	5	5
sintering	5	12	2	5	4	0
sisal fabric	2	0	0	0	0	0
sisal fiber	2	6	2	5	11	10
sisal	1	1	0	2	4	0
size	0	0	0	0	0	3
sizing agent	28	13	14	21	17	13
sizing machine	7	8	11	12	8	5
sizing	41	45	36	49	28	20
skin (human)	2	3	5	9	10	3
skirt	4	2	4	5	7	3
skiwear	3	3	0	2	8	5
sleeping bag	2	9	3	2	2	1
sleepwear	0	3	1	0	1	0
sleeve	19	18	22	24	27	12
sley	0	0	0	0	0	2
slip resistance	10	11	12	5	11	13
slit	0	2	0	0	0	0
slitting	7	9	5	1	3	3
sliver can	5	3	1	2	3	0
sliver guide	2	2	0	1	0	0
sliver	52	40	34	21	30	12
slurry	0	0	0	0	0	1
smart fabric	0	3	19	44	44	42
smart textiles	0	0	1	0	0	0
snakeskin	0	0	0	0	1	0
social history	0	0	0	13	14	6
social impact	0	0	1	15	18	7
sock	25	40	19	33	31	19
sodium hydroxide	0	0	0	0	1	0
softening agent	20	28	40	32	38	26
software	5	86	97	103	89	72
soil burial test	0	0	0	0	0	2
soil reinforcement	8	22	45	24	26	15
soil release	17	21	20	18	13	24
soil resistance	9	14	15	14	12	14
soil test	0	2	2	8	7	1
sol-gel process	0	0	0	1	0	0
solubility	22	31	61	71	58	27
solvent	30	44	64	53	35	13

	1999	2000	2001	2002	2003	2004
sorption	15	12	19	27	36	13
sound proofing	0	0	0	0	1	0
sourcing	21	23	17	30	18	20
soybean protein fiber	0	0	0	0	0	1
space dyeing	2	2	2	7	2	2
spacer fabric	11	19	20	25	24	26
spandex fabric	6	3	3	1	1	1
spandex fiber	12	11	7	8	5	5
spandex yarn	2	6	3	1	2	1
spandex	7	5	14	8	7	16
Spanish broom fiber	0	1	0	0	0	0
special precaution	0	0	0	0	0	1
spectrometry	0	0	0	0	1	0
spectrophotometry	10	14	22	17	9	11
spectroscopy	18	15	62	78	49	16
speed	0	0	1	4	45	56
spider silk	2	7	4	11	7	10
spill	0	0	0	0	1	0
spin-drawing	5	6	4	1	0	4
spin dyeing	0	0	0	0	1	0
spin finish	17	35	24	21	16	12
spindle	31	37	24	26	17	12
spinnability	8	15	16	7	14	14
spinneret	29	15	13	22	24	17
spinning industry	43	28	17	24	23	17
spinning machine	162	171	155	128	138	108
spinning solution	45	35	19	24	18	15
spinning speed	0	0	0	0	3	0
spinning	154	138	178	118	136	87
splicing	9	20	13	13	16	14
splint	0	2	1	2	0	2
spool	0	15	20	10	9	7
sports goods	37	61	43	29	19	23
sports surface	4	13	4	1	8	15
sportswear	169	148	100	90	80	77
spray application	0	0	1	1	8	12
spray deposition	0	0	0	0	1	0
spray dyeing	3	0	0	4	2	2
spraying	0	0	0	0	1	0
spreading machine	8	4	3	1	4	1
spreading	8	10	6	11	11	9
spunbonding machine	6	6	1	2	1	5
spunbonding	47	25	29	30	29	43

	1999	2000	2001	2002	2003	2004
sputter etching	0	1	1	0	0	0
squeegee	0	0	1	1	2	2
stab resistance	9	7	3	2	5	5
stability	0	0	0	0	0	3
stabilizing agent	14	16	19	11	5	10
stain removal	9	21	25	22	11	16
stain resistance	11	22	23	18	15	27
stainless steel	0	0	0	0	2	0
standard	68	106	110	140	124	85
staple fiber	32	35	31	42	36	24
starch	13	7	15	27	34	33
static electricity	18	28	25	6	29	18
steam	10	23	17	19	23	9
steaming machine	9	5	6	8	5	5
steaming	11	19	17	17	11	8
steel	0	0	0	0	0	3
stencil	0	1	2	4	3	4
stentering machine	14	17	4	11	4	5
stentering	7	6	3	4	2	4
sterilization	0	4	2	4	4	4
stickiness	9	5	4	7	2	3
stiffness	16	23	28	38	38	32
stitch-bonding machine	6	4	4	5	5	2
stitch-bonding	19	14	14	15	20	8
stitch	33	68	90	78	73	48
stonewashing	8	7	4	6	1	2
stop motion	3	11	0	4	5	5
storage	51	62	24	26	38	21
strain	11	19	29	24	27	22
streakiness	0	1	0	0	0	2
strength	104	99	153	143	182	131
strengthering	1	0	0	0	0	0
stress-strain property	36	26	41	34	19	27
stress	33	37	67	59	62	47
stretch	65	79	85	87	87	64
stripping	0	1	0	0	0	0
structural analysis	69	81	86	56	56	52
structural property	92	98	133	120	106	63
structure	1	0	0	0	0	0
stuffer box	0	0	1	3	1	3
stuffing	4	3	3	1	1	1
substrate	13	9	16	12	11	8
suction device	22	22	16	22	15	5

	1999	2000	2001	2002	2003	2004
suit	9	15	14	8	11	6
suiting	0	0	1	1	0	0
sulfur dye	11	10	13	6	8	5
sulphonic acid	0	0	0	0	1	0
superabsorbency	2	0	0	0	0	0
superabsorbent material	4	22	21	29	18	24
supercritical carbon dioxide dyeing	15	7	10	7	8	4
supercritical carbon dioxide	0	1	12	15	18	10
supercritical fluid	0	0	0	0	0	1
supplier	56	69	44	40	36	25
supply chain	0	6	2	43	51	54
supply	0	39	43	52	31	17
support garment	0	41	30	22	17	8
support	0	2	0	0	0	0
surface-active agent	48	42	58	78	74	46
surface modification	58	82	70	111	107	53
surface property	72	79	52	43	49	33
surgical product	73	71	51	34	44	24
sustainability	0	8	7	17	24	18
sustainable resource	2	8	0	1	0	0
suture	26	24	10	7	10	13
sweater	4	5	2	4	2	0
swelling agent	4	4	4	4	1	1
swelling	0	3	4	27	13	10
swimwear	15	27	19	16	17	14
symmetry	0	0	0	0	0	1
synthesis	15	28	260	396	340	158
synthetic-organic dye	1	0	0	0	0	0
synthetic dye	2	4	5	4	4	1
synthetic fiber fabric	35	40	21	16	18	21
synthetic fiber industry	9	21	14	9	6	13
synthetic fiber yarn	28	31	18	14	15	19
synthetic fiber	195	60	73	83	91	54
synthetic latex	0	0	0	0	0	1
synthetic polyamide	0	0	0	0	0	12
synthetic resin	0	0	0	0	0	2
synthetic rubber	0	0	0	0	0	9
T-shirt	3	1	5	3	3	5
tafetta fabric	0	0	0	1	2	0
tailoring	0	0	0	0	2	0
take-down mechanism	4	5	3	4	7	3
take-up mechanism	16	21	8	17	13	8

	1999	2000	2001	2002	2003	2004
tanglelacing	9	2	2	2	0	1
tank	0	0	0	0	1	0
tanning	0	0	0	0	1	0
tape	20	25	20	13	19	15
tapestry	5	4	14	17	16	12
tariff structure	0	0	1	21	27	16
tarpaulin	6	4	5	5	2	1
tartan	0	0	0	0	1	0
tasar silk	1	4	7	12	7	8
tax system	0	0	0	0	1	0
teaching	0	0	0	0	0	1
tear resistance	5	8	7	9	8	12
teasel	0	0	1	0	0	0
technical textile	130	137	229	196	250	185
technology	0	0	0	0	0	1
temperature control	15	19	8	17	25	9
temperature effect	15	22	54	97	114	45
temperature	31	42	62	39	35	8
temple	2	2	0	1	2	1
tenacity	21	32	36	31	28	26
tensile property	103	164	167	180	175	76
tension control	42	54	36	31	32	32
tension	77	79	82	62	63	38
tensioner	21	15	13	6	10	7
tent	25	23	19	34	37	19
terminology	3	5	3	16	7	3
terpolymer	0	0	0	0	3	3
terrorism	0	0	0	0	0	2
terry fabric	14	11	26	9	11	7
testing instrument	47	68	84	70	76	44
testing	330	348	345	321	298	172
textile-clothing industry	190	238	217	201	217	179
textile art	29	22	100	129	158	181
textile conservation	0	11	5	5	1	3
textile history	146	126	142	216	178	139
textile industry	266	276	269	355	333	392
textile machinery industry	187	127	135	100	147	140
textile machinery	0	0	0	0	2	0
textile technology	0	0	0	0	1	0
texturing machine	39	29	32	28	22	33
texturing	64	46	47	51	43	33
theft	0	0	0	0	0	1
theoretical analysis	39	57	70	80	40	19

	1999	2000	2001	2002	2003	2004
theory	2	30	57	47	40	20
thermal analysis	0	0	19	40	20	11
thermal anaysis	0	0	1	0	0	0
thermal conductivity	2	18	8	10	25	9
thermal expansion	0	0	0	3	0	2
thermal insulation	33	43	24	28	28	17
thermal phenomena	0	0	0	0	0	1
thermal property	38	49	105	127	158	65
thermal stability	1	11	43	58	55	16
thermal treatment	19	13	10	2	0	0
thermobonded	0	1	0	0	0	0
thermochromic dye	0	2	3	1	1	2
thermodynamics	0	0	0	11	22	9
thermogravimetry	0	3	24	39	19	6
thermomechanics	0	0	0	0	0	1
thermoplastic fiber	49	55	55	36	25	27
thermoplastic polymer	0	0	0	0	1	9
thermoplastic property	66	101	92	83	42	42
thermosetting property	0	0	0	1	0	1
thermosetting resin	0	0	0	0	1	0
thermosetting	0	0	0	0	1	0
thickening agent	12	15	17	16	12	7
thickness	0	0	3	29	66	29
thinning	1	1	2	0	1	3
threading	21	18	11	13	2	9
three-dimensional fabric	0	0	0	0	1	0
three-dimensional textile	3	44	61	62	57	77
tie dyeing	0	0	3	2	3	1
tie	5	16	10	2	1	1
tin oxide	0	0	0	0	0	1
tire cord	9	22	17	15	11	16
tire	12	12	13	6	9	7
tissue engineering	0	11	16	15	8	13
titanium chloride	0	0	0	0	0	1
titanium dioxide	0	0	0	0	4	4
titanium nitride	0	0	0	0	0	1
titanium oxide	0	0	0	0	0	2
titanium	0	0	0	0	4	4
toe closing machine	4	5	1	5	4	1
toe closing	9	9	4	9	11	10
top dyeing	0	0	3	1	0	0
top making	10	12	20	13	5	4
torque	5	3	3	7	12	6

	1999	2000	2001	2002	2003	2004
toughness	0	0	0	0	17	17
tourniquet	0	0	0	2	0	0
tow conversion	6	3	8	4	3	2
tow	11	5	8	12	7	9
towel	8	9	6	6	7	7
toxicity	27	30	22	30	27	8
toxicology	0	0	0	0	0	2
toy	0	4	5	6	5	4
tracking	0	0	0	0	0	1
trade agreement	37	68	66	55	69	37
trade association profile	0	0	0	0	0	1
trade organization	0	0	0	0	6	7
trade union	0	0	0	0	2	1
trade	0	0	0	0	0	1
traditional textile	28	27	40	26	22	50
train	2	0	1	3	6	10
training pant	0	0	0	0	1	0
training	38	47	63	71	58	51
transfer printing	14	15	13	13	11	9
transmission electron microscopy	2	1	1	1	1	0
transparency	0	5	8	15	12	8
transparent property	1	4	0	0	0	0
transport	45	34	39	52	38	8
transportation vehicle	0	0	0	0	0	3
trash	29	19	11	10	13	6
traveler	7	10	4	4	15	11
traverse mechanism	5	13	1	8	6	2
triacetate fabric	0	1	1	0	0	1
triacetate fiber	0	0	0	1	0	0
triacetate yarn	0	0	0	1	0	0
triacetate	0	0	1	0	0	0
tribology	6	5	11	5	7	5
trichromatic dye	1	8	0	1	3	1
trichromatic dyeing	1	0	0	1	0	0
tricot machine	0	0	0	0	1	0
trim	9	3	2	1	2	3
trimming	1	11	3	6	10	11
trousers	12	24	12	9	10	9
tube	26	22	13	5	15	17
tubular knitting machine	0	2	4	1	0	3
tubular material	0	0	25	24	39	31
tuft	6	6	13	0	7	9

	1999	2000	2001	2002	2003	2004
tufting machine	15	20	21	13	11	22
tufting	18	26	24	17	10	28
turbine	0	0	0	0	0	3
turnover	0	4	6	24	39	62
tussah silk	2	3	2	0	0	0
tweed	0	0	0	0	1	1
twill	0	0	0	0	1	1
twist	26	45	42	39	37	20
twisting machine	45	27	16	21	22	18
twisting	55	51	37	38	33	30
two-dimensional textile	0	0	0	1	1	0
ultrasonics	21	37	42	38	42	20
ultraviolet radiation	27	34	52	45	49	33
ultraviolet resistance	42	52	32	26	34	23
umbrella	1	1	3	0	0	1
underwear	67	71	41	55	29	31
uniform	15	18	9	15	16	15
uniformity	24	34	26	27	25	16
union dyeing	2	1	0	0	1	0
unwinding	2	8	6	9	6	6
upholstery	34	37	32	45	57	45
urban design	0	1	1	8	10	7
urea	0	0	0	0	0	1
use	119	25	8	5	10	0
vacuum	16	18	18	13	10	12
valve	0	0	0	1	0	0
vapor	3	7	6	8	10	10
vascular graft	0	0	0	0	1	0
vat dye	8	13	20	15	6	17
vat dyeing	8	6	8	7	0	1
vegetable dye	1	1	7	5	11	4
vegetable fiber fabric	0	1	1	0	0	0
vegetable fiber yarn	1	0	0	0	0	0
vegetable fiber	9	4	3	10	3	7
vehicle component	0	0	0	0	0	3
vehicle	0	0	0	2	0	6
velocity	0	0	0	0	0	3
velour fabric	0	8	6	4	4	6
velvet fabric	3	5	8	5	2	3
ventilation	6	8	13	16	12	9
vessel	7	8	8	22	27	19
vibration damping	7	10	14	16	14	16
vibration	24	28	32	40	27	27

	1999	2000	2001	2002	2003	2004
vicuna fiber	1	3	0	1	0	0
vinyl coating	0	2	3	0	0	0
vinyl polymer	0	0	0	0	0	1
virus resistance	0	1	2	2	2	3
viscoelasticity	8	24	37	40	37	16
viscoplasticity	0	0	0	0	0	1
viscose rayon fabric	3	7	9	3	5	5
viscose rayon fiber	12	22	17	25	24	14
viscose rayon yarn	2	12	6	12	3	5
viscose rayon	3	6	18	15	11	7
viscosity	13	29	60	71	66	28
viscous rayon	1	0	0	0	0	0
visibility	2	2	1	4	9	10
volatile compound	0	0	0	0	1	0
volatile organic compound	0	0	0	2	2	8
vortex spinning	0	0	0	0	1	6
wadding	4	1	5	3	2	3
wader	0	0	0	0	0	1
wage	6	16	14	12	18	9
waistband	4	13	4	3	5	3
wall covering	5	8	9	8	2	3
walnut bark	0	0	0	0	1	0
warehouse	0	0	0	0	0	1
warp knitting machine	25	25	20	23	11	6
warp knitting	54	68	84	57	66	54
warp preparation	12	7	10	27	17	16
warp yarn	41	75	61	37	17	8
warp	27	25	32	20	17	9
warping machine	20	19	18	36	20	12
warping	18	21	17	31	18	10
washfastness	32	31	39	37	32	23
washing machine	72	101	97	108	135	154
washing	43	54	38	35	38	21
waste containment	7	8	20	4	8	13
waste disposal	15	18	9	23	14	4
waste management	45	51	39	24	20	25
waste	32	19	16	29	39	13
wastewater	5	41	52	56	43	25
water-jet loom	6	8	3	7	4	8
water conservation	16	32	20	17	20	6
water consumption	0	0	2	1	6	1
water jet	6	14	11	28	22	17
water pollution	23	8	2	5	3	3

	1999	2000	2001	2002	2003	2004
water resistance	71	87	72	57	86	80
water treatment	3	2	25	22	24	20
water vapor	16	11	12	12	21	9
water	25	24	30	33	37	17
waterproofing	35	15	7	0	3	1
wave-shed loom	4	0	0	1	0	0
wax	6	8	5	13	3	7
wear resistance	20	23	20	17	18	11
wear	19	21	29	18	25	19
weather resistance	17	21	22	26	21	19
weavability	1	15	13	10	4	10
weave	23	37	46	46	54	24
weaving industry	39	31	12	25	20	22
weaving	204	287	344	291	243	267
web forming machine	9	5	0	0	3	0
web forming	18	34	21	50	54	58
web	119	46	53	44	26	20
webbing	7	7	5	7	4	2
website	0	0	0	0	1	0
weed control	0	0	2	1	2	1
weft insertion	55	54	52	30	24	36
weft knitting	30	27	18	35	33	25
weft	41	47	42	32	24	21
weight	22	45	43	58	85	46
welding	7	9	18	21	20	15
welfare provision	0	0	0	0	3	1
wet finishing	4	7	1	3	6	4
wet laying	14	8	7	12	8	3
wet processing machine	17	7	5	4	11	2
wet processing	40	20	20	19	33	21
wetcleaning machine	1	0	0	0	0	0
wetcleaning	1	0	0	1	0	1
wettability	0	0	0	0	1	0
wetting	16	27	25	29	15	13
whisker	6	6	7	5	4	3
whiteness	0	0	0	0	1	0
whitening	0	0	0	2	10	6
whole-garment knitting	6	17	8	8	9	5
wholesaling	0	0	0	0	1	0
wicking	8	18	18	19	18	7
wig	0	0	0	1	0	0
winch dyeing	0	0	0	1	1	0
wind power	0	0	0	0	1	1

	1999	2000	2001	2002	2003	2004
wind turbine	0	0	0	0	0	4
winding machine	67	91	49	56	48	30
winding	79	94	84	93	59	39
wing	0	0	0	0	1	0
wipe	17	19	35	32	37	61
wire	0	0	1	15	13	13
womens employment	0	0	0	0	4	3
womenswear	69	51	48	50	41	31
wood	0	0	0	0	0	2
wool fabric	81	80	78	76	63	40
wool fiber	100	80	80	74	75	37
wool grease	2	3	9	1	1	2
wool industry	74	60	84	167	128	90
wool yarn	20	33	21	29	25	10
wool	113	70	75	81	90	36
working conditions	0	0	0	1	7	8
workwear	47	30	26	33	37	25
World Trade Organisation	0	0	1	0	1	1
worsted fabric	24	19	10	8	13	12
worsted yarn	10	20	24	8	10	11
woven fabric	352	361	364	285	298	267
wrap spinning	9	9	10	9	4	3
X-ray analysis	5	20	20	21	22	10
yak wool	0	2	0	0	0	0
yarn brake	4	17	18	8	11	3
yarn breakage	23	25	16	20	15	6
yarn construction	0	0	0	0	9	5
yarn diameter	9	5	14	10	6	1
yarn length	0	0	0	0	1	1
yarn manufacture	70	186	136	142	180	243
yarn manufacturing	0	1	0	0	0	0
yarn modification	1	27	15	15	18	16
yarn preparation	0	0	0	0	0	1
yarn property	173	289	244	156	171	146
yarn	230	103	100	71	71	22
yellowing	8	7	4	9	4	5
yield	216	108	24	17	17	3
Young modulus	3	8	13	14	8	7
Z-twist yarn	0	1	0	0	0	0
z-twist	0	1	5	2	0	1
zein	0	0	0	0	2	1
zeta potential	0	0	1	1	2	0
zinc oxide	0	0	0	0	0	4

	1999	2000	2001	2002	2003	2004
zipper	0	7	6	5	7	6
zirconium dioxide	0	0	0	0	0	1

USE OF MANUFACTURER INDEXING PER YEAR

	1999	2000	2001	2002	2003	2004
3C	0	0	1	0	0	0
3M	0	4	3	8	5	35
3Tex	0	1	5	1	0	1
3X	0	0	0	0	0	1
62 Group	0	0	0	1	0	0
A and A Material	0	0	0	1	0	1
A Celli	1	0	3	1	0	1
A Deschamps et Fils	0	1	0	0	0	1
A Fem	0	1	1	0	0	0
A H Meyer	0	0	0	0	0	1
A W Hainsworth and Sons	0	0	0	1	0	1
A P I Saldatrici	1	0	0	0	0	0
A1 Embroidery	1	0	0	0	0	0
AAA Wholesale Fabrics	0	0	0	0	1	0
Aachen Composite Alliance	1	0	0	0	0	0
Abaqus	0	0	0	0	0	1
Abbey Event	0	1	0	0	0	0
Abercrombie and Fitch	0	0	1	0	0	1
Abhishek	0	0	0	1	0	0
Abington	0	0	1	0	0	0
Absecon Mills	0	0	0	0	1	0
Absorbent Products	0	1	0	0	0	0
Acabados Tintoretto	0	0	0	0	0	1
Academy Awnings	0	0	1	0	0	0
AccuTrack Automation	0	0	0	0	1	0
ACF Tarp and Awning	0	0	0	0	1	0
ACIMIT	0	0	2	1	0	0
Ackermann	0	0	0	0	2	1
Acme McCrary	0	0	1	1	0	0
Acoltex	1	0	1	0	0	0
Acordis Speciality Fibres	0	0	0	3	5	0
Acordis	20	18	12	6	7	9
Acorn	0	0	1	0	0	0
AcryMed	0	0	0	0	1	1
Acushnet	0	0	0	0	0	1
Adams Carpets	0	1	0	0	0	0
Addax	1	0	0	0	0	0
Adidas Salomon	0	0	0	1	0	0
Adidas	1	3	3	0	0	2
Aditya Birla Group	0	0	0	2	0	1
ADMA	1	0	0	0	0	0
ADO	0	0	0	0	0	1
Adria	0	0	0	0	1	0
Advance	0	0	0	0	1	0
Advanced Ceramics	0	0	0	0	0	1

WORLD TEXTILES Manufacturer Indexing

	1999	2000	2001	2002	2003	2004
Advanced Composite Structures	0	0	0	0	1	0
Advanced Composites Group	1	0	0	0	0	1
Advanced Design Concepts	0	0	0	0	0	1
Advanced Elastomer Systems	0	0	0	0	0	1
Advanced Extrusion Technologies	0	0	0	2	1	0
Advanced Glassfiber Yarns	0	0	1	1	0	0
Advanced Medical Solutions	0	0	0	1	0	0
Advanced Polymer Engineering	0	0	0	0	0	1
Advanced Structures	0	0	0	0	1	0
Advanced Technical Products	1	0	0	0	0	0
Advanced Tent Rental	0	0	0	0	0	1
Advanced Tissue Sciences	0	0	1	0	0	0
Advantage Manufacturing Group	1	0	0	0	0	0
Aegis	0	0	0	1	1	1
Aerocomposit Occitane	0	0	0	0	0	1
Aerolaminates	0	1	0	0	0	0
Aeromode	1	0	0	0	0	0
Aerorider VOF	0	0	0	0	0	1
Aerospace Fabrication and Materials	0	0	1	0	0	0
Aerospatiale	1	0	0	0	0	0
Aerovac	0	1	0	0	0	0
Afrodite Mode Collection	0	0	0	1	0	0
AGI	0	0	1	0	1	0
AgION Technologies	0	0	0	1	0	0
AGM Jactex	0	0	0	0	0	1
Agro Fibres Technologies Plasturgie	0	0	0	0	0	1
Agrotechnology and Food Innovations	0	0	0	0	0	2
Agteks	0	1	0	1	0	1
Ahlers	0	0	0	0	0	1
Ahlstrom	0	0	1	0	1	4
Ahlstrom Corporation	0	0	0	1	1	0
Aigis	0	0	0	1	0	0
Aigle	1	0	0	0	0	0
Air Dimensional Designs	0	0	1	0	0	0
Air Logistics	0	0	0	1	0	0
Air Products	0	0	0	0	2	2
Air Set	0	0	0	1	0	0
Airbus	1	0	0	0	0	3
Aircast	0	0	0	0	0	1
Airtech Advanced Materials	0	0	0	0	0	1
Airventures Uninsured	0	0	1	0	0	0
Aisin Seiki	0	0	0	1	0	0

	1999	2000	2001	2002	2003	2004
Aisin	0	0	1	0	0	0
Ajinomoto	1	0	0	0	0	0
Akab	1	0	0	0	0	0
Aker	0	1	1	0	0	0
Akinal Sentetik	0	0	0	1	0	0
AKSA	0	0	0	1	0	0
Akzo Nobel	6	3	0	2	0	0
Al Mirkaz	0	0	0	0	0	1
Alabama Textile Manufacturers Association	0	0	1	0	0	0
Alacantra	0	0	1	0	0	0
Alan Del Rosario	1	0	0	0	0	0
Alaze	1	0	0	0	0	0
Albany International	0	1	1	0	1	20
Albert Dormanns	0	0	1	0	0	0
Alberta Research Council	0	0	0	0	0	2
Albini	0	2	0	0	0	1
Albis	0	0	1	0	0	2
Albon Chemie	0	1	0	1	0	0
Albrecht	0	0	0	0	0	1
Albright and Wilson	2	0	0	0	0	0
ALC Computertechnik	1	0	0	1	0	1
Alcantara	0	0	0	0	1	3
Alcare Company	0	0	0	0	1	0
Alceru Schwarza	0	0	0	2	0	0
Alcocertex	0	1	0	0	0	0
Alcore Brigontine	0	1	0	0	0	0
Aldan Industries	1	0	0	0	0	0
Aldeco	0	0	0	1	1	0
Aldi	0	0	0	0	1	2
Aldila	2	1	0	0	0	0
Alex Begg and Company	1	1	0	0	0	0
Alexander Drew	0	1	0	0	0	0
Alexander Machinery	0	1	0	0	0	0
Alexanders of Scotland	0	0	0	1	0	0
Alexandra Workwear	1	1	0	2	0	2
Alfred Tenzler	0	0	0	1	0	0
Alfredo Grassi	0	0	0	0	1	0
Alfsen og Gunderson	0	0	0	0	0	1
Alice Manufacturing Company	0	0	1	0	0	0
Alichem	0	0	0	1	0	0
All About August	0	1	0	0	0	0
All Nippon Airways Company	0	0	0	0	1	0
All Nippon Nonwovens Association	0	0	0	1	0	0
Allchem	0	0	0	0	0	1
Allders	0	1	1	0	0	0

WORLD TEXTILES Manufacturer Indexing

	1999	2000	2001	2002	2003	2004
Allegiance Healthcare	2	0	1	0	0	1
Alliance Russian Textiles	0	0	0	0	0	1
Alliance	0	1	0	1	0	0
Alliant Techsystems	0	0	0	0	0	1
Allibert Buckhorn	0	0	0	0	0	1
Allied Fiber Systems	0	1	0	0	0	0
Allied Textile Companies	1	0	1	0	0	0
AlliedSignal	7	2	0	0	0	3
Allma	0	0	1	1	0	1
Almafuerete	0	0	0	0	0	1
Alnet	0	0	0	0	0	1
Alpe Adria Textil	0	0	0	0	1	0
Alpek	0	0	1	0	0	0
AlpenMode	0	0	0	1	0	0
Alpha Engineering Company	0	0	0	0	0	1
Alpha Scientific	0	0	0	0	0	1
Alpha Shirt Company	0	0	0	0	0	1
Alpha Tops	0	1	1	1	0	0
Alpha Trims	0	1	0	0	0	0
Alphabex	0	0	0	0	1	0
Alphatex Beijing Knitting Company	0	0	0	0	1	0
Alps Industries	0	1	2	0	1	0
Alstom	0	0	0	0	0	1
Alt Group	0	0	0	0	0	1
Altamira Colour	1	0	0	0	0	0
Alterfil	0	0	0	1	0	0
Alto Yarn	1	0	0	0	0	0
Alucarbon	0	0	0	0	1	0
Alusuisse Airex	0	0	1	0	0	0
Alvaro Cunha	0	1	0	0	0	0
Alvis Boniver Conte	0	0	0	0	0	1
Amann	2	0	1	0	0	2
Amarnaathh Enterprises	0	0	0	0	1	0
Ambar	0	1	0	0	0	0
Ambassador Clothing Company	0	0	0	0	0	1
AMC Guatemala	0	0	0	0	1	0
Amcor	0	0	1	0	0	0
America Textil	0	0	1	0	1	0
American Agrifabrics	0	0	0	1	0	0
American and Efird	0	0	1	0	0	0
American Apparel Producers Network	0	1	1	0	0	0
American Association of Textile Chemists and Colorists	0	0	0	0	1	0
American Button	0	1	0	0	0	0
American Canopy and Equipment	0	0	1	0	0	0

	1999	2000	2001	2002	2003	2004
American Composite Materials Engineering	0	0	0	0	0	1
American Composites Manufacturers Association	0	0	0	0	1	0
American Eagle Outfitters	0	0	1	0	0	1
American Furniture Manufacturers Association	0	0	0	1	1	0
American Materials and Technology	0	1	0	0	0	0
American Medical Systems	0	0	0	1	0	0
American Metric	0	0	0	1	0	0
American National Textile Center	0	0	0	0	0	2
American Non-Slip Products	0	0	0	0	0	1
American Nonwovens	1	0	0	0	0	0
American Oil Comp	1	0	0	0	0	0
American Printing Company	0	0	1	0	0	0
American Sahm	0	1	0	0	0	0
American Silk Mills	0	0	0	0	1	0
American Society for Testing and Materials	0	0	0	0	1	0
American Synthetic Fiber	0	0	1	0	0	0
American Tandem Medical	0	0	0	1	0	0
American Textile Manufacturers Institute	1	1	1	1	0	0
Americhem	0	0	0	0	1	0
Ameritex	0	0	1	0	0	0
Ames	0	1	0	0	0	0
AMF Reece	0	1	0	0	1	0
Amfibe	0	1	0	0	0	0
Amiantit	0	0	0	0	0	1
Amicus	0	0	0	0	1	0
Amietex	0	0	0	1	0	0
Amital	0	0	1	1	0	0
Ammar Textiles	0	0	0	1	0	0
Amoco Fabrics	0	2	0	0	0	1
Amtek	0	0	0	1	2	0
Amtex	0	0	0	1	1	0
Anateks Group	0	0	0	0	1	0
Anchor	0	0	1	0	0	0
Andar	1	0	0	2	2	4
Andreas Kufferath and Company	0	0	0	0	0	1
Andrew Textile Industries	0	0	0	1	0	0
Anglo American Sewing Machine Company	0	0	1	0	0	0
Anglo Belgian	1	0	0	0	0	0

WORLD TEXTILES Manufacturer Indexing

	1999	2000	2001	2002	2003	2004
Anglo Felt	0	0	0	1	0	1
Angus Handloom Weavers	0	0	1	0	0	0
Anita Haas	0	0	0	0	0	1
Ann Sutton Foundation	0	0	0	0	1	0
Ann Taylor	1	0	0	0	0	1
Anne Fontaine	0	0	0	0	0	1
Anne Somoff	0	0	1	0	0	0
Ansell	0	1	0	0	0	0
Antex-Angels	0	0	0	0	1	0
AOC	0	1	0	0	1	1
Apacinti	0	0	1	1	0	0
Apollo	1	0	0	0	0	0
Apparel International	1	0	0	0	0	0
Appleton Papers	0	0	0	0	1	2
Applied Color Systems	0	0	0	0	0	1
Applied Extrusion Technologies	0	0	1	0	0	0
Applied Materials and Technologies	0	0	0	1	0	0
Applied Therapies	0	0	0	1	0	0
Apres Party and Tent Rental	1	0	0	0	0	0
Aprion	0	1	2	0	3	0
Aquafil	1	1	3	4	1	3
Aquarius	0	0	1	0	0	0
AquaTex Industries	0	2	0	0	0	0
Ara	0	0	0	0	1	0
Aracruz Celulose	0	0	0	1	0	0
Aramido Corporation	0	0	0	0	1	0
Arca	2	0	0	0	0	0
Arcadia Group	1	0	0	0	0	0
Arcelik	0	0	0	0	0	2
Arcelor	0	0	0	1	0	0
Archetype Solutions	0	0	0	0	0	1
Architectural Fabric Structures	0	0	0	1	0	0
Archroma Global Services	0	0	0	0	1	0
Arctronic	0	0	1	0	1	0
Ardekan Textile Company	0	0	0	1	0	0
Arena	0	1	2	0	0	1
Argelich Termes y Cia	0	2	1	0	2	0
Arhar Wool	0	0	0	1	0	0
Arianne	0	0	0	0	1	0
Ariella	0	0	1	0	0	0
Arioli	2	5	0	2	1	0
Aristo Pharmaceuticals	0	1	0	0	0	0
Arkwright Incorporated	0	0	0	0	0	1
Armfield Passano	0	0	1	0	0	0
Armotec	0	0	0	0	0	1
Armstrong	0	1	0	0	0	1

	1999	2000	2001	2002	2003	2004
Arnold Webbing	1	0	0	0	0	0
Arnotts	0	1	0	0	0	0
Artereria Daya Mulia	0	0	0	0	0	1
Arteva	0	0	0	0	0	1
Artextil	0	0	1	0	0	0
Arthur Bell	1	1	1	0	0	0
Articulos Higienicos	0	1	1	0	0	0
Artratex	1	0	0	0	0	0
Arvind Mills	0	0	0	0	2	0
As Seen on Screen.com	0	0	0	0	1	0
Asahi Glass	0	0	0	0	0	2
Asahi Kasei	8	9	5	1	1	22
Asche Masche	0	0	0	1	0	0
Asea Brown Boveri	0	0	1	0	0	0
Asglawo	0	0	0	0	1	1
Ash City	0	0	0	0	0	1
Ashima	0	0	1	0	0	0
Ashland	2	1	0	0	0	0
Ashworth	0	0	0	0	0	1
ASICS	1	0	1	0	0	0
ASON	0	1	0	2	0	1
Asota	1	3	4	3	1	0
Aspen Aerogels	0	0	0	0	0	1
Aspide Medical	0	0	1	0	0	0
Aspo	0	0	0	1	0	0
Asselin	0	1	3	1	0	0
Associated Hygiene Products	0	0	1	0	0	1
Association of German Textile Finishing Experts	0	0	1	0	0	0
Association of Light Industry Enterprises of Lithuania	0	0	1	0	0	0
Association of Universities for Textiles	0	1	1	0	1	0
Associazione Tessile e Salute	0	0	0	0	2	0
Associazione Tessile Italiana	0	0	2	0	0	0
Assofibre	0	0	0	3	1	0
Assyst Bullmer	0	0	1	0	0	2
Assyst	1	3	0	1	1	0
Astar	0	1	0	0	0	0
Astechnologies	0	0	0	2	0	0
AstenJohnson	1	0	0	0	0	3
Asutex	0	1	1	0	0	0
Atateks	0	0	0	1	0	0
ATD Corporation	0	0	0	1	0	1
Ateliers de Belmont	0	0	0	0	0	1
Athanor	0	0	1	0	0	0
Athena Nordic	0	0	0	1	0	0
ATIRA	0	0	0	0	1	0
Atiyeh	0	0	1	0	0	0

WORLD TEXTILES Manufacturer Indexing

	1999	2000	2001	2002	2003	2004
ATK	0	0	0	0	1	0
Atlanta Attachment Company	0	1	0	0	0	1
Atlantic Research	1	0	0	0	0	0
Atlantis Carpets	0	0	0	0	0	1
Atlas Copco	0	0	0	0	1	2
Atlas	0	2	0	0	1	0
Atofina	0	0	0	3	1	4
Atrium Medical	0	0	0	0	0	1
August Herzog	0	0	1	0	0	1
Augustine	0	0	1	1	0	0
Aunde Group	0	0	0	1	0	0
Austin Group	1	0	0	0	0	0
Austin Reed	1	1	0	1	0	0
Australian Aerospace	0	0	0	0	0	1
Australian Alpaca Association	0	0	0	0	0	1
Australian Alpaca Cooperative	0	0	0	2	0	0
Australian Composites	0	0	0	0	0	1
Australian Dyeing Company	0	1	0	0	0	0
Australian Merino Spinners	0	1	0	0	0	0
Australian Topmaking Services	0	1	1	0	1	0
Australian Wool Exchange	0	1	1	1	0	2
Australian Wool Handlers	0	0	1	0	0	0
Australian Wool Innovation	0	0	0	1	2	6
Australian Wool Network	0	0	1	0	0	0
Australian Wool Research and Promotion Organisation	0	1	0	0	0	0
Australian Wool Services	0	1	0	0	0	0
Australian Wool Testing Authority	0	0	1	0	0	1
Austrim	0	2	2	0	0	0
Autefa	1	2	1	0	3	1
Authentic Fitness	0	0	0	1	0	0
Autoconcept	0	0	1	3	1	0
Autoliv	0	0	1	0	0	0
Automazioni Tessili Frigerio	0	0	0	0	0	2
Automold	0	0	0	0	1	0
Autotype	0	0	0	0	0	1
Avecia	0	1	1	1	1	7
Avenue Data Systems	0	0	0	0	1	0
Avery Dennison	0	0	0	0	0	3
Avgol	0	0	0	0	0	2
Avoca Handweavers	1	0	0	0	0	0
Avocet	0	0	0	0	0	1
Avol	0	0	1	0	0	0
Avondale Mills	0	2	1	7	0	0
Axapta Fashion	0	0	0	0	1	0
Axento Ellefeld	0	0	1	0	1	0

	1999	2000	2001	2002	2003	2004
Axson	0	0	0	0	0	1
Ayka	0	0	0	1	0	0
Aytemizler	0	0	0	0	1	0
Azeta	1	0	0	0	0	0
Azimut Yachts	0	0	0	0	1	1
Aznar Group	1	0	0	0	0	0
Aztex	0	0	0	0	0	1
B and T Textilia	0	0	0	1	0	0
Babcock	1	2	2	0	0	0
BAE Systems	0	0	1	0	0	2
Bafatex	0	0	0	0	1	0
Bahariye	0	0	0	0	0	1
Bainbridge	0	0	1	1	0	0
Baisch	0	0	0	1	0	0
Baja Marine	0	0	0	1	0	0
Baku Australia	0	0	1	0	0	0
Baldwin	0	0	0	0	0	1
Balijana Anchalik Janseva Samiti	0	0	0	1	0	0
Ballantyne Cashmere	1	0	1	0	0	0
Bally Ribbon Mills	0	1	0	0	0	1
Balta Group	2	0	0	2	0	0
Bandex	0	0	0	0	0	1
Bandolera+A407	1	0	0	0	0	0
Bangkok Weaving Mills	0	1	0	1	0	0
Bangladesh Garment Manufacturers and Exporters Association	0	0	0	0	1	0
Bangladesh Textile Mills Association	0	0	0	0	1	0
Banner and Beau Brummel	0	0	0	1	0	0
Banner Creations	0	1	0	0	0	0
Bar	0	0	0	0	0	1
Barbarian Rugby Wear	0	0	1	0	0	0
BarcoVision	2	2	1	2	3	0
Bard	0	0	0	1	1	1
Barella	0	0	0	1	0	0
Barmag	5	14	15	18	13	12
Barnhardt Manufacturing	2	0	0	0	0	0
Barracuda	0	0	0	1	0	0
Barrett Enclosures	1	0	1	1	0	0
Barrymores	0	0	1	0	0	0
Barsmark	0	1	0	0	0	0
Barudan	0	0	0	0	1	4
Basaltex	0	0	0	0	1	0
Basell	0	0	0	1	0	0
BASF	7	13	33	23	17	21
Basofil Fibers	0	0	0	0	0	1
Bassett Furniture	0	0	0	1	1	0

WORLD TEXTILES Manufacturer Indexing

	1999	2000	2001	2002	2003	2004
Bastian Wickeltechnik	0	1	0	1	0	1
Bateks	0	0	0	0	0	1
Batliboi	0	0	1	1	0	1
Bauerhin	0	0	0	0	2	0
Baugur Group	0	0	0	0	0	2
Baumler	0	0	2	1	0	1
Baxenden Chemicals	0	1	0	0	0	0
Baxenden	0	1	0	0	0	0
Baxter	0	0	0	0	0	1
Bay Mills	1	0	1	1	0	0
Bay Side Canvas	0	0	0	1	1	0
Bayer	1	6	10	9	12	24
Bayex	0	1	0	0	0	0
Baysaller Tekstil	1	0	0	0	0	0
Baytex	0	0	0	1	0	1
BBA Group	7	2	8	9	6	11
Beacon Power	0	0	0	0	0	1
Beal Group	0	0	0	1	0	0
Beam Dynamics	0	0	1	0	0	0
Beaulieu	0	2	1	0	0	0
Beaver Industries	0	1	0	0	0	0
Beba	0	1	0	0	0	0
Becatron	0	0	0	0	0	1
Beck	0	0	1	0	0	0
Beckwith Bemis	0	0	0	1	0	0
Beechfield	0	0	1	0	0	0
Behr	0	0	0	0	1	0
Beiersdorf	0	0	1	2	5	9
Beijing Topnew	0	0	0	0	0	1
Beijing Yanshan Petrochemical Company	0	0	0	0	1	0
Beka	0	1	0	0	0	0
Bekaert Australia	0	0	0	0	0	1
Bekaert Group	0	0	1	3	1	6
belChem	0	0	1	0	0	0
Bell Horn	0	0	0	0	0	1
Bell New Ceramics	0	0	0	0	0	1
Bella Bustiere	0	1	0	0	0	0
Belle Fabrics	0	0	0	0	0	1
Bellora	0	0	0	0	0	1
Belltex	1	0	0	0	0	0
Bellview Hosiery	0	0	0	1	0	0
Belmont	1	1	0	0	0	0
Beluga Tauchsport	0	0	0	1	0	1
Bemberg	0	1	0	0	0	0
Benetton	0	0	1	0	2	2
Benninger	7	8	10	7	5	15
Benq	0	0	0	0	0	1
Bentley Harris	0	0	2	0	0	0

	1999	2000	2001	2002	2003	2004
Beogradska Konfekcija	0	0	0	1	0	0
Berger Seiba Technotex	0	0	0	0	0	3
Bergers	0	1	0	0	0	1
Berghaus	0	0	1	0	0	0
Berkeley Apparel	0	0	1	0	0	1
Bernado fashions	0	0	0	0	1	0
Berning and Sohne	0	1	0	0	0	0
Bertex	1	0	0	0	0	0
Bertheas and Cie	0	1	0	0	0	0
Berwick Delaware	0	0	0	0	0	1
Berzel Group	0	0	0	0	0	1
Besserman	0	0	0	0	1	0
Best Air Systems	0	0	0	0	0	1
Best Cutting Die	0	0	0	1	0	0
Best Technologies Study and Research	0	1	1	1	2	1
Betty Barclay	0	0	0	0	0	1
BFC	0	0	0	1	0	0
BFF Nonwovens	1	0	1	0	0	0
BFGoodrich	1	0	0	0	1	1
BGB Stockhausen	0	0	0	3	0	0
BHT Bicma	0	0	0	1	0	0
BI Group	0	0	0	0	0	1
Bianca	0	0	1	0	0	0
Biancalani	2	6	2	7	2	1
Bianco	0	0	0	2	2	0
Biax Fiberfilm	1	0	0	1	0	0
BIC Carpets	0	0	0	1	0	0
Bidim	0	0	0	0	0	1
Biella Shrunk Process	0	0	0	0	0	1
Bierrebi	1	0	0	1	0	0
Big W	1	0	0	0	0	0
Bigagli	0	1	0	0	0	0
Bikoma	0	2	0	0	0	0
Billon Freres	0	0	1	0	0	0
Binder	1	0	0	0	0	0
Binsfeld Engineering	0	0	0	0	0	1
Biocell	0	0	1	0	0	0
BioClip	0	0	0	1	0	0
Biocompatibles International	1	0	0	0	0	0
Biocompatibles	0	0	0	1	1	1
BioLargo	0	0	1	0	0	0
Biotrace International	1	0	0	0	0	0
Biotronik	1	0	0	0	0	0
Bird and Cronin	0	0	0	1	0	0
Birdair	0	3	2	1	0	0
Birds Portchmouth Russum	0	0	0	1	0	0
Bischoff Textil	0	0	0	2	0	0
Bisco	0	0	0	0	1	0

WORLD TEXTILES Manufacturer Indexing

	1999	2000	2001	2002	2003	2004
Bisio	1	1	0	2	0	0
Biteam	0	0	0	0	0	1
Bitexma	0	2	0	0	0	0
Bittante	0	0	0	0	1	0
BKB	0	0	0	1	0	1
BKI	0	0	0	0	0	3
BKW Group	0	0	0	0	0	1
BL Color	0	0	0	1	0	0
Black Peony Group	0	0	0	1	0	0
Blacks Leisure Group	0	0	0	1	2	0
Blackwood Carpet Yarns	1	0	0	0	0	0
Blank Textiles	0	0	0	0	0	1
Blaver	0	0	0	0	0	1
Bloch and Behrens Wool	0	0	0	0	0	1
Blucher	0	0	0	0	1	0
Blue Generation	0	0	0	0	0	1
Blue Ridge Industries	0	0	1	0	0	0
Blue Sky Designs	0	0	0	0	1	0
Bluefly	0	0	0	1	0	1
Blundstone	0	0	0	0	0	1
BMB Group	0	0	0	0	0	1
BMP	0	0	1	0	0	1
BMW Group	0	0	0	0	1	2
Boag	1	0	0	0	0	0
Bob Timberlake Apparel	1	0	0	0	0	0
Boco	0	1	0	0	0	0
Bodet and Horst	0	1	0	1	0	0
Boehme	0	1	0	1	1	2
Boeing Company	0	1	0	0	3	4
Bogner Jeans	0	0	0	1	0	0
Bolton Institute	0	0	0	0	0	1
Bombardier	0	0	0	0	0	1
Bombay Textile Research Association	0	0	3	4	0	2
Bonanza	0	0	1	0	0	0
Bonas	1	0	0	1	0	1
Bonazzi	1	0	2	1	0	0
Bond Laminates	0	1	0	0	0	0
Bond Worth	0	0	0	0	0	1
Bondwise Reliant	0	1	0	0	0	0
Bonino	0	1	0	0	0	0
Bonneterie Dupe	1	0	0	0	0	0
Bonnie Strauss	1	0	0	0	0	0
Boos Group	0	0	0	1	0	0
Boos Textile Elastics	0	2	0	1	0	0
Borchert and Moller	1	0	0	0	0	0
Borden	0	0	0	0	2	1
Border Combing Company	0	0	0	0	1	0

	1999	2000	2001	2002	2003	2004
Borealis	1	1	0	0	2	3
Borujerd Textile	0	1	0	0	0	0
Bosch und Siemens Hausgerate	0	0	0	0	0	17
Bosideng Group	0	0	0	0	0	1
Bossi	1	0	0	0	0	0
Bostik	1	0	0	0	0	0
Boston Scientific	0	0	0	0	1	1
Botto Poala	0	0	3	0	1	0
Bower Roebuck	1	1	1	1	1	1
Boyce Spars	0	0	0	0	0	1
Boyteks	0	0	0	0	0	1
Bozzetto	0	0	0	0	0	1
BP Amoco	2	0	1	0	0	0
Bracker	0	0	0	1	0	0
Bracks	0	1	0	0	0	0
Bradford Dyers Association	2	0	0	0	0	0
Bradford Industries	0	0	0	1	0	3
Bradley Industrial Textiles	0	2	0	0	0	0
Bradmill Outdoor Fabrics	0	0	0	0	1	0
Bradmill Undare	2	1	0	1	1	0
Bradsal	0	0	0	0	1	0
Braiconf	0	0	0	0	1	0
Brainchild Design	0	0	0	0	1	0
Brandix	0	0	0	0	0	1
Brawd Manufacturing	0	0	0	1	0	0
Brax	0	0	0	0	0	1
Brayton	0	0	0	1	1	0
Brazzoli	0	6	2	0	3	0
Breakaway	1	0	0	0	0	0
Bremer Woll Kammerei	0	2	3	0	0	2
Brenig	0	0	0	0	0	1
Brenner	0	0	0	0	0	1
Brennet	0	0	0	2	0	0
Bretschneider	0	0	1	0	0	0
Brice	0	0	0	1	0	0
Bridgehaugh	0	1	1	1	0	0
Bridgestone	0	0	0	2	1	0
Brierley Brothers	0	0	0	1	0	1
Brilen and Nurel	0	0	0	1	0	0
Brilen	0	0	0	1	0	0
Brimar	0	0	0	0	0	1
Brintons	0	1	1	0	2	3
Brisay	1	1	0	1	0	1
Bristol Myers Squibb	0	2	2	3	2	1
Brite Ideas	0	0	0	0	0	1
British Alpaca Fine Fiber Company	0	0	0	1	0	0
British Millerain	0	0	0	0	1	0

WORLD TEXTILES Manufacturer Indexing

	1999	2000	2001	2002	2003	2004
British Petroleum	0	2	0	2	0	0
British Polyethylene Industries	0	0	0	1	0	0
British Wool Marketing Board	0	0	0	0	0	1
British Wool Textile Export Corporation	0	0	0	0	1	2
Brock	0	0	1	0	0	0
Brockway	0	1	0	0	0	0
Broderie Suizo	1	0	0	0	0	0
Brodrene	0	0	0	0	1	0
Broll and Company	0	0	0	0	1	0
Brookhouse Composites	1	0	0	0	0	0
Brookline	0	1	1	0	0	0
Brooks Brothers	0	0	0	0	1	0
Brooks Sports	0	0	0	0	0	1
Brookstone	0	0	1	0	0	0
Brother	2	3	0	1	2	18
Brown Shoe	0	0	0	0	2	0
Brownell	0	1	0	0	0	0
Bruck	0	1	0	0	1	2
Bruckner	0	3	0	3	6	2
Brugmann	0	0	0	0	1	0
Bruin Plastics+A653	0	1	0	0	0	0
Bruno Banani	0	0	0	1	0	0
Bruno Kleine	0	0	0	0	0	1
Brunswick Technologies	3	0	0	0	0	0
Bryte	1	0	0	0	0	0
BSB	0	0	0	0	0	1
BSN	0	0	0	1	1	3
BTC	0	0	0	0	1	0
BTP	1	0	0	0	0	0
Buckeye	2	3	2	1	4	0
Buckfast Spinning Company	0	0	0	0	1	0
Bug-Out Outdoor Wear	0	0	0	0	0	1
Buhler	1	0	0	0	0	0
Building Materials Investment	0	0	0	0	0	2
Bulk Molding Compounds	0	0	1	1	0	0
Bulkolot	0	0	0	1	0	0
Bullmer	1	0	0	0	0	0
Bulmer and Lumb Group	0	0	0	0	0	1
Bultel	0	0	0	0	0	1
Bundesverband	0	0	1	0	0	0
Burberry	0	2	0	1	0	0
Burke Mills	1	1	0	0	0	0
Burley Sekem	0	1	0	0	0	0
Burlington Coat Factory	0	0	1	0	0	0
Burlington Industries	4	6	2	1	2	1
Burmatex	0	0	1	0	0	0

	1999	2000	2001	2002	2003	2004
Bursel	0	0	0	0	0	1
Burtco	0	0	0	1	0	0
Busi Giovanni	0	0	0	0	0	1
Buteks	0	0	0	0	1	0
Butyl Products	1	0	0	0	0	0
BWK Elders	0	0	0	1	0	0
C A Kennedy	0	1	0	0	0	0
C and A	0	0	0	0	0	1
C and C Metal	0	1	0	0	0	0
C and C Originals	1	0	0	0	0	0
C and D Aerospace	0	0	0	1	0	0
C B Goldner	0	1	0	0	2	0
C Bel for Awnings	0	0	0	1	0	0
C Bruhl	0	0	0	1	0	0
C F Baker and Sons	0	0	0	2	0	0
C J Grenier	0	0	1	0	0	0
C P Company	0	0	0	0	1	0
C F Hathaway	1	0	0	0	0	0
Cabela	0	0	0	0	1	0
Calator Ruckh	0	0	1	0	0	0
Calcoup	1	0	0	0	0	0
Calder Group	0	0	0	0	0	1
Calida	1	0	0	0	0	0
Caligen	0	0	0	1	1	0
Calitri	0	0	1	0	0	0
Calze	0	0	0	2	0	0
Calzeat	0	1	0	0	0	0
Camber	1	2	0	0	0	0
Camdeboo Mohair	0	0	0	0	0	1
Campen	0	0	0	1	0	0
Campi y Jove	0	1	0	0	0	0
Canadian Textile Institute	0	0	0	1	0	0
Canalair	0	0	0	0	0	2
Canali	0	0	0	0	1	0
Candlewick Yarns	0	1	0	0	0	0
Candy	0	0	0	0	1	4
Canesis Network	0	0	0	0	1	3
Canfab	0	0	0	1	0	0
Cannibal	0	0	0	0	0	1
Cannon Tecnos	0	0	0	1	0	0
Canobbio	2	0	0	1	0	0
Canon	0	0	0	0	0	3
Cantel	0	0	0	0	0	1
Canvas and Cushions	0	0	0	0	2	0
Canvas Chic	0	0	0	0	0	1
Cape Mohair and Wool	0	0	0	1	0	1
Cape Mohair Holdings	0	0	0	1	0	0
Cape Mohair International	1	0	0	1	0	0
Cape Wools	0	0	0	1	0	0

WORLD TEXTILES Manufacturer Indexing

	1999	2000	2001	2002	2003	2004
Caperdoni	2	0	0	0	0	0
Capron Carter	0	0	1	0	0	0
Carapace	0	0	0	1	0	0
Carbon Everflow	0	1	0	0	0	0
Card Monroe	0	0	0	0	1	3
Cardinal Health	0	0	0	0	1	0
CardioTech	0	0	0	1	0	0
Cargill Dow	0	12	10	11	12	1
Cargill	0	1	0	1	0	2
Carhartt	1	0	0	0	0	0
Carilon	1	0	0	0	0	0
Carl Kohl	0	1	0	0	0	0
Carl Merz	0	0	2	0	0	0
Carlee Corporation	0	1	0	0	0	0
Carli Gry	0	1	0	0	0	0
Carlin International	0	0	0	1	0	0
Carlo Barbera	0	0	0	1	0	0
Carlo Colucci	0	0	1	0	0	0
Carnegie Fabrics	0	0	1	0	0	0
Carolina CoverTech	0	0	0	1	0	1
Caroll	1	0	0	0	0	0
Carolon Company	0	0	0	0	0	1
Carpet and Rug Institute	0	0	1	0	0	0
Carpet Express	0	1	0	0	0	0
Carpetright	1	1	0	1	1	0
Carpets International	0	2	0	0	0	0
Carpets of Kidderminster	0	1	1	1	1	0
Carpets of Worth	0	0	1	0	0	1
Carr Reinforcements	1	0	0	0	0	0
Carrefour	0	0	0	1	0	0
Carrington Career and Workwear	0	0	0	0	1	0
Carrington	0	1	1	0	0	0
Carthage	0	1	0	0	0	0
Caru	0	1	0	0	0	0
CASA	0	0	0	1	0	0
Casaveen	0	1	0	0	0	0
Cashmere and Camel Hair Manufacturers Institute	0	1	0	0	1	0
Cashmere Fibres International	0	1	0	0	0	0
Cashmerefine	0	1	1	0	0	0
Caspian Custom Canvas	0	0	0	0	0	1
Castro Model	0	0	0	0	0	1
Catalana de Polimers	0	0	0	0	1	0
Catalytic Materials	0	0	0	0	0	1
Cavalier Bremworth	1	0	2	0	0	0
Cavalier Corporation	1	1	0	2	0	0
Cavalltec	0	0	0	0	0	1

	1999	2000	2001	2002	2003	2004
Cavisa	0	0	0	2	1	0
Cavitec	0	0	0	2	2	4
Ceccato	1	0	0	0	0	0
Celanese	0	4	0	0	0	1
Celebrity Lab	0	0	0	1	0	0
CellTran	0	0	0	1	0	2
Cellulose Converting Equipments	1	1	0	0	1	0
Celsum Technologies	0	0	0	0	0	1
Centa Antriebe Kirshey	0	0	1	0	0	0
Centenari e Zinelli	0	1	1	0	1	1
Centexbel	0	0	0	0	3	1
Cento Tec	0	0	0	0	1	0
Central Institute for Research on Cotton Technology	0	0	1	3	0	0
Central Silk Board	0	0	0	0	0	1
Central Silk Technological Research Institute	0	0	0	0	1	0
Central Tasar Research and Training Institute	0	0	0	0	0	1
Central Textile Machine Company	0	0	0	0	1	0
Central Textiles	0	0	0	0	1	0
Centre for Objective Measurement and Innovation Technologies	0	1	0	0	0	0
Centre for Textile Technology	0	0	1	0	0	0
Centrepulse	0	0	0	0	1	0
Centro Lino Italiano	0	0	0	0	1	0
Centro Ricerche Fiat	0	0	0	0	0	1
Centrocot	0	0	1	1	0	0
Century Dyeing	0	0	1	0	0	0
Century Investment Group	0	0	1	0	0	0
Century Mills	0	1	0	0	0	0
Cerex	1	1	1	1	0	1
Certain Teed	0	1	1	1	0	0
Cetex	0	0	2	0	0	1
Chaine e Trame	0	0	1	0	0	0
Chamatex	0	1	1	0	0	0
Champion Products	0	1	0	0	0	0
Chantelle	1	0	0	0	0	0
Chapelthorpe	1	2	1	2	2	1
Chapman Thermal Products	0	0	0	0	0	1
Charcoal Cloth International	0	0	0	1	2	0
Chargeurs	0	0	7	3	3	2
Charles Crossley and Company	0	0	1	0	0	0
Charles Parsons Group	0	0	0	1	0	1

WORLD TEXTILES Manufacturer Indexing

	1999	2000	2001	2002	2003	2004
Charles River	0	0	0	1	0	0
Charles Vogele Group	0	0	0	0	0	1
Charm Young	0	0	0	0	0	1
Chas W House and Sons	0	0	1	0	0	0
Chase Machine and Engineering	0	0	0	1	0	0
Chavanoz Industrie	0	0	0	0	0	1
Cheil Industries	0	0	0	0	2	0
Chelest	0	0	1	0	0	0
Chemax Multikotes	0	0	0	0	0	1
Chemfab	1	0	0	0	0	0
Chemical Process Equipment	0	1	0	0	0	0
Chemlon	0	1	0	0	0	0
Chemtex	1	0	2	3	0	1
Chengzhou Fei Shen	0	0	0	1	0	0
Chesfields	1	0	0	0	0	0
Chester Barrie	0	1	0	0	0	0
Chicl	0	0	0	1	0	0
Chico FAS	0	0	0	1	0	0
Chiemsee and Company	0	0	0	0	0	1
China Cocoon and Silk Exchange	0	0	1	0	0	0
China National Offshore	0	1	0	0	0	0
China National Textiles Imports and Exports Corporation	0	0	0	0	2	1
China Petroleum and Chemical	0	0	0	0	0	1
China Silk	0	0	1	0	0	0
China Textile Institute	0	0	0	0	1	1
China Textile Machinery	0	0	1	0	0	0
China Textile Network	0	0	0	0	0	1
China Wool Textile Association	0	0	0	0	0	1
China Worldbest	0	0	1	1	0	1
Ching Chi Machine Company	0	0	0	0	0	4
Chisso	3	2	3	1	2	4
Chiva Carpet	0	0	0	0	1	0
Chomarat Composites	0	0	0	0	1	1
Chori	2	1	0	0	0	0
Chris Kay	0	0	1	2	0	0
Christian Eschler	1	1	4	3	10	3
Christian Sandler	1	0	5	3	2	2
Christian Schmidt-Prestin	0	1	0	0	0	0
Christory	0	0	0	0	1	0
Chroma	0	1	0	0	0	0
Chromarat	0	0	0	0	0	1
Chromatech	0	1	0	0	0	0
CHT R Beitlich	1	1	0	3	4	1

	1999	2000	2001	2002	2003	2004
Chubu Chelest	0	0	1	0	0	0
Chul Thai	0	0	0	1	0	0
Chun Au Knitting Factory	0	0	0	1	0	0
Cia Industrial Abayan	0	1	0	0	0	0
Ciba Corning	0	0	0	0	1	2
Ciba Specialty Chemicals	7	22	25	19	18	26
Cibitex	0	4	1	0	1	1
CiDRA	0	0	0	0	1	0
CIEL Textiles	0	0	0	1	0	0
Cimi	2	2	1	0	1	0
Cincinnati Machine	0	1	0	0	0	0
Ciocca	0	0	0	0	0	1
CIRAD	3	0	0	0	0	0
Circa 1801	0	1	0	0	0	0
Circon Corcle	0	0	0	0	1	0
CIT Group	0	0	1	0	0	0
Citroen	0	0	0	0	1	0
Clariant	5	6	10	7	12	11
Clark and Gallagher Technologies	0	0	0	0	1	0
Clark Cuttler McDermott Company	0	1	0	0	0	0
Classic Concepts Company	0	0	0	0	0	1
Classic Exports	0	0	0	1	0	0
Clea	0	1	0	0	0	0
Clemson University	0	0	0	0	1	1
Clinton Grosshandels	0	0	0	0	0	1
Clissold	0	0	0	0	1	0
Clopay	1	0	2	1	0	1
Clorox	0	0	0	0	0	1
Clover Manufacturing Company	0	0	0	0	0	2
Cloverbrook	0	0	0	1	0	0
ClozeX	0	0	0	0	0	2
Club Amici del Lino	0	0	0	0	1	0
CMT	2	0	1	0	0	0
Coatema	1	1	2	3	4	4
Coates Brothers	0	0	0	0	0	1
Coating Applications Group	0	0	1	0	0	0
Coats Viyella	3	3	1	0	0	0
Coats	1	2	1	3	2	4
Cobble	0	3	1	1	0	1
Cobra International	0	0	0	0	0	1
Codi	0	1	0	0	0	0
Coevin Licensing	0	0	0	0	0	2
Cogetex Well	1	0	0	0	0	0
Cognatex	0	0	0	1	0	0
Cognesint	1	0	0	0	0	0
Cognetex	0	0	2	3	2	1

WORLD TEXTILES Manufacturer Indexing

	1999	2000	2001	2002	2003	2004
Cognex	0	0	0	1	0	0
Cognis	0	0	2	2	2	11
Coimbatore Air Control Systems	0	0	0	0	0	1
Colbond	1	0	2	2	0	2
Cole Haan	0	0	0	0	1	0
Colefax and Fowler	1	1	0	0	0	0
Colefax Group	0	0	0	1	1	0
Colgate Palmolive	1	0	2	2	0	5
Colibri	0	0	1	0	0	0
Collagen Matrix	0	0	0	0	1	0
Collections Magi	0	1	0	0	0	0
Collins and Aikman	1	1	0	0	0	3
Coloplast	0	0	0	2	0	0
Color Chem	0	0	0	1	0	0
Color Marketing Group	0	0	0	0	1	0
Color Service	0	3	0	0	0	0
ColorSpan	0	2	1	0	0	0
Colortronic	0	0	1	0	0	0
Colsman	1	0	0	0	0	0
Columbia Insurance Company	0	0	0	0	0	1
Columbia Sportswear	0	0	0	0	1	2
Comerama	0	0	0	1	0	0
Comercial Catalana de Maquinaria Textil	0	0	0	1	0	0
Comercial Julia Ingles	0	0	1	0	0	0
Comerio Ercole	0	1	0	0	0	0
Comet	0	1	0	0	0	0
Comeureg	0	0	0	1	0	0
Comex	0	0	0	1	0	0
Comez	2	2	1	1	2	3
Comfortage	0	0	0	1	0	0
Commonwealth Scientific and Industrial Research Organisation	1	4	2	1	1	2
Comoli Fermo	0	0	0	2	0	0
Compagnie d'Importation de Laines	0	1	0	0	0	0
Compania Industrial de Parras	2	1	0	0	0	0
Compania Industrial Nuevo Mundo	0	1	0	0	0	0
Compass	0	0	0	0	1	0
CompMess	0	0	0	0	1	0
Composite Integration	0	0	0	0	1	0
Composite Machines Company	0	1	0	0	0	0
Composite Materials	1	0	0	0	0	0
Composite Solutions	0	1	0	0	1	0

	1999	2000	2001	2002	2003	2004
Composite Systems	0	0	0	0	0	1
Composites Consulting	0	0	0	1	0	0
Composites Distribution	1	0	0	0	0	0
Composites Jareno	0	1	0	0	0	0
Composites One	1	0	0	0	0	0
Composittrailer	0	0	0	0	1	0
Compotech	0	0	0	0	1	0
Compression Garment Technologies	0	0	0	0	1	0
Computer Integrated Textile Design Association	0	1	0	0	0	0
Comsa Comercial	0	0	0	1	0	0
ComTec	1	0	0	0	0	0
Concert	0	1	0	1	3	2
Cone Mills	0	3	0	0	1	0
Confecciones Colombia	0	1	0	0	0	0
Conoco	1	1	3	0	0	1
ConocoPhilips	0	0	0	0	1	0
Consejo Intertextil	0	0	0	0	1	0
Consip	0	0	0	0	1	0
Consolidated Ecoprogress Technology	1	2	2	0	0	0
Consolidated Midland Corporation	1	0	0	0	0	0
Consolidated Minerals	0	0	0	0	0	1
Consolitex	0	0	0	1	0	0
Consorcio Textil del Pacifico	1	0	0	0	0	0
Consumer Product Enterprises	1	0	0	0	0	0
Conterra	0	1	0	0	0	0
Conti Complett	3	0	0	0	0	0
Contifibre	0	0	0	1	1	0
Contifix	0	1	0	0	0	0
Continental Fibres	0	1	0	0	0	0
Continental Teves	0	0	1	0	0	0
Convatec	0	0	0	2	0	0
Conyplex	0	0	0	0	1	0
Cookson and Clegg	0	0	0	0	1	0
Cookson Fibres	1	0	0	0	0	0
Cool Options	0	0	0	0	0	1
Cooley	0	0	0	1	0	0
CoolZs	0	0	0	0	0	1
Coop	0	0	0	0	1	0
Cooperative Research Centre for Advanced Composite Structures	2	0	0	0	1	0
Coorstek	0	0	0	0	1	0
Copack	0	0	1	0	0	0
Copitex Group	0	0	1	0	0	0

WORLD TEXTILES Manufacturer Indexing

	1999	2000	2001	2002	2003	2004
Coppley Apparel	0	0	0	0	1	0
Coprolan	0	0	0	1	0	0
Coramtex	0	0	0	0	0	2
Corbex	0	1	0	0	0	0
Corcoy	0	0	2	0	0	0
Cordant	0	0	2	0	0	0
Cordenka	0	0	1	0	0	0
Core Materials	1	0	0	0	0	0
Corino	0	1	0	0	0	1
Corman	0	1	1	0	0	0
Corn Products	0	0	0	0	1	0
CorNova	0	0	0	0	0	1
Cornwell Parker	0	1	1	1	0	0
COROMEX	0	0	0	1	0	0
Coronet Werke	0	0	0	0	0	1
Coroplast Fritz Muller	0	0	0	0	0	1
Corovin	0	0	0	0	0	1
Corsadi	0	0	0	0	0	1
Cortese	1	1	0	1	0	0
Cortex	0	0	3	0	1	0
Cortman	0	0	1	0	0	0
Cosmopolitan Textiles	1	0	0	0	0	0
Costruzione Macchine Tessili	0	1	0	0	0	0
Costruzioni Apparecchiature Elettroniche Nucleari	0	0	0	0	1	0
Cotonificio Oltlina	0	1	0	0	0	0
Cotton Council International	0	0	0	1	1	0
Cotton Group	0	0	0	0	1	0
Cotton Incorporated	0	6	3	8	1	0
Cotton Knit	0	0	1	0	0	0
Cotton Moisture	0	0	0	0	0	1
Cotton Traders	0	0	0	0	0	1
Country Casuals	0	1	0	0	0	0
Country Road	0	1	0	0	0	0
Couristan	1	0	0	0	0	0
Courtaulds	5	6	2	3	1	0
Courts	0	1	0	2	0	1
Cousin Biotech	0	0	0	0	0	1
Cousin Filterie	1	0	0	0	0	0
Covess	0	0	0	0	1	0
Coville	0	1	0	0	0	0
Crabtree	0	1	0	0	0	0
Craghoppers	0	1	0	0	0	0
Crane and Company	0	0	1	0	0	0
Cray Valley	0	1	0	0	0	0
Creare	0	0	0	0	0	1
Creation Baumann	0	0	0	0	0	1
Creative Carder	0	0	0	0	1	0
Creative Inflatables	0	0	0	0	0	1

	1999	2000	2001	2002	2003	2004
Creavis	0	0	0	0	0	1
Creditex	0	0	0	0	1	0
Crestell	0	0	0	0	0	1
Creswick Woollen Mill	0	0	0	0	1	0
Croll-Reynolds	0	0	0	0	1	0
Cromos Tintoria Nueve Fibre	1	0	0	0	0	0
Crompton and Knowles	0	2	0	0	0	0
Crompton Corporation	0	0	0	0	0	2
Crown Crafts	0	1	0	0	0	0
Crystal Group	0	0	1	2	0	1
CSS Technology	0	0	0	0	0	1
Cubotex	0	1	0	2	0	0
Culp	4	1	3	0	0	0
Cultor	1	0	0	0	0	0
Cuno	0	0	0	0	0	2
Curioni	0	0	1	0	0	0
Curt G Joa	0	0	0	1	2	1
Curtis Wool Direct	0	0	0	0	0	1
Custom Chenille Embroidery	0	0	0	0	0	1
Custom Marine Canvas	0	1	0	0	0	0
Custom Woolen Mills	0	0	0	0	0	1
Custom Industries	0	0	1	0	0	0
Cutter and Buck	0	0	0	1	0	0
CV International Brand Licensing	1	0	0	0	0	0
CVC Capital Partners	2	0	2	0	0	0
Cyberware	0	0	1	0	0	0
Cyclics	0	0	0	2	0	1
Cytec	2	0	0	0	0	3
D'Hersigny	0	0	1	0	0	0
D and E	0	1	0	0	0	0
D Avenza	0	1	0	0	0	0
d3o Lab	0	0	0	0	0	1
Da Kong	0	1	0	0	0	0
Da Rue	0	1	0	0	0	0
Dachstein	1	0	0	0	0	0
Daido Metal Company	0	0	0	0	0	2
Daikin	0	1	1	0	0	4
Daikyo Chemical Company	0	0	0	0	0	1
DaimlerChrysler	0	3	1	1	6	3
Dainippon	0	0	0	0	1	1
Daisy	0	0	0	0	0	1
Daiwa	0	1	0	1	0	0
Daiwabo	2	5	0	0	0	1
DAK Americas	0	0	1	0	1	0
Dakahlaya	0	0	0	1	0	0
Dakota Coatings	0	0	0	0	0	1
DAKS Simpson	1	0	0	0	0	0
Dalteks	0	0	0	1	0	0

WORLD TEXTILES Manufacturer Indexing

	1999	2000	2001	2002	2003	2004
Dalyplastic	0	0	0	0	0	1
Damart	0	1	0	0	0	0
Dambi	0	0	0	0	0	1
Damei	0	0	0	1	0	0
Dan Company	0	0	0	0	0	2
Dan River	0	1	0	0	1	0
Dan Web	0	0	0	2	1	4
Dana Corporation	0	0	0	1	1	0
Danalco	1	0	0	0	0	0
Dancotex	0	0	0	0	1	0
Danis	0	0	0	0	0	1
Danisco	1	0	0	0	0	0
Danish Bestseller	0	0	0	0	0	1
Danksin	0	1	0	0	0	0
Danley	0	0	0	0	0	2
Danti Paolo	1	0	0	0	0	0
Daramic	0	0	1	0	0	0
Dariotex	0	0	1	1	0	1
Dassault	1	0	0	0	0	2
Datacolor	5	6	3	3	1	3
Datatex	0	1	0	1	1	0
Dauphin	0	0	0	1	0	0
David Bentley	0	1	0	0	0	0
David Crabtree and Son	0	0	1	0	0	0
David Rigby Associates	0	0	0	0	1	0
David Wine Tents	0	1	0	0	0	0
Dawson International	0	1	3	1	2	0
Dazian	0	0	0	1	0	0
DBW	0	0	0	0	0	1
De Boer	1	3	2	1	0	0
De Poortere Freres	0	0	2	0	0	0
Deb Shops	1	0	0	0	0	0
Decathlon	0	0	0	0	0	1
Deckers Outdoor	0	0	1	0	0	0
Declercq and Declercq	0	1	0	0	0	0
Decor Rest	0	0	0	0	0	1
Decoray	0	0	0	0	0	1
Defab	0	0	0	1	0	1
Defence Research	0	0	1	0	0	0
Degussa	0	0	0	0	0	2
Dehnco	0	0	0	0	1	0
Deichmann Umwelttechnik	0	0	0	0	0	1
Deimo	0	0	1	0	0	0
Deko-Weberei Falkenstein	0	0	1	0	0	0
Dell'Orco	0	0	0	0	0	1
Delphi	0	1	0	0	1	2
Delpiano	0	0	0	0	0	1
Delta Apparel	0	0	0	0	1	0

	1999	2000	2001	2002	2003	2004
Delta Converting Equipment	0	1	0	0	0	0
Delta Galil	0	0	0	0	1	1
Delta Tie	0	0	0	1	0	0
Denali	1	0	0	0	0	0
Dentelles Darquer	0	1	1	0	0	0
Derby Associates	0	1	0	0	0	0
Desco von Schulthess	0	0	1	1	0	0
Design Solutions	0	0	0	0	0	1
Design Spun	0	0	0	1	0	0
Designer Textiles International	0	0	0	0	0	1
Designtex	0	1	0	0	0	0
Designworks	0	0	1	0	0	0
Desseilles+A1124	0	0	1	0	0	0
Desso	0	1	0	0	0	1
Detect Sensory Fabrics	0	0	0	0	1	0
Dethmers	0	0	1	0	0	0
Deutsche Bank Industrial Holdings	0	0	1	0	0	0
Deutsche Institute fur Textil- und Faserforschung	0	0	0	0	0	2
Deutsches Textilforschungszentrum Nord West	0	0	1	0	0	1
Devan	0	0	0	1	0	1
Devi Polymers	0	1	0	0	0	0
Devold AMT	0	0	1	0	1	0
DeWal	0	1	0	0	0	0
Dewavrin	0	1	2	2	1	0
Dewhirst	2	1	0	0	1	0
Dexter	1	1	0	0	0	0
DIAB Technologies	0	1	0	1	0	0
Diagram Design	0	0	0	1	0	0
Diatec	0	1	0	1	0	0
Dickson Constant	0	0	0	1	0	0
Dienes	1	0	1	0	0	1
Diesel	1	0	0	0	0	2
Dietrich Wetzel	0	0	2	0	0	0
Difco	0	1	1	1	1	0
Digital Printing Systems	0	3	0	0	0	0
Digital Weaving Norway	0	0	0	1	1	0
Dillards	0	0	0	1	0	0
Dilo	0	5	5	6	4	5
Diolen	0	0	0	0	1	1
Disposable Soft Goods	0	1	0	0	0	0
Dixie Group	1	2	0	0	0	0
DLR Institut fur Strukturmechanik	0	0	0	0	0	2
DM and E	0	0	0	1	0	0

WORLD TEXTILES Manufacturer Indexing

	1999	2000	2001	2002	2003	2004
DMC	0	0	1	0	0	0
DNP	0	0	0	0	1	0
DO CERAM	0	0	1	1	0	1
Dockers	0	0	0	1	0	0
Dogi	0	0	0	1	0	0
Doko Spinning	0	2	0	0	0	0
Doku	0	0	0	1	0	0
Dolina	0	1	0	0	0	0
Domino Printing Sciences	0	0	0	0	0	1
Dommer	0	0	1	0	0	0
Domo	0	0	0	0	0	2
Domtex	0	0	0	0	1	0
Don and Low	3	0	1	0	0	0
Donisthorpe	0	0	0	2	0	0
Donna Karan	0	1	0	0	0	0
Dormeuil	0	0	0	1	0	0
Dornier	9	13	14	5	11	13
Double Visions	0	0	0	0	1	0
Doubletex	0	0	0	0	0	1
Dovecot Studios	0	0	0	2	0	0
Dover Marquees	0	0	0	1	0	0
Dow	1	2	9	10	11	18
Doyle Fraser	0	0	0	0	0	1
Dracup	0	0	0	0	0	1
Drahtcord Saar	0	0	0	0	0	2
Draper Knitting Company	0	0	1	0	0	0
Dream Weaver	0	0	0	0	1	0
Dritex	0	0	0	1	0	0
Drummond Group	0	0	2	0	0	0
Drypers	0	0	1	0	0	1
DSM	3	5	13	7	2	14
Dua	0	0	0	1	0	1
DuBay Polymer	0	0	0	0	0	2
Ductel	0	0	0	0	1	0
Duflot	0	2	0	3	0	1
Dunlop Slazenger Group	0	0	1	0	0	0
Dunlop Standard Aerospace Group	0	1	0	0	1	1
DuPont	32	58	60	55	52	81
Durkopp Adler	1	3	2	1	2	7
Dye-Namix	0	0	0	0	0	1
Dyejet Printing Systems	0	0	0	0	0	1
Dyersburg Fabrics	1	0	0	0	0	0
Dyestuffs Manufacturers Association of India	0	0	0	0	1	0
Dynatex	0	0	0	0	1	1
Dyneon	0	0	0	1	0	0
Dyson	0	0	0	1	0	3

	1999	2000	2001	2002	2003	2004
DyStar	6	11	13	17	18	19
E Boselli and Company	0	0	0	0	1	0
E S Originals	0	0	0	0	0	2
E Sallis	0	0	0	1	0	0
Eagle Nonwovens	0	1	0	0	0	0
Eagle Ottawa	0	0	0	0	0	2
East India Company	0	0	0	1	0	0
Eastern Garment Manufacturing Company	0	0	0	0	1	0
Eastman Chemical	0	0	0	1	2	1
Eastman Kodak	1	2	1	1	2	5
Eastman Machine	1	0	0	1	0	0
Eastman Systems	1	0	0	0	0	0
Ebara	0	0	0	0	0	1
Ebay	0	0	0	0	0	1
Ebo	0	0	0	1	0	0
Ecco Gleittechnik	0	0	0	0	1	0
Ecolab	0	0	0	0	1	3
Ecolog	0	0	1	0	0	0
Ecotextil	1	1	0	0	0	1
Eddie Bauer	0	0	1	0	0	0
Edel	0	0	0	0	1	0
Edgeworks	0	1	0	0	0	0
Edilfloor	0	0	1	0	0	0
Edmund Erdmann	1	0	0	0	0	0
EDV	0	0	0	0	1	0
Edward Hall	1	0	0	0	0	0
Edward Keller	0	0	0	1	0	0
Edward Lifesciences	0	0	1	0	0	0
Edwards Garment Company	0	0	0	0	0	1
eFashion Solutions	0	0	0	0	0	1
EGO Group	0	0	0	0	0	1
Ehmke Manufacturing Company	0	0	0	0	1	0
Eide	0	0	0	1	0	0
Eing	1	0	0	0	0	0
Eisele Apparate - und Geratebau	1	1	0	0	0	0
Ekamat	0	0	1	0	0	0
Eke Tekstil	0	0	0	0	0	1
Ekkon	0	0	0	0	0	1
Elana	1	0	0	0	0	0
Elastic Corporation of America	0	0	0	0	1	0
Elbit Vision Systems	0	0	2	0	0	0
Elco Direct	1	0	0	0	0	0
Elders	0	2	0	0	2	2
Electrical Systems	0	1	0	0	0	0
Electro-Jet	0	0	0	0	2	1

WORLD TEXTILES Manufacturer Indexing

	1999	2000	2001	2002	2003	2004
Electro Textiles	0	0	2	0	0	0
Electrolux	1	0	0	2	1	8
Electronik Fabricmakers	0	0	0	1	0	0
Electrotextiles	0	0	1	0	0	0
Electrovac	0	0	0	0	1	0
Elf Atochem	2	0	0	0	0	0
Elgi Electric	0	0	0	0	0	1
Elite Group	0	1	0	0	0	0
Elize	0	2	0	0	0	0
Elk Corporation	0	0	0	1	1	1
Elkoteks	0	0	0	1	0	0
Ellenberger and Poensgen	0	0	0	0	0	1
Elliot Musgrave	0	0	1	0	0	0
Ellis Developments	0	1	0	0	0	0
Ellis	0	0	0	0	0	1
Elsan	0	1	0	0	0	0
ELTEK	0	0	0	0	0	1
Eltex	0	0	0	0	0	1
Elyon	0	0	1	0	0	0
EM Fiberglas	0	0	0	0	0	1
Embroiderers Guild	0	0	0	0	0	1
Emerson and Renwick	2	1	1	0	0	0
Emil Jager	0	0	0	0	1	0
Emil Muhlmann	1	0	0	0	0	0
EMM	0	0	0	1	0	0
Empresa Textil Nortenha	0	0	1	0	0	0
EMS Chemie	1	2	2	0	0	0
EMS Griltech	0	0	4	1	0	1
Encad	0	0	1	0	0	0
Eneas Informatica	0	0	0	0	0	1
Energy Science Laboratories	0	1	0	0	0	0
Engineered Filtration Products	0	0	0	1	0	0
Engineering Recycling Maschinen und Anlagen	0	0	1	2	1	2
Engineering Technology	0	1	0	0	0	0
English Card Clothing Company	0	0	1	0	0	0
Engraw	0	0	0	1	0	0
Enhanced Retail Solutions	0	0	0	0	0	1
Enka Tecnica	0	1	0	0	1	1
Enka	0	1	3	3	0	0
Enkador	0	0	0	0	1	0
Enshu Senko	1	0	0	0	0	0
Entec Composites Machines	0	1	0	0	0	0
Entra Iberica	0	1	0	0	0	0
Enviro Pac	0	0	0	0	0	1
Environmental Language	0	0	0	0	0	1

	1999	2000	2001	2002	2003	2004
EPC Industrial Engineering	0	0	0	0	1	0
Epic Group	0	0	1	0	0	0
Epson	0	0	0	0	1	1
Equi-Fibers	0	0	0	0	0	2
Equistar	0	0	0	0	0	1
Erbatech	0	0	0	0	1	0
Erez Thermoplastic Products	0	0	1	0	0	0
Erhardt and Liemer	1	1	2	0	1	0
ERKO	0	0	1	0	0	1
Ermenegildo Zegna	0	1	0	1	0	0
Ernst Jacobi	0	0	0	0	0	1
Ernst Keller	0	0	0	1	0	0
Erwotex Erzgebirgische Wolltextilien	0	0	1	0	0	0
Esch	1	1	0	0	0	0
Esfil	1	0	0	0	0	0
Esgyrn Guanaco	0	0	1	0	0	0
ESI Group	0	0	0	1	0	1
ESMA	0	0	0	0	1	0
Esmery Caron Structures	0	0	0	0	1	0
Espace Textile	0	0	0	1	0	1
eSpin Technologies	0	0	0	1	1	0
Espirit	1	0	0	0	0	1
Essential Machines	0	1	0	0	0	0
Etam	0	0	1	0	0	1
Eterna Mode	1	0	0	0	0	0
ETH	3	0	0	0	0	0
Ethicon	2	0	0	0	1	8
Ettlin	0	0	0	1	0	0
Ettrick Yarn Dyers	0	0	2	1	1	0
Eureka Verpackungs	1	0	0	0	0	0
Euro-Composites	0	0	0	0	0	1
Eurocopter	0	0	0	0	1	0
Eurojersey	0	0	0	1	0	1
Eurolube	0	0	2	0	0	0
Europa Aircraft	0	0	0	1	0	0
European Aeronautic Defence and Space Company	0	0	0	1	0	1
European Apparel and Textile Organization	0	0	0	0	1	1
European Association of Textile Polyolefins	0	0	0	1	0	0
European Colour	1	0	1	1	1	1
European Composites Industry Association	0	0	0	0	1	0
European Composites Recycling Service Company	0	0	0	0	0	1
Europrotect	0	0	2	0	0	1

WORLD TEXTILES Manufacturer Indexing

	1999	2000	2001	2002	2003	2004
Euroresins	0	0	1	1	0	1
Event Labor Works	0	0	0	0	0	1
Event Rentals	0	0	0	0	0	1
Evergreen Nylon Recycling	0	2	0	0	0	0
Everlight	0	0	0	0	0	2
eWarna.com	0	0	0	1	0	0
Exacta	1	0	0	0	0	1
Excelpress	0	0	0	1	0	0
Executive Garment Processors	0	1	0	0	0	0
Exel	0	0	0	0	0	1
Exeq	0	0	0	0	0	1
Extremeties	1	0	0	0	0	0
Extrusion Systems	1	0	0	0	0	0
Exxon	0	0	1	0	0	1
Ezio Ghiringhelli	0	0	1	0	0	0
F A Kumpers	0	0	0	0	1	0
F Drake	0	5	0	0	0	0
Fabelta	0	0	1	0	0	0
Faber Group	0	0	0	1	0	0
Faber Pro Motion	0	0	0	1	0	0
Fabric Images	0	0	0	1	0	1
Fabric Ireland	0	1	1	0	0	0
Fabrica International	0	1	0	0	0	0
Fabricato Tejicondor	0	0	0	0	1	1
Fabricato	0	0	0	0	1	0
FabriTec	0	0	0	0	1	0
Fabritex	0	0	0	0	0	1
Fagor	0	0	0	0	0	1
Fahilos	1	0	0	0	0	0
Falke G+A1417roup	0	0	0	0	1	0
Falmac	0	2	0	0	0	1
Fameccanica	1	1	0	0	0	0
Fan Everyday Wear	0	0	1	0	0	0
Fanco	0	2	0	0	0	0
Fanificio Becagli	0	0	0	0	1	0
Fanuc Robotics	0	1	0	0	0	0
Far Eastern Textiles	0	0	0	0	2	0
Fardis	0	1	0	0	0	0
Fare	1	3	1	0	0	1
Farr West Fashions	1	0	0	0	0	0
Fashion Box	0	0	0	0	0	2
Fashion for Men	0	0	0	0	0	1
Fashion Group International	0	0	0	1	0	0
Fashion Institute of Technology	0	0	0	0	1	0
Fashionchain	0	1	0	0	0	0
Fasionizer	1	0	0	0	0	0
Fast Retailing	0	0	1	1	0	0

	1999	2000	2001	2002	2003	2004
Faupel	0	1	0	1	1	1
Faytex	0	0	0	0	0	1
FD Alliance	0	0	0	0	1	0
Federal Mogul	0	0	0	0	0	2
Fehrer	0	4	0	3	1	7
Fein Elast	0	0	1	0	0	0
Feinjersey	0	0	0	0	1	0
Felina	0	0	2	0	0	0
Felters Group	0	1	0	0	0	0
Feltex	1	0	2	0	1	0
Feltrifici Internazionali Riuniti	0	0	1	0	0	0
Fempro	0	1	2	0	0	0
Ferber	0	0	1	0	0	0
Ferrari	0	2	1	1	1	0
Ferrer Dalmon	1	0	0	0	0	0
Ferris Pharmaceuticals	0	0	0	0	0	1
Feyecon	0	0	0	0	0	1
Fibaflo	0	0	0	1	0	0
Fiber and Finishing	0	0	0	0	0	1
Fiber Innovation Technology	0	1	1	0	0	1
Fiber Innovations	0	0	1	0	0	1
Fiberex	0	1	0	0	1	0
Fiberline Composites	0	0	0	0	1	1
Fiberlok	0	0	0	1	0	1
FiberMark	1	0	2	1	2	1
Fiberstruct	0	0	0	0	0	1
Fibertech	0	1	1	0	0	1
Fibertex	2	0	2	1	0	1
FiberVisions	3	7	5	8	4	2
Fiberweb	0	1	0	0	0	0
Fibras Quimicas	0	1	0	0	0	0
Fibre All	0	1	0	0	0	0
Fibre Reinforced Products	0	1	0	0	0	0
Fibre Resources Group	0	0	1	0	0	1
Fibreguide	0	1	1	1	0	0
Fibreloft	0	0	0	0	1	0
Fibrevision	0	0	1	0	0	0
Fibro Organic	0	0	0	0	1	0
Ficamp	1	0	0	0	0	0
FiDiVi Tessitura Vergnano	0	0	0	0	0	1
Fields	0	0	1	0	0	0
Fieldturf	0	0	0	1	1	0
Fife Corporation	0	0	0	1	0	0
Figleaves.com	0	0	0	0	0	1
Fii	1	1	1	0	0	0
Fil Man Made	0	0	0	0	2	0
Fila	0	0	0	1	1	1
Filasur	0	0	1	0	0	1

	1999	2000	2001	2002	2003	2004
FilaTech	0	0	1	0	0	0
Filati Buratti	0	0	0	0	1	0
Filati di Ziche	0	0	0	0	1	0
Filati Drago	0	0	0	0	1	0
Filatura di Pollone	1	0	0	0	0	0
Filatura e Tessitura di Puglia	0	1	0	0	0	0
Filatura e Tessitura di Tollegno	0	1	0	0	0	0
Filatura Paolo Paschetto	0	0	0	1	1	0
Filatures et Tissage de Saulxures	2	0	0	0	0	0
Fillattice	0	2	0	0	0	0
Filligent	0	0	0	0	0	1
Filodoro	0	0	0	1	0	1
Filpucci	0	0	1	0	0	0
Filteco	1	0	0	0	0	0
Filzfabrik Fulda	0	0	1	0	0	0
Fimatex	0	0	0	0	0	1
Fina Technology	0	0	0	0	2	1
Fincarde	0	0	0	0	1	1
Findlay Industries	0	0	0	0	0	1
Fineline	0	1	0	0	0	0
Finlane Group	0	0	1	1	0	0
Finot Group	0	0	0	1	0	0
Fipipe	0	0	0	1	0	0
Firplak	0	0	0	0	0	1
Firsan	0	0	0	1	0	0
First Quality Nonwovens	0	0	0	0	2	2
Fischer Advanced Composite Components	0	0	0	0	0	1
Fischer and Paykel	0	0	0	0	0	2
Fisipe	0	0	0	1	0	0
FK Group	0	0	0	0	1	0
FLA Orthopaedics	0	0	0	0	0	1
Flag Shop	0	0	0	0	1	0
Flags Unlimited	0	0	0	0	1	0
Flainox	0	1	0	0	0	1
Flanders Tapestries	0	0	0	1	0	0
Fleissner	9	13	27	16	11	17
Flemings	1	1	0	0	0	0
Fletcher International Exports	1	1	1	1	1	0
Fletcher International Products	1	0	0	0	0	0
Fletcher Jones	0	0	0	1	0	0
Flexform Technologies	0	0	0	0	0	1
Flexible Alternatives	0	0	0	0	1	0
Flexitec Structures	0	0	0	0	1	0
Flokser	0	0	1	1	0	0
Floreal Group	0	0	0	1	0	0
Flude	0	0	1	0	0	0

	1999	2000	2001	2002	2003	2004
Flying Colours	0	0	0	0	1	0
FMJ ChemBio	0	0	0	1	0	0
Foam Matrix	0	0	0	2	0	0
Fokker	1	1	0	0	0	0
Fongs	1	4	2	1	2	3
Ford Motor Group	1	1	2	1	0	1
Forest Line	0	0	0	0	0	1
Forever	0	0	0	0	0	1
Formia Nuova	0	0	0	0	0	1
Formosa Plastics	1	0	0	0	0	0
Formosa Taffeta Company	0	0	0	0	0	1
Forschungskuratorium Textil	0	0	0	1	0	0
Forsell and Littleton	0	1	0	0	0	0
Forster Rohner	0	0	1	0	0	0
Forster Textile Mills	0	0	1	0	0	0
Fort James	0	0	1	0	0	1
Forta	0	0	0	0	0	1
Fortafil	0	0	0	0	0	1
Fortex	0	0	0	0	0	1
Foss	0	0	1	1	0	3
Foster Miller	0	0	0	0	1	1
FotoWear	0	0	0	0	0	1
Fountain Set	0	0	0	1	1	3
Four Leaf Textiles	0	1	0	0	0	0
Fourne	0	1	1	0	0	0
Fowler	0	0	0	1	0	1
Fox and Lillie	1	2	2	0	1	0
Framis	0	0	0	0	2	0
Frank Kochmann Gruppe	0	0	1	0	0	0
Frank Usher	0	0	0	0	1	0
Franz Gahlert Steppwaren	0	0	0	1	0	0
Franz Lasser	1	0	0	0	0	1
Franz Schafer	0	0	0	0	1	0
Franz Ziener	0	0	0	0	0	1
Franzoni Filati	0	1	1	4	0	0
Fratelli Citterio	0	0	0	0	1	0
Fratelli Robustelli	0	0	0	0	0	1
Fred Lawton and Son	1	0	1	0	0	0
Free Spirit	0	0	0	0	0	1
Freesky	0	1	0	0	0	0
French Dressing Jeanswear	0	0	0	0	1	0
French	0	0	0	1	0	0
Frenzelit	0	0	2	0	0	0
Fresco Plastics	0	0	0	0	0	1
Fresenius	0	0	0	0	0	1
Fresh Produce	0	0	0	0	1	0
Freshtex	0	0	0	0	0	1
Freudenberg	7	7	10	5	2	17

WORLD TEXTILES Manufacturer Indexing

	1999	2000	2001	2002	2003	2004
Freysinnet	0	0	0	0	1	0
Friction Free Technologies	0	1	0	0	0	1
Frisby	0	0	1	1	0	0
Fritz Blanke	0	2	0	0	0	0
Fritz Gegauf	0	0	0	0	0	2
Fritz Landolt	0	0	0	0	1	0
Frottana	0	1	1	0	0	0
Frowein	0	0	0	1	0	0
Fruit of the Loom	2	0	3	0	1	0
Fuji Heavy Industries	0	0	0	0	0	5
Fuji Spinning Company	2	1	1	0	1	2
Fuji Xerox Company	0	0	0	0	0	1
Fujii Keori	0	0	0	1	0	0
Fukaki Woollen Textile Company	0	0	0	1	0	0
Fukusen	2	0	0	0	0	0
Fulcrum	0	0	0	0	0	1
Fulda Gruppe	0	1	0	0	0	0
Furmanite	0	0	0	0	0	1
Furukawa Electric Company	0	0	0	0	0	1
G and J Awning and Canvas	0	0	0	0	0	1
G III	0	0	0	1	0	0
G Modiano	0	0	1	0	0	0
Gabriel Chemie	0	0	0	0	1	0
Gadoon Textile Mills	0	0	0	0	2	0
Gadzooks	1	0	0	0	0	0
Gafitex	0	0	0	0	0	1
Galan	1	0	0	3	1	0
Gale Pacific	0	0	0	0	1	0
Galey and Lord	0	1	2	0	0	0
Galvanin	0	1	0	0	0	0
Gamma Holding	0	0	1	0	0	1
Gap Tekstil	0	0	0	0	0	1
Gap	0	0	0	1	0	0
Garan	0	0	1	0	0	0
Garant	0	0	0	0	0	1
Gardella	0	0	0	1	0	0
Gardeur	0	0	0	0	1	0
Gardiner of Selkirk	1	2	1	0	0	2
Garland Company	0	0	0	0	2	0
Garnett Controls	0	2	1	1	0	0
Garnett Group	0	0	0	1	0	0
Garnveredlungswerke	0	0	0	2	0	0
Gartner	0	0	1	0	0	0
Garware Wall Ropes	0	0	0	2	0	0
Gaskell	2	1	0	1	0	0
Gaston County Dyeing Machine	1	3	0	0	0	0
Gates Unitta	0	0	0	0	0	1

	1999	2000	2001	2002	2003	2004
Gavazzi	0	0	0	0	1	0
Gayatri	0	0	1	0	0	1
Gazal	0	1	0	1	0	0
GE Advanced Materials	0	0	0	0	0	2
Geelong Textiles	1	1	1	0	1	0
Gehring Textiles	0	3	1	1	3	0
Geislinger	0	0	1	0	0	0
Gematex	1	3	0	0	0	0
Gemeinschaft unweltfreundlicher Teppichboden	0	0	1	0	1	0
Gemmill and Dunsmore	1	0	0	1	0	0
Gemsan	0	0	0	1	0	1
Genencor	3	0	4	0	0	2
General Electric	0	0	1	0	2	8
General Motors	0	1	0	1	3	1
Genesta	0	0	0	0	0	1
Gentle Fit Lingerie	1	0	0	0	0	0
GEOfabrics	0	0	0	0	0	1
Geoff Bade	0	0	0	0	1	0
George Medical	0	0	0	0	0	1
Georgia Composites	0	0	0	0	1	1
Georgia Pacific	0	0	0	1	1	0
Georgia Tech Research Institute	0	0	0	0	1	2
Geox	0	0	0	0	0	1
Gerber	4	6	8	4	2	3
Gerflor	0	0	0	0	1	0
German Plant and Machinery Manufacturers Association	0	0	1	0	0	0
German Wool Research Institute	0	0	0	0	1	0
Getzner	0	0	1	1	0	0
Gevas	1	0	0	1	0	0
GfE Medizitechnik	0	0	0	0	3	1
Gibson Group	0	0	1	0	0	0
Giesse	0	1	0	1	1	0
Gigliotti e Gualchieri	0	0	0	0	0	1
Gilbos	0	1	0	1	0	0
Gild the Lily	0	0	0	1	0	0
Gildan	0	0	0	1	0	1
Gill	0	0	1	0	0	0
Gillette	0	0	0	1	0	0
Gimbel Glove Company	0	0	0	0	1	0
Gina	0	0	0	1	0	0
Giordano	0	0	0	0	0	1
Giovanelli	0	1	0	1	1	0
Giovanni Bozzetto	0	0	0	1	0	0
GiS	0	0	0	0	1	0

WORLD TEXTILES Manufacturer Indexing

	1999	2000	2001	2002	2003	2004
Gividi	0	0	0	0	1	1
GKN	1	1	1	0	0	1
Glasform	1	0	1	0	0	0
Gleistein	1	0	0	0	0	0
Glen Raven	1	7	1	4	1	4
Glenanne Jacquards	0	1	0	0	0	0
Glenro	0	0	0	1	0	0
Glo-Tech	0	1	0	0	0	0
Global Brand Marketing	0	0	0	0	0	1
Global Corte	0	0	0	2	0	0
Global Textile Partnership	0	0	0	4	0	0
Global Sewing	0	2	0	0	0	0
Globe Manufacturing	1	2	1	0	0	0
Globe Worsted	1	2	0	0	0	0
Globetec Nonwovens	1	0	0	0	0	0
Glocon	0	0	0	0	1	0
Glowyarn	0	0	0	0	0	2
GMP	0	0	0	0	0	1
Gneuss	0	2	4	2	4	0
Gobi	1	0	0	0	0	0
Godfrey Hirst	0	0	0	1	0	0
Gokuldas	0	0	0	1	0	0
Gold Shield Indiana	0	0	0	0	0	1
Gold Toe	0	0	0	1	0	1
Golden Lady	1	1	0	1	1	3
Goldschmidt	0	0	0	0	1	1
GoLite	0	1	0	0	0	1
Goller	0	5	0	0	0	0
Goodacre	1	0	0	0	1	0
Goodyear	0	1	0	0	2	8
Gorlitzer Hanf und Drahtseilerei	0	0	0	0	1	0
Gosen	1	0	0	0	0	1
Goulburn Wool Scour	0	0	1	1	1	0
Goulston	0	0	0	0	3	2
GPMI	0	0	0	0	1	0
Grand National Apparel	0	0	0	0	0	1
GrandStand	0	1	0	0	0	0
GranStand	0	0	1	0	0	0
Graph to Graphics	0	0	0	0	0	1
Graphtec	0	0	1	0	0	0
Grasim	1	0	0	0	0	0
Great Lakes	0	0	0	0	2	1
Great Plains Buffalo Products	0	0	0	1	0	0
Great Universal	0	1	1	0	1	0
Green Bay Nonwovens	0	1	1	1	0	0
Green Label Company	0	0	0	0	1	0
Green Marine	0	0	1	0	0	0
Greenstone Automation	0	0	0	0	1	0

	1999	2000	2001	2002	2003	2004
Greiner	0	0	0	0	0	1
Gretag Macbeth	0	1	1	0	0	2
Greystone	0	0	0	0	1	0
Griffith Textile Machines	0	1	1	0	1	1
Grindi	0	0	0	0	0	1
Grob Horgen	0	0	0	1	1	1
Grossenhainer	0	0	0	0	1	0
Grotto	0	0	0	0	0	1
Groupe Socota	0	0	0	1	0	0
Groz Beckert	3	7	5	8	4	8
Grupo Antolin Ingenieria	0	0	0	1	0	0
Grupo El Asturiano	0	0	1	0	0	0
Grupo Inca	1	1	0	1	0	0
Grupo Industrial Miro	0	1	0	0	0	1
Grupo M	0	0	0	0	0	1
Grupo Sedamil	0	0	0	1	0	0
Grupo Textil Providencia	0	0	0	0	1	0
Gruppo Colle	0	0	0	0	0	1
Gruppo Drago	0	0	0	0	1	0
Gruppo Fedora	0	0	0	1	0	0
Gruppo Linificio	0	0	0	0	0	1
GSE	0	0	1	0	0	0
GTC Technology	0	0	1	0	0	0
GTN	0	0	0	3	1	0
GTZ	0	0	0	1	0	0
Guabello	0	1	0	0	0	0
Guardian Protective Clothing	0	0	0	0	0	1
Guardian	0	0	0	0	0	1
Guelph Twines	0	0	1	0	0	0
Guengual Group	0	0	0	0	0	1
Guess	0	1	0	0	0	0
Guido Unger	0	1	0	0	0	0
Guilford Mills	1	4	1	1	0	3
Gundle/SLT Environmental	0	0	0	1	0	0
Gunn and Inggs	0	1	0	0	0	0
Gunold and Stickma	1	0	5	0	0	0
Gunold	0	0	1	2	0	2
Gunze	0	0	0	0	0	1
Gustav Daiber	1	0	1	0	0	1
Gutermann	0	0	0	0	1	1
Guth und Wolf	1	0	0	0	0	0
Guy and O'Neill	0	0	0	0	1	0
Gwlan Teifi	0	0	1	0	0	0
Gygli	0	0	0	2	1	0
Gymboree	0	1	0	0	1	0
H and G Hegmanns	0	0	0	0	0	1
H Dawson Sons	0	0	0	1	0	0
Habasit	0	0	0	0	1	2

WORLD TEXTILES Manufacturer Indexing

	1999	2000	2001	2002	2003	2004
Hacoba	0	1	0	0	0	0
Haggar	1	0	0	0	1	0
Hagihara Industries	0	0	1	0	0	0
Hainsworth	0	0	0	1	0	1
Halbmond	0	1	1	0	0	0
Halmatic	0	0	0	1	0	0
Hamby Textile Research Laboratories	0	0	0	0	0	1
Hamel	0	0	1	1	0	0
Handi Quilter Company	0	0	0	0	0	1
Hanes Printables	0	0	0	0	0	2
Hangzhou Xinhua Group	1	1	0	0	0	0
Hanil E Hwa	0	0	0	0	0	1
Hansa	1	0	1	1	0	0
Harbor Healthcare	0	0	0	0	0	1
Hardcore Composites	1	1	0	0	0	1
Hardwire	0	0	0	0	1	0
Harish	0	0	2	0	0	2
Harley Davidson	0	0	0	0	0	2
Harodite	0	0	0	0	1	2
Harold Watson	1	0	0	0	0	0
Harriet and Henderson	1	1	0	1	0	0
Harris Fabrics	0	0	1	0	0	0
Harris Speciality Chemicals	1	0	0	0	0	0
Harrison Gardner	0	0	1	0	0	0
Harry Lucas	1	1	0	0	0	0
Harry Teague	0	1	0	0	0	0
Hart Stores	0	0	0	0	1	0
Hartmarx	0	0	1	1	0	0
Hartzell	0	0	1	0	0	0
Harveys	1	1	0	1	0	0
Hasbro	0	0	0	1	0	1
Hashima Company	0	0	0	0	0	1
Hass Mondomix	0	0	0	0	0	1
Hassan Group	0	0	1	2	0	0
Hatch Corporation	0	0	0	0	1	0
Hatley The Little Blue House	0	0	0	0	0	1
Hauser	1	1	0	0	0	0
Havencrown	1	0	0	0	0	0
Hawick Cashmere	0	1	1	1	0	0
Hawkeye	0	0	0	0	0	1
Hawtin	0	1	1	0	0	0
Hay and Robertson	0	0	1	1	0	0
HB Engineering	0	0	0	1	0	0
HDO	0	0	1	0	1	0
HealthShield	1	1	0	0	0	0
Healthtex	0	0	0	0	0	1
Hebbecker	1	0	0	0	1	0
Heberlein	1	2	3	3	6	7

	1999	2000	2001	2002	2003	2004
Hecking Deotexis	0	0	0	1	0	1
HEH	0	0	0	0	0	1
Heidemann	0	0	1	0	0	1
Heimbach Group	0	0	0	0	0	2
Heliot	1	1	0	3	0	0
Hellmann	0	0	0	0	0	1
Helly Hansen	1	0	0	0	0	1
Hemcon	0	0	0	0	1	0
Hempline	0	0	0	1	0	0
Hemptown Clothing	0	0	0	0	1	0
Henderson Machinery	0	0	0	0	0	1
Heng Yuan Xiang	0	0	0	0	0	1
Henkel	1	0	0	0	0	4
Hennes and Mauritz	0	0	0	0	1	2
Henredon	0	0	0	0	1	0
Henriksen	0	0	1	0	0	0
Heraeus	0	0	0	0	0	2
Hercules	0	0	0	0	1	2
Heritage Textile Company	0	0	0	1	0	0
Hermitage Holdings	1	0	1	0	0	0
Herzog und Brauer	0	0	0	0	0	1
Heusker	0	0	0	0	0	1
Hewlett Packard	0	0	0	0	0	2
Hexcel	3	1	8	1	2	2
Hickory Springs Manufacturing Company	0	0	0	1	0	0
Hicks Tufting Machine Service	0	0	0	0	0	2
High Modulus	0	0	0	1	0	0
High Voltage Graphics	0	0	0	0	0	1
Highstil	0	0	0	0	1	0
Hilanderias Andinas	0	0	0	0	1	0
Hilasal	0	0	0	0	1	0
Hilaturas Llaudet	0	2	0	0	0	0
Hilden Manufacturing	1	0	0	0	0	0
Hills Incorporated	0	0	1	1	1	5
Hills Industries	0	0	0	0	0	1
Hilos and Telas	0	0	0	0	1	2
Hilton Fabrics	0	0	1	0	0	0
Himatsingka Seide	2	0	0	0	0	0
Hinspergers	0	0	0	0	0	1
Hirano Tecseed	0	0	1	0	0	0
Hiraoka and Company	0	0	0	0	1	0
Hirdaramani	0	0	0	0	0	1
Hitachi Chemical Company	0	0	0	0	1	0
Hitachi	0	0	0	0	0	1
HMZ Fashion Group	0	0	0	0	1	0
Hobas	0	1	0	0	2	0
Hodgson of Scotland	0	0	0	0	0	1

	1999	2000	2001	2002	2003	2004
Hoechst and Reitex Hydraulik	0	0	0	0	1	0
Hoechst	1	1	0	0	0	0
Hoff and Korbel	0	0	1	0	0	0
Hoffman	0	0	0	1	1	1
Hohensteiner Institute	0	0	0	1	0	0
Hokkaido University	0	0	0	0	0	1
Holeproof	0	1	0	0	0	1
Hollinee	0	0	0	0	0	1
Hollingsworth and Vose	0	0	0	0	0	1
Hollingsworth	1	2	0	0	0	1
Hollister Incorporated	0	0	0	0	0	1
Holly Bra	1	0	0	0	0	0
Holofiber Enterprises	0	0	0	0	1	0
Home Products	0	0	0	0	0	1
Home Source	0	0	0	0	0	1
Homestyle	0	0	0	1	1	0
Honda	0	0	0	0	0	2
Honeywell	1	4	14	4	2	26
Hornung Mills	0	0	0	0	0	1
Horrockses Fashions	0	0	0	1	0	0
Hospital Specialty Products	0	1	1	0	0	0
Hot Topic	0	0	0	0	1	1
House of Monatic	0	0	0	1	0	0
Howa	0	0	0	0	0	2
Howard Showers	0	1	0	0	0	0
Howies	0	0	0	0	0	1
HP Chemie Pelzer	0	0	0	1	0	2
HTP Unitex	0	4	0	2	1	1
HTS	0	0	0	0	0	1
Hualon	0	1	1	0	0	0
Huber and Suhner	0	0	4	0	0	0
Huber Manufacturing Company	0	0	0	0	0	1
Hubtex	0	0	0	0	0	1
Huddersfield Fine Worsteds	1	1	1	1	0	1
Huddersfield Textile Centre of Excellence	0	0	1	0	0	0
Huddersfield Textile Society	0	0	0	0	1	0
Hudson Awning and Sign Company	0	0	0	1	0	0
Hudson Industries	0	0	0	0	0	1
Huesker	0	0	0	0	2	4
Hugh Mackay	0	0	0	0	0	1
Hugo Boss	0	1	1	2	1	10
Huish	0	0	0	0	0	1
Huizen Industrial Fibres	0	0	0	0	1	0
Humatro	0	0	0	0	1	0
Hummelsheim	1	0	0	0	0	0

	1999	2000	2001	2002	2003	2004
Hummer Turfgrass Systems	0	0	0	0	1	0
Hun Kun Enterprise	0	0	0	0	0	1
Hunter Douglas	0	0	1	0	0	3
HunterLab	1	2	0	0	0	0
Hunters of Brora	1	1	1	0	0	0
Huntingdon Mills	0	0	0	0	1	1
Huntsman	0	0	0	0	0	2
Hurell-Hispano	0	0	0	0	0	1
Huyck Australia	0	0	1	0	0	0
Hy-Tec	0	0	0	0	1	0
Hygienic Industries	0	0	1	0	0	0
Hyosung	0	1	0	1	0	4
Hypercar	0	0	1	3	0	0
Hyperion Catalysis	1	3	0	0	0	2
Hysport	0	1	0	0	0	0
Hytex Group	0	0	0	0	1	0
Ibena	1	0	0	0	0	0
ICBT	6	8	3	0	0	0
Ichikawa Company	0	0	0	0	1	9
ICI	1	0	0	0	0	0
Icomatex	0	0	0	0	1	0
Icon Laser Technologies	0	0	0	0	1	0
Icopal	0	0	0	0	1	0
Icotec	0	0	0	0	1	0
ICS Group	0	1	0	0	0	0
Ideas to Market	0	0	0	0	1	0
Idego	1	0	0	0	0	0
Idemitsu Unitech	0	0	0	0	1	1
Idrosistem	0	0	1	0	0	0
Ihsan Sons	0	0	0	1	0	0
IIshin	0	0	1	0	0	0
IKO Industries	0	1	0	0	0	0
Ilford Imaging	0	0	0	0	0	1
Illawarra Alpacas	0	0	0	1	0	0
Ilpea	0	0	0	0	0	2
Imatex	0	0	1	0	0	0
Imax	1	0	0	0	0	0
Impact Composite Technology	0	0	1	1	0	0
Impak	1	0	1	0	0	0
Implant Sciences	0	0	0	0	0	1
In Stitches	0	1	0	0	0	0
INC	0	0	1	0	1	1
Inca Tops	0	1	1	0	0	0
Incorporated Society of London Fashion Designers	0	0	1	0	0	0
Indas	0	0	0	0	1	0
Indian Cotton Mills Federation	0	0	0	1	0	1

WORLD TEXTILES Manufacturer Indexing

	1999	2000	2001	2002	2003	2004
Indian Institute of Technology	0	0	0	0	0	1
Indian Jute Mills Association	0	1	0	0	0	0
Indian Terrain	0	0	0	0	1	0
Indian Textile Accessories and Machinery Manufacturers Association	0	0	0	0	0	1
Indian Woollen Mills Federation	0	0	0	1	0	0
IndianOil	0	0	1	0	0	0
Inditex	0	0	0	2	0	0
Indorama	0	1	2	1	1	0
Industria Ambrosiana Filati	0	1	0	0	0	0
Industria Bergamasca Rifrangenti	0	0	0	0	0	1
Industrial Air	0	0	1	0	0	0
Industrial Copolymers	0	1	1	0	0	0
Industrial Fabrics Association International	0	0	0	0	1	1
Industrias Campins	0	0	0	0	1	0
Inex Ukus	0	0	0	1	0	0
Infineon	0	0	0	7	0	0
Inflatable Images	0	0	1	0	0	0
Inflate Design	0	0	0	0	0	1
Informatica Textil	0	0	0	1	0	0
Inmetess	0	0	0	1	0	0
Innothera Topic	0	0	0	1	3	0
Innovative Global	0	0	0	0	0	1
Innovative Technologies	0	1	0	0	0	0
Inoplast	0	0	0	0	0	3
Inotex	0	0	0	0	1	0
Inova	0	1	0	0	0	0
Inspec Fibres	0	1	0	0	0	0
Instituform	0	0	0	0	0	1
Institut Francais du Petrole	0	0	0	0	1	0
Institut Francais Textile Habillement	0	0	4	1	0	0
Institut fur Polymerforschung Dresden	0	0	0	0	0	1
Institut fur Textil und Bekleidungstechnik	0	0	0	0	0	1
Institut fur Textil und Verfahrenstechnik	1	0	0	4	0	5
Institute for Manufacturing Automation	0	0	0	1	0	0
Institute for Soldier Nanotechnologies	0	0	0	1	2	0
Institute of Chemical Fibres	0	0	1	2	0	0
Instrumar	0	1	0	0	0	0

	1999	2000	2001	2002	2003	2004
Intangible	0	1	0	0	0	0
Integra	0	0	0	1	0	0
Intel	0	0	0	0	0	1
Intelligent Textile Technological Alliance	0	0	0	0	0	1
Intera	2	0		0	0	0
Intercontinental Polymers	0	0	2	0	0	0
Interface	3	5	0	0	1	4
Interloop	0	0	0	0	1	0
Internacional de Comercio	0	0	0	1	0	0
International Fashion Machines	0	0	0	0	0	1
International Fibre Centre	0	0	2	0	0	2
International Finance Corporation	0	0	0	0	1	0
International Flavors and Fragrances	0	0	0	0	1	0
International Forum for Cotton Promotion	0	0	0	0	1	0
International Mohair Association	0	1	0	0	0	0
International Rayon and Synthetic Fibre Comittee	0	0	0	1	0	0
International Silk Association	0	1	0	0	0	0
International Surface Design Association	0	0	0	0	0	1
International Taly	0	0	0	1	0	0
International Textile Manufacturers Federation	0	0	0	1	0	0
International Wool Textile Organisation	0	1	0	0	1	1
Interspe Hamann Group	0	0	0	0	0	1
Intertape Polymer Group	0	0	0	0	1	1
Intertex	0	0	0	0	0	1
Intex	0	1	0	0	0	1
Intier	0	0	0	0	0	1
Inuisia	0	0	0	0	0	1
Inventa Fischer	1	6	6	3	4	3
Inventio	0	0	0	1	0	0
Investa	0	0	1	0	1	0
Investronica	1	0	0	2	0	0
Invibio	0	0	1	0	0	0
Invista	0	0	0	0	3	21
Iran Sericulture Company	0	0	0	1	0	0
Irmac	0	1	0	0	0	0
IRO	1	0	0	0	1	1
Iropa	0	0	0	0	0	10
Ishikawajima Harima Heavy Industries	0	0	0	0	1	1

WORLD TEXTILES Manufacturer Indexing

	1999	2000	2001	2002	2003	2004
Ishizuka	0	0	0	0	0	1
Isko	0	0	0	0	1	1
Isra Vision Systems	0	0	0	0	0	1
Issey Miyake	0	0	1	1	0	0
Istituto Commercio Esterno	0	0	1	0	0	0
Italian Textile Association	0	0	1	0	0	4
Italpaca	0	0	0	0	0	1
Italplastic	0	0	3	0	0	0
ITC	0	0	0	0	1	0
Itema Group	0	0	3	4	2	1
ITL Software	0	0	0	1	0	0
Itochu	1	3	0	1	1	0
ITT Flygt	0	0	1	2	1	0
IVM	1	0	0	0	0	0
IZWool	0	0	0	0	0	1
J and J Dyson	1	0	0	0	0	0
J and J Industries	0	1	1	0	0	0
J Barbour and Sons	1	1	0	0	1	0
J C Rennie	0	0	1	0	0	0
J F Muller	0	0	0	1	1	0
J G Boswell Company	0	0	0	1	0	0
J L de Ball	0	0	0	1	0	0
J S Brooksbank	1	0	1	0	1	0
J Schneider	1	0	0	0	0	0
Jack Laurie	0	0	1	0	0	0
Jackson of Geelong	0	1	0	0	0	0
Jacob Holm	1	0	0	0	0	1
Jadran Hosiery	0	0	0	2	0	0
Jaeger	0	0	0	1	0	1
Jaeggli	0	0	0	0	0	1
Jaguar	0	0	0	0	1	0
Jaipur Polo Club	0	0	1	0	0	0
Jakob Muller	2	4	8	4	4	3
Jakob Schlaepfer	0	0	0	0	0	1
Jam Session	0	0	0	0	1	0
James Beattie	0	0	0	0	1	1
James H Heal	1	1	3	0	0	1
James Hardie Research	0	0	0	0	1	1
James Holdsworth and Brothers	1	0	0	0	0	0
Jandakot Wool Washing	1	1	0	1	1	0
Jangyin Shengai Industrial Company	0	0	0	0	1	1
Janney Daniel	0	0	0	0	0	1
Janome	0	0	0	0	1	3
Japan Absorbent Technology Institute	0	0	0	0	0	1
Japan Aircraft Development	0	0	0	0	0	1
Japan Matex	0	0	0	0	0	1

	1999	2000	2001	2002	2003	2004
Japan Science and Technology	0	0	0	0	0	2
Japan Textile Machinery Association	0	0	0	1	0	0
Japan Vilene	1	0	1	0	0	0
Jarmain	1	0	0	0	0	0
Jason Incorporated	1	0	0	0	0	1
Javor	0	0	0	1	0	0
JCPenney	0	0	0	1	2	1
Jeanly	0	0	0	3	0	0
Jeantex	0	0	1	0	0	0
Jemtex	0	1	0	0	0	0
Jensen	0	0	0	0	0	2
Jentex	0	1	0	0	0	0
Jentschmann	0	0	1	0	0	0
Jeremiah Ambler	0	0	0	1	0	0
Jergens	0	0	1	0	0	0
Jerzees	0	0	0	0	0	2
Jet Ice	0	0	0	0	0	1
Jewel Power Company	0	0	0	1	0	0
Jiangsu Sunshine	0	0	0	1	0	1
JNCO	1	0	0	0	0	0
Jockey	1	1	0	0	0	0
Johann Borgers	0	1	0	0	0	0
Johannes Zimmer	0	1	1	3	2	1
John Charles Designs	0	0	0	0	1	0
John D Hollingworth	0	0	1	1	0	1
John David Sports	1	0	1	1	1	0
John Deere	0	0	0	1	0	0
John England	0	0	1	0	0	0
John Foster	0	0	0	1	0	1
John Hanly and Company	0	0	1	1	0	0
John Heathcoat	1	1	0	1	0	1
Johns+A2052 Manville	0	0	2	0	2	6
John Marshall and Company	0	0	0	1	0	0
John McNutt	0	0	1	0	0	0
John Smedley	0	0	0	0	0	1
Johnson and Johnson	1	1	2	1	1	5
Johnson Controls	0	1	0	0	0	1
Johnson Matthey	0	0	0	0	0	1
Johnston	0	0	0	1	0	0
Johnstons of Elgin	2	2	2	2	1	2
Jomac Wells Lamont	0	0	1	0	0	0
Jones Apparel	1	1	1	1	1	0
Jordan and Sons	0	0	0	1	0	0
Jordan Lovatt and Jones	0	0	0	1	1	0
Jos A Bank	0	0	0	0	1	0
Josef Kanz	0	0	0	2	0	0
Josef Otten	0	1	0	0	0	0

WORLD TEXTILES Manufacturer Indexing

	1999	2000	2001	2002	2003	2004
Joslin Orthopedic Gear	0	0	0	0	0	1
Jossi	1	0	0	0	1	1
JUKI Union Special	0	1	0	0	0	0
Juki	0	0	0	2	1	4
Jules Tournier and Fils	0	0	1	0	0	0
Kaesor	0	0	1	0	0	0
Kafus	2	2	0	0	0	0
Kalamazoo	0	0	0	0	1	0
Kanebo Gohsen	6	2	3	0	2	0
Kanebo	6	6	1	2	1	3
Kaneka	0	0	1	1	0	2
Kanematsu	0	0	0	0	0	1
Kaneta	0	1	0	0	0	0
Kang Na Hsuing	0	1	1	0	0	0
Kannegiesser	1	1	1	0	0	1
Kao	0	0	0	0	0	7
Kappler	0	0	0	0	0	1
Kaputone	0	0	1	0	1	0
Karibe Industria Comercio	0	0	1	0	0	0
Karl Mayer	11	15	32	43	34	43
Karla Colletto	0	0	0	1	0	0
Karma Kumas	0	0	0	0	1	0
Karnataka Handloom Development Corporation	0	0	0	0	0	1
Karrimor	0	1	0	0	0	0
Karstadt-Quelle	0	1	0	0	0	1
Karsten	0	0	0	0	1	0
Katag	0	0	0	0	0	2
Kathy Ireland Worldwide	0	0	0	1	0	0
Kaufhof	0	0	0	0	0	2
Kaztek	1	0	0	0	0	0
KBN Group	0	0	0	0	0	1
Kea	0	0	1	0	0	0
Keilmann	0	0	0	0	1	0
Keisokki Kogyo	1	0	0	0	1	0
Kelheim Fibres	0	0	0	0	0	2
Keller	0	0	0	1	0	0
Kellwood	1	0	1	0	1	0
Kelly and Windsor	0	0	0	0	0	2
Kelt	0	0	0	0	1	0
Kelty	0	1	0	0	0	0
Kem-Wove	1	0	0	0	0	0
Kemcor	1	0	0	0	0	0
Kendall Confab	0	1	1	0	0	0
Kenex	0	0	0	1	0	0
Kenneth Cole	0	1	0	0	0	0
Kenvelo	0	0	0	0	0	1
Keraplast	0	0	0	1	0	0
Keratec	0	0	0	0	0	1

	1999	2000	2001	2002	2003	2004
Kermel	0	0	0	0	1	2
Kern Liebers	2	1	3	0	0	0
Kettelhack	0	0	0	0	0	1
Keum Yong	1	0	0	0	2	0
KG Fibers	0	0	0	1	0	0
Khalke	0	0	0	1	0	0
KI	0	0	0	0	1	0
Kidou Company	0	0	1	0	0	0
Kikkawa Weaving	0	1	0	0	0	0
Kikuchi Kogyo	0	0	0	0	0	2
Kilgour French Stanbury	0	0	0	0	0	1
Kimberly Clark	8	4	4	13	24	92
Kinetic Concepts	0	0	0	0	1	0
Kingfisher	0	1	0	0	0	1
Kingsmead Carpets	0	0	0	1	0	0
Kirloskar Toyoda	0	0	0	0	1	1
Kissel and Wolf	1	0	0	0	0	0
Kiu Hung Industries	0	0	0	0	0	1
KKA Kleinwefers Anglagen	0	0	2	0	0	0
Klaus Steilmann and Company	0	0	0	1	0	0
Kleinewefers	4	2	1	0	0	0
Klieverik Heli	0	0	2	3	1	1
Klopman	1	0	0	0	3	1
Kluz	0	0	0	1	0	0
KM Fabrics	0	1	0	0	0	0
Kmart	0	0	0	0	0	1
Knauf	0	0	0	0	0	1
Knit-Rite	0	0	0	0	0	2
Knoll	0	1	1	1	0	0
Knotex	1	0	0	0	1	0
Knowaste	0	0	0	0	0	1
Knowlton Nonwovens	0	0	0	0	0	1
Kobe	0	0	1	0	0	0
Kobelco	1	0	0	0	0	0
Kobest Company	0	0	0	0	0	2
Koch Industries	0	0	0	0	1	2
Kockums	0	0	0	0	0	1
Koenig und Bauer	0	0	0	0	0	2
Koenig	0	0	0	0	2	0
Kohls	0	1	0	0	0	0
KolnMesse Group	0	0	0	1	0	0
Kolon	0	1	0	0	1	7
Komatsu Seiren	3	3	0	0	0	3
Komer	0	1	0	0	0	0
Kondo	0	1	0	0	0	0
Konen	0	0	0	1	0	0
Kookai	0	1	0	0	0	0
Koppermann	0	1	0	0	0	0

WORLD TEXTILES Manufacturer Indexing

	1999	2000	2001	2002	2003	2004
Koramsa	0	0	0	1	0	0
Kordarna	0	1	0	0	0	0
Korea Tarpaulin	0	0	0	1	0	0
Korma	0	1	0	0	0	0
Kortec	0	0	0	1	0	0
Korteks	0	0	0	0	2	0
KoSa	0	5	13	2	4	3
Kotka	1	0	0	0	0	0
Krantz	0	5	0	0	0	0
Kraton Polymers	1	0	0	0	0	1
Kraus Carpet Mills	0	0	1	0	0	0
Krauss Maffei	1	0	0	0	0	2
Krelus	0	0	0	0	1	0
Krettek	1	1	0	0	0	0
Kreyenborg	0	0	1	0	0	0
Kronflie Spundale	1	0	0	0	0	0
Kroy	2	0	2	0	1	0
Kuaerner Oilfield Products	1	0	0	0	0	0
Kufner	2	1	0	1	0	0
Kunal Organics	0	0	0	0	0	1
Kunert	0	0	0	2	0	0
Kunststofftechnik Sachsen	0	0	0	0	1	0
Kurabo	1	3	1	1	1	0
Kuraray	5	7	8	2	1	13
Kuris-Wastema	0	2	0	0	0	0
Kurray	0	0	0	0	1	0
Kusters	6	16	7	7	5	3
KVG	0	0	0	0	0	2
Kyung-In	0	0	0	0	0	1
L'air Liquide	0	0	0	0	0	1
L'etendard	0	0	0	0	1	0
L-Fashion Group	0	0	0	0	0	1
L and P Property Management	0	0	0	0	1	5
L F Brands	0	0	0	0	3	0
L J Specialities	0	0	1	0	0	0
L S International	0	0	0	0	1	0
La France Industries	0	0	1	0	0	0
La Meccanica	1	0	0	1	1	1
La Perla	0	0	0	1	0	0
Laboratoire Hydrex	0	0	0	0	1	0
Laboratoires d'Hygiene et de Dietetique	0	0	0	0	0	1
Laboratorios Indas	0	1	0	1	0	0
Labortex	0	1	0	0	0	0
Lacent	0	1	0	0	0	0
Lacom	0	1	0	0	0	0
Lacoste	0	0	1	0	0	1
LaCrosse Furniture Company	0	0	0	0	1	0

	1999	2000	2001	2002	2003	2004
Lafer	2	6	2	2	1	2
Laib Yala Tricot	1	0	0	0	0	0
Laidlaw and Fairgrieve	0	1	0	0	0	0
Lainiere de Picardie	1	0	0	1	0	0
LAIP	0	0	0	0	1	0
Lakeland Industries	0	0	0	1	0	3
Lako	0	0	1	0	0	0
Lakshmi Mills	2	1	2	1	5	6
Lamberti	0	0	1	2	1	0
Lamiflex	0	1	0	0	0	1
Lammertz	0	1	0	0	0	0
Lamperti	1	1	0	0	1	0
Lana Lee Fashions	1	0	0	0	0	0
Lanas Trinidad	0	0	0	1	0	0
Lanasur	0	0	0	1	0	0
Lancair	0	0	0	1	0	0
Lancaster Fibre Technology	0	1	0	0	0	0
Landec	0	0	0	0	0	1
Landrell Fabric Engineering	1	0	0	0	0	0
Landrell	0	0	0	1	0	0
Lands End	1	1	1	1	0	0
Landskroon	0	0	0	1	0	0
Lane Bryant	0	0	0	1	0	0
Langholm Dyeing	2	0	0	2	1	0
Lanieri Italiana	0	0	0	0	1	0
Lanificio Botto Guiseppe	0	0	1	0	0	0
Lanificio Carlo Barbera	0	1	0	0	0	0
Lanificio di Schio	0	0	1	1	1	0
Lanificio di Tollegno	0	0	1	0	1	0
Lanificio Fratelli Cerruti	0	1	1	0	1	0
Lanificio Luigi Botto	0	0	0	1	0	0
Lanificio Luigi Colombo	0	0	0	1	0	0
Lano	1	2	0	0	0	0
Lanolin Trading Company	0	0	1	0	0	1
Lantal	0	0	0	0	0	1
Lantor	0	2	0	2	0	0
Lanxess	0	0	0	0	0	1
Lanzhou	0	0	1	0	0	0
Laresa	0	0	0	0	1	0
Laroche	1	3	2	1	3	2
Las Vegas Awnings	0	0	0	0	0	1
Latex Equipment Sales and Service Company	0	1	1	0	0	0
Latexia	0	0	0	1	0	0
Laufa	0	0	1	0	0	0
Lauffenmuhle	0	0	0	0	0	1
Laura Ashley	0	0	0	0	1	0
Laurastar	0	0	0	0	0	1
Laurence Pierce	0	0	0	1	0	0

WORLD TEXTILES Manufacturer Indexing

	1999	2000	2001	2002	2003	2004
Lavorazione Commercio Tessile	0	0	0	1	1	0
Lawer	0	2	0	2	0	0
Lawrence Schiff	0	1	0	0	0	0
Lawrence Technological University	0	0	0	0	0	1
Lawson Hemphill	0	1	0	1	4	0
Laycock	0	0	0	1	0	0
Lazzati Franco	0	0	0	0	0	1
Leading Lady	0	0	2	0	0	0
Leading Spinning	0	0	1	0	0	0
Lear	1	1	0	0	0	1
Leavitt and Parris	0	0	0	0	1	0
Lebanon Apparel	0	1	0	0	0	0
LecTec	0	0	0	1	0	0
Lectra	2	10	9	2	7	3
Lee Industries	0	0	0	0	1	0
Lee Weitzman Furniture	0	0	0	1	0	0
Leeds Group	0	0	2	0	0	0
Lees Carpets	0	1	0	1	0	0
Leffers	0	0	1	0	0	0
Legend Care	0	0	0	0	0	1
Leggett and Platt	0	0	0	1	0	0
Lempiere	0	1	0	0	0	0
Lendell Manufacturing	0	0	0	1	0	0
Lentex	0	1	0	0	0	0
Lenzi Egisto	0	0	1	0	0	0
Lenzing	17	12	17	23	17	11
Leopoldo Pozzi	0	6	0	0	0	1
Lessona	1	0	0	0	0	0
Levante	0	0	0	1	0	0
Levi Strauss	1	3	3	3	1	5
Lexington	0	1	0	0	0	0
Leyer SynTex	0	0	0	0	1	0
LG Chemical	0	0	1	0	1	2
LG Electronics	0	0	0	0	2	61
LGL Electronics	0	0	0	0	0	1
LHTaylor	0	0	0	0	0	1
Li and Fung	0	0	0	0	0	2
Li Peng	0	0	0	1	3	0
Liba	2	4	2	4	5	2
Liberty	1	2	0	1	0	0
Life Science	0	0	0	0	0	1
LifeTex	0	0	0	0	0	1
Liftport	0	0	0	0	1	0
Light Line	0	0	0	0	1	0
Lilly Pulitzer	0	0	0	0	0	1
Lincoln Composites	0	0	0	1	0	0
Lincoln Lustre	0	0	0	1	0	0

	1999	2000	2001	2002	2003	2004
Linda Lundstrom	0	1	0	0	0	0
Lindenfarb	0	1	0	0	0	0
Lindstrand	0	0	0	2	1	0
Linea Clothing	0	0	0	0	0	1
Lineapiu	0	0	0	0	1	0
Linear Composites	0	0	0	0	0	1
Linificio Canapificio Nazionale	0	0	0	2	0	0
Linotec Development	0	0	0	1	0	0
Linq	0	0	1	1	0	0
Lintrend	0	0	0	1	0	0
Linz Textil	0	0	1	0	0	0
Lion Apparel	0	0	0	1	0	0
Lithium Technology	1	0	0	0	0	0
Little Jumbucks	1	0	0	0	0	0
Littlewood and Son	0	1	0	0	0	0
Liwaco	0	0	1	0	0	0
Liz Claiborne	3	0	1	1	1	1
LL Bean	0	1	0	0	0	0
LM Glasfiber	0	0	0	1	1	2
LNP	0	0	0	0	1	1
Loane Brothers	0	0	0	0	0	1
Lochcarron of Galashiels	0	1	1	0	0	0
Lockheed Martin	0	0	1	0	0	4
Locks	0	0	1	0	0	0
Loepfe	0	2	0	0	1	0
Loffler	0	0	0	1	2	0
Loftex	0	0	0	0	0	1
Logan Textiles	0	0	2	0	0	0
Logotech	1	0	0	0	0	0
Lohmann	0	0	1	0	3	0
Lojigma	0	1	0	0	0	0
Loma Company	0	0	1	0	0	0
Lonati	1	6	1	1	1	7
Longclose	1	2	0	0	0	0
Longina Phillips Design	0	1	0	0	1	0
Loomdata Systems	0	1	0	0	0	0
Loptex	0	0	1	0	0	0
Lora Piana	0	0	1	0	0	0
Lorber	0	1	0	0	0	0
Loris Bellini	2	5	0	0	1	1
Loro Piana and Company	0	0	0	0	2	1
Lorz	0	0	0	0	0	1
Lothian Coated Fabrics	0	0	1	0	0	0
Louis de Poortere	0	2	1	0	0	0
Louis Vuitton	0	0	0	0	0	1
Low and Bonar	0	2	0	1	0	1
Lows of Dundee	0	0	1	0	0	0
Loyalty Partner	0	0	0	0	0	1

	1999	2000	2001	2002	2003	2004
Luckenhause Technische Textilien	0	0	0	0	1	0
Lucky Brand	0	0	0	1	0	0
Ludlow Textiles	0	1	0	0	0	0
Ludwig Wissenbach	0	0	1	0	0	0
Luftedersysteme	0	0	0	0	1	0
Lui-Jo	0	0	0	0	1	0
Lumax	0	0	0	1	0	0
Lummus	1	0	0	0	0	1
Lunatex	0	0	0	0	0	1
Lunn Fabrics	0	0	0	1	0	0
Lurex Company	0	1	0	0	0	0
Lurgi Zimmer	4	5	2	0	0	0
Luwa	0	1	2	1	0	2
Luxilon	0	1	0	0	0	0
Lydall	0	0	0	0	1	2
Lydney Vehicle Systems	0	0	1	0	0	0
Lyle and Scott	1	0	0	2	0	0
Lysac	0	1	0	1	0	0
M-Base	1	0	0	0	0	0
M-Tec	3	4	0	1	0	2
M and R Printing	0	0	0	0	3	0
M Dohmen	0	0	3	0	0	0
M J Soffe Company	0	0	1	1	0	0
M Recht	0	0	0	0	0	1
Maag	0	1	0	2	1	1
Mabel	0	0	0	1	0	0
MacDermid ColorSpan	0	0	1	0	1	0
Mackays	0	0	0	0	0	1
Mackintosh	0	1	0	0	0	0
MacLean Vehicle Systems	1	0	0	0	1	0
Macpi	0	0	0	2	0	0
Macquarie	0	0	1	1	3	0
Madan Trading Company	0	0	1	0	0	0
Maddison Filter	0	1	0	0	0	0
Madeira Freiburg	1	0	0	0	0	1
Madison Filter	0	0	1	0	0	0
Madival	0	1	0	0	0	0
Madura Garments	0	1	0	0	1	0
Mageba	1	0	1	2	0	2
Magee Rieter	0	0	0	0	0	1
Magee Weaving	0	0	1	1	0	0
Magellan Systems	0	0	0	1	0	0
Magenta Master Fibers	0	0	1	0	0	0
Maglificio Dalmine	0	0	0	0	0	1
Magna	1	0	0	0	1	0
Magnolia Manufacturing	0	0	0	0	0	1
Mahindra and Mahindra	0	0	0	0	0	1
Mahlo	1	3	5	5	1	0

	1999	2000	2001	2002	2003	2004
Maibo	1	0	0	1	0	0
Maidenform	0	1	0	0	0	0
Maier Unitas	0	1	0	0	1	0
Maile	0	0	0	0	0	1
Main	0	0	0	0	0	1
Mainelab	0	0	0	0	1	0
Mainettech	0	0	0	0	1	0
Maino	0	0	1	0	1	3
Maison Lesage	0	0	0	0	0	1
Maison Prelle	0	1	0	0	0	0
Maje	0	1	0	0	0	0
Malden Mills	1	6	4	1	3	6
Malimo	0	1	1	0	0	0
Malwa Cotton Spinning Mills	0	0	0	0	0	1
Mambo	0	1	0	0	0	0
Mamiliano Dini	0	0	1	0	0	1
Mammut	0	0	1	0	0	2
Manabe Yuichi	0	0	0	0	1	0
Manfred Huck	0	1	0	0	0	1
Mango	0	0	1	0	0	1
Manifattura Crespi	0	0	0	0	1	0
Manifattura d'Olmina	0	1	0	0	0	0
Manifattura di Albiate	0	1	0	0	0	0
Manifattura di Legnano	0	0	1	0	0	0
Manifattura di Valle Brembana	0	1	0	0	0	0
Manifattura Ferno e Borgomaneri	0	1	0	0	0	0
Mantegazza Antonio Arti Grafiche	0	0	1	0	0	0
Mantero	0	1	1	0	1	0
Manual Woodworkers and Weavers	0	0	0	0	1	1
Marantech	0	0	0	0	1	0
Marcart	0	0	1	0	0	0
Marcel Liebaert	1	0	0	0	0	0
Marco Maranghi	0	0	0	0	0	1
Margola	0	1	0	0	0	0
Marimekko	0	0	0	1	0	0
Marine Composites	0	0	0	0	1	0
Marine Custom Canvas	0	0	0	0	1	0
Marinella Napoli	0	1	0	0	0	0
Mario Crosta	1	2	0	0	0	0
Marioboselli	0	0	0	1	0	0
Marion Composites	0	0	1	0	0	0
Marisol	0	0	0	0	1	0
Markische Faser	2	0	0	0	0	0
Marks and Spencer	3	4	1	5	8	2
Marlow Ropes	1	0	0	1	0	0

WORLD TEXTILES Manufacturer Indexing

	1999	2000	2001	2002	2003	2004
Marmot	0	1	0	0	1	0
Marquillas Fast	0	0	0	1	0	0
Marshfield Furniture	0	0	0	1	0	0
Martel	0	1	0	1	0	0
Martex	0	0	0	1	0	0
Martin Color Fi	0	0	1	0	0	0
Martin International	2	1	0	1	1	1
Martin Marietta	1	0	0	1	0	0
Maru	0	1	0	0	0	0
Marusan	0	0	0	0	0	1
Marutaka Sangyo	0	0	1	0	0	0
Marzoli	3	2	1	3	1	4
Marzotto	0	1	2	1	2	0
MAS Holdings	0	0	1	0	0	0
Mascioni	0	0	1	1	0	0
Maskara Industries	0	1	0	0	0	0
Masland	0	0	1	0	0	0
Massachusetts Institute of Technology	0	0	0	1	0	0
Massen	1	0	3	3	0	0
Massimo Guarducci	0	0	0	0	0	1
Massimo Osti	0	1	0	0	0	0
Mast Mode	0	0	0	0	0	1
Master Italia	0	0	1	0	0	0
Master Pharma	0	0	0	1	0	0
Mastercraft Fabric	0	0	0	0	1	0
Masters of Linen	0	1	1	2	0	0
Masureel	0	0	0	0	0	1
Masurel Direct	0	0	0	0	1	0
Masurel Fils	0	0	1	1	0	0
Matalan	0	1	0	1	1	0
Matec	1	3	1	6	8	4
Material ConneXion	0	0	0	1	0	0
Materials Resources	1	0	0	0	0	0
Mathec	0	0	0	1	0	0
Mathis	0	0	0	1	0	1
Matsushita	0	0	0	0	0	1
Mattes and Ammann	1	1	0	0	0	0
Maurer	1	0	0	0	0	0
Maurizio Peruzzo Industries	2	0	0	0	0	0
Mautom	0	0	0	1	0	0
Mavani Group	0	0	0	1	0	0
Max Goller	0	1	0	0	0	0
Max Planck	0	0	0	0	0	1
May Company	0	0	0	1	0	0
Mayborn	0	0	0	0	1	0
Mayer and Cie	4	4	8	9	16	6
Mayford Group	0	0	0	0	1	0
Maytag	0	0	0	0	2	3

	1999	2000	2001	2002	2003	2004
Mazda Motor Corporation	0	0	0	1	1	0
Mazzini ICI	0	0	0	0	1	0
MBI International	0	0	0	0	0	1
McAirlaid	0	1	1	1	0	0
McClean Anderson	0	0	1	1	0	0
McDonnell Douglas	1	0	0	0	0	1
McKinnon Land Moran	0	0	0	1	0	0
McNaughton	0	0	1	0	1	0
McNeil	0	0	0	0	0	4
MCS	3	0	4	2	0	2
MD Viola	0	0	0	1	0	0
Meadowbrook Inventions	0	3	0	0	0	0
Meadstone	0	1	0	0	0	0
MeadWestvaco	0	0	0	0	0	1
Mebal	0	0	1	0	0	0
Mec-Mor	0	0	0	0	1	0
Meca Provence	0	0	0	1	2	3
Mecanique Application Tissus Mecatiss	0	0	1	0	0	0
Meccaniche Moderne	1	0	0	1	0	0
Mecoplast	0	0	0	0	1	0
MedArray	0	0	0	0	0	1
Medico	1	1	0	0	0	0
Medicorp	0	0	0	1	0	0
Medima	1	0	0	0	0	0
Medline	0	0	0	0	0	1
Megatrex	0	0	0	0	0	1
Mei	0	0	1	1	0	0
Meias Scalina	0	0	0	0	1	0
Meijer	0	0	0	0	1	0
Melco	1	0	0	1	1	1
Memminger-IRO	0	5	2	4	0	3
Memteks	0	0	0	0	0	1
Men's Warehouse	1	0	0	1	0	0
Menegatto	1	0	0	0	0	0
Mensa	0	0	1	0	0	0
Menzel	0	0	1	0	0	0
Menzolit Fibron	0	0	0	0	1	0
Merchant Taylors Company	0	0	0	0	1	0
Meraklon	1	1	0	0	0	0
Merida Meridian	0	0	0	0	1	0
Meridian Automotive Systems	0	0	0	1	0	0
Merino Gold	0	0	1	0	1	0
Merino New Zealand	0	0	1	0	0	0
Meritas	0	2	0	0	0	0
Merkalon	0	0	0	0	1	0
Merloni	0	0	0	0	0	3
Mertz	0	0	1	0	1	0

WORLD TEXTILES Manufacturer Indexing

	1999	2000	2001	2002	2003	2004
Meshpack	1	0	0	0	0	0
Messe Frankfurt	0	0	0	0	0	1
Messier Bugatti	1	0	0	1	0	6
Metabolix	0	0	1	0	0	0
Metag	0	0	2	0	0	0
Metallized Engineering	1	0	0	0	0	0
Metris	0	0	0	0	0	1
Metro Group	0	0	0	0	0	1
Metro Sportswear	0	0	0	1	0	0
Metso Paper	0	0	0	2	0	0
Mettler Toledo	1	0	0	0	0	0
Mexx	0	0	1	0	1	1
Meyer and Meyer	0	0	0	0	0	1
Mezzera	0	1	0	0	0	0
MFG	1	0	0	0	0	0
MG Rover Group	0	0	0	0	0	1
MHM	1	0	0	2	0	0
Miami Awnings	0	0	2	1	0	0
MiCell	0	0	0	0	0	2
Michele Le Palmer	0	0	0	0	1	0
Michele Ratti	0	0	0	1	0	0
Michelin	1	0	0	0	1	5
Michell and Cia	0	1	0	2	0	0
Michell Group	0	2	4	2	1	0
Michell	0	0	0	0	0	1
Micrex	0	0	0	0	0	1
Micro D	0	0	0	1	0	0
Microban International	0	1	0	0	0	1
Microfaser Repro	0	0	0	0	1	0
Microfibres	0	0	0	0	0	1
MicroMuscle	0	0	0	0	0	1
Microtek	0	0	0	0	0	2
Microzine	0	0	0	0	0	1
Mid Micron	0	0	1	0	0	0
Midwest	0	1	0	0	0	0
Mike's Marine Custom Canvas	0	0	0	0	1	0
Miller Canvas	0	0	0	1	0	0
Milliken	4	4	4	6	7	57
Milyan	0	1	0	0	0	0
Mimaki	0	0	0	0	0	1
Mingazzini	1	0	0	0	0	0
Minolta	0	1	2	0	0	0
MIRALab	0	0	0	1	0	0
Miro Radici Group	0	2	0	0	0	0
Miroglio	5	1	2	4	1	0
Missoni	0	0	0	0	1	0
Mitloedi	0	1	0	0	0	0
Mitsu Tec	0	0	0	0	1	1

	1999	2000	2001	2002	2003	2004
Mitsubishi Chemical	0	0	0	1	0	2
Mitsubishi Materials	0	0	0	0	0	1
Mitsubishi Rayon	2	4	0	0	3	4
Mitsubishi	1	3	0	1	0	4
Mitsui	0	1	0	0	0	4
MK	0	0	0	0	1	0
mnemoScience	0	0	0	0	1	0
Mobiltex	0	0	0	0	1	0
ModaCAD	1	0	0	0	0	0
Modern Petrofils	0	0	0	0	1	0
Modipon	0	0	1	0	0	0
Moenus	1	1	0	0	0	0
Mogul	0	0	1	1	0	1
Mohair Council of America	0	0	0	0	0	1
Mohair Promotion Cape	0	0	1	0	0	0
Mohair South Africa	0	0	1	1	1	0
Mohair Spinners South Africa	0	1	0	1	0	0
Mohawk	3	1	0	1	0	2
Moire	0	0	1	0	0	0
Moley	0	1	0	0	0	0
Molnlycke	1	1	1	1	1	0
Momentos	0	0	0	1	0	0
Monarch	0	3	1	2	3	7
Mondetta Clothing	1	0	0	0	0	0
Mondo	0	0	0	0	0	1
Mondomix	1	0	0	0	0	0
Monforts-Koenig	1	0	0	0	0	0
Monforts	1	8	5	11	4	12
Monkwell	0	1	0	0	0	0
Monotub	0	0	0	0	1	0
Monsoon Accessorize	0	1	0	2	1	0
Mont Blanc Technology	0	0	0	0	1	0
Montefibre	0	1	5	5	1	3
Montell	1	0	0	0	0	1
Monterey	0	0	0	0	1	0
Monti Antonio	0	1	1	0	0	0
Monti Australia	0	0	0	0	1	0
Monti Group	0	1	0	0	1	1
Montreal Centre for Contemporary Textiles	0	0	1	0	0	0
Montrose Textile Machinery	1	0	0	0	0	0
Mony Industrial	0	0	0	0	1	0
Moorhouse and Brook	0	0	0	1	0	0
Moorpark	0	0	0	0	1	0
Moran Canvas Products	0	0	0	1	0	0
Morikawa Nenshi	0	0	0	0	0	1
Morning Pride Manufacturing	0	0	0	0	0	1

WORLD TEXTILES Manufacturer Indexing

	1999	2000	2001	2002	2003	2004
Morris	0	0	0	0	0	1
Morrison Textile Machinery	0	2	1	0	0	0
Morton, Young and Borland	0	0	1	0	0	0
Moss Bros	1	1	0	0	0	0
Moss	0	3	0	0	0	0
Mossley	0	1	0	0	0	0
Mothercare	0	0	0	1	0	0
Motohiro and Company	0	0	1	1	0	1
Motz Group	0	0	0	0	1	0
Mouchel	1	0	1	2	1	0
Mount Vernon Mills	0	9	0	0	1	0
Mountain Equipment Coop	0	0	1	0	0	0
Mountain Hardware	0	1	0	0	0	0
Mountain Spinning Mills	0	0	0	0	1	0
MOVEngineering	0	0	0	1	0	1
Mowbrays	0	1	0	0	0	0
Moxon	1	0	1	0	0	0
Moygashel Mills	0	0	1	0	0	0
MS Machinery and Systems	1	2	0	0	0	0
MSC	0	0	0	0	1	0
MTC Medical Fibers	0	0	0	0	1	0
Mulberry	0	0	0	1	0	0
Muller	0	0	0	1	0	3
Multifabs Survival	1	1	0	0	0	1
Multiflo Instruments	0	0	0	0	0	1
Multiplast	0	0	0	0	0	1
Munsinger	0	0	0	0	1	0
Munzinger	0	0	1	0	0	0
Mura	0	0	0	1	0	0
Murata	7	3	7	6	2	10
Muratec	0	1	0	5	0	0
Musto	0	0	1	0	0	0
Muvan	0	1	0	0	0	0
Muzzi	0	1	1	0	0	0
Myer Grace Brothers	0	1	0	0	0	0
Mykal	0	0	1	0	0	0
Mykrolis	0	0	0	0	0	1
Mylene	0	3	0	0	0	0
Mytrex	0	0	1	0	0	0
N and MA Saville	0	0	0	0	0	1
N B International	0	0	0	0	1	0
Nabro	0	0	0	0	0	1
Nagao Shoji	0	0	0	0	1	0
Nagata Seiki	0	1	0	1	0	0
Nagler Design	0	0	0	1	0	0
Nakagoshi	0	1	0	0	0	0
Nallis	0	1	0	0	0	0
Nan-Woel	1	0	0	0	0	0

	1999	2000	2001	2002	2003	2004
Nan Ya Plastics	0	0	0	0	0	3
Nan Yang Textile Group	0	0	0	1	0	1
Nanning Sugar	0	0	0	0	0	1
Nano Tex	0	0	2	1	3	6
Nanotec	0	0	0	0	0	1
NanoTechniques	0	0	0	1	0	0
Nanoval	0	0	0	1	0	0
Naomoto	0	1	0	1	0	0
Napier	2	0	0	0	0	0
Naqeeb Khalid	1	0	0	0	0	0
Narricot	0	1	0	0	0	0
Nash	0	1	0	0	0	0
Nassimi	0	0	0	0	0	1
Nastrificio Gavazzi	0	0	1	1	0	0
Natesh	0	0	0	0	0	1
Nationaal Lucht- en Ruimtevaartlaboratorium	0	0	0	0	0	2
National Aeronautics and Space Administration	0	3	4	1	3	6
National Composite Center	0	0	0	0	1	1
National Cotton Council	0	0	0	1	0	0
National Institute of Design	0	0	0	0	0	1
National Institute of Fashion Technology	0	0	1	0	0	0
National Nonwovens	0	1	0	1	0	0
National Physical Laboratory (India)	0	1	0	0	0	0
National Sorbents	0	0	1	0	0	0
National Starch	0	1	0	0	0	2
National Textiles	1	0	3	0	0	0
Natracare	0	0	2	0	0	0
Naue Group	0	0	0	1	2	0
Nautilus Skin Touch	0	0	0	1	0	0
Naylor Jennings	1	0	0	0	0	0
NDT Solutions	0	0	0	0	0	1
Nederlandse Organisatie voor Toegepast	0	1	0	0	0	0
NedGraphics	0	0	0	0	1	1
Neill Johnstone	1	1	0	0	0	1
Nejdek Wool Combing	0	1	0	0	0	0
Nellcor	0	0	0	1	0	0
Nelson Laboratories	1	0	0	0	0	0
Nemesis	0	1	0	0	0	0
Neopress	0	0	0	0	1	0
Neptco	0	0	0	1	0	0
Nester	0	0	0	1	0	0
Netti	0	0	0	0	0	1
Neuenhauser	1	2	1	0	1	1
Neumag	1	5	3	6	4	7

	1999	2000	2001	2002	2003	2004
Neumann	0	0	1	0	0	0
Neuss	0	1	0	0	0	0
New Balance	0	0	1	0	0	1
New Dimensions	0	0	0	0	0	1
New England Wool	0	0	0	0	1	0
New Long	0	0	0	0	0	1
New Look	0	1	0	1	1	1
New World Environmental Products	0	0	0	0	0	1
New Zealand Fabrics	0	0	1	0	0	0
New Zealand Merino Company	0	0	0	0	1	1
New Zealand Wool Board	1	0	0	2	0	0
New Zealand Wool Testing Authority	0	1	1	1	0	1
Nexia Biotechnologies	0	1	1	3	2	3
Next	0	0	0	0	0	2
Nextec	0	0	3	1	0	0
Neyret	0	1	0	0	0	0
Nicast	0	0	0	0	0	1
Niche Digital Print	0	0	0	0	0	1
Nichiban	0	0	0	1	0	0
Nihon Balluff	0	0	0	0	0	1
Nike	0	2	2	4	4	4
NikWax	0	0	0	1	0	0
Nile Linen Group	0	0	1	0	0	0
Nilit	0	0	0	1	1	0
Nina Campbell	0	1	0	0	0	0
Nine West	0	1	0	0	0	0
Nippon Dom	1	0	0	0	0	0
Nippon Electric Glass	0	0	0	1	0	0
Nippon Filicon Company	0	0	0	0	0	1
Nippon Graphite Fiber	0	0	0	0	1	0
Nippon Kayaku	0	1	0	0	1	0
Nippon Keori Kaisha	0	0	0	1	0	1
Nippon Mayer	0	0	0	1	0	1
Nippon Mitsubishi Oil	0	0	0	0	1	1
Nippon Nozzle Company	0	0	0	1	0	1
Nippon Petrochemicals	0	2	0	0	1	0
Nippon Pillar Packing	0	0	0	0	0	1
Nippon Sheet Glass	0	0	0	0	0	2
Nippon Shokubai	1	0	0	0	0	1
Nippon Soda	0	0	0	0	1	0
Nippon Steel	1	0	0	1	1	1
Nishi Nenshi	0	0	0	1	0	0
Nissan	1	0	0	0	0	0
Nisshinbo	3	5	1	2	0	0
Nissho Iwai	0	1	1	0	0	0
Nitex	0	0	0	1	0	0

	1999	2000	2001	2002	2003	2004
Nittetsu Mining Company	0	0	0	0	1	0
Nitto Boseki	0	2	2	1	0	0
Nitto Glass Fiber Manufacturing	0	0	0	1	0	0
NKD Vertriebs	0	0	0	0	1	0
No 2 Worsted Ace Textile Industrial Company	0	0	1	0	0	0
No Mulligens	0	0	1	0	0	0
Noble	0	0	0	2	0	0
Nordenia	0	0	0	0	0	1
Nordifa	0	1	0	0	0	0
Nordson	1	0	1	2	0	8
Norman Ritchie	0	0	0	0	0	1
Norman Shatz	0	1	0	0	0	0
Norsea Natural Products	0	0	0	0	1	0
Norsel	0	0	0	0	0	1
NorSud Orientteppiche	0	0	1	0	0	0
North American Bus Industries	0	0	0	0	0	1
North Beach Marine Canvas	1	0	0	0	0	0
North Carolina State University	0	0	1	0	1	3
North Country Canvas	0	0	1	0	0	0
North Marine Group	0	0	0	0	0	1
Northern Elastomeric	0	0	0	0	1	0
Northern Group Retail	0	0	0	0	0	2
Northern India Textile Mills Assosiation	0	1	0	0	0	0
Northern India Textile Research Association	0	0	1	0	1	1
Northern Textiles	0	0	0	1	0	0
Northrop Grumman	0	0	2	0	0	2
Norwalk Furniture	0	1	0	1	0	0
Noseda	0	1	2	1	0	0
Nottinghamshire Sports and Safety Systems	0	0	0	0	0	2
Nova Mosilana	0	0	1	0	0	0
Novaceta	1	0	0	0	0	0
Novakust	1	0	0	0	0	0
Novartis	0	1	0	1	0	0
Novelty Textiles Mills	0	1	0	0	0	0
Noveon	0	0	0	2	0	0
Novibra	1	0	0	0	0	0
Novo Nordisk	1	0	0	0	0	0
Novocut	0	1	0	0	0	0
Novoteks	0	0	0	1	0	0
Novozymes	0	0	3	2	0	5
Nowo	0	0	0	0	1	0
Noyon	0	0	1	0	0	0
NoyVallesina	0	2	1	0	0	0

WORLD TEXTILES Manufacturer Indexing

	1999	2000	2001	2002	2003	2004
NRB Bearings	0	1	0	0	0	0
NTM Obermaier	0	0	1	1	0	0
Nu-kote	0	0	0	0	0	1
NUBAR	0	1	0	0	0	0
Numarine	0	0	0	0	0	1
Nuno	0	0	0	0	1	0
Nuova Isotex	1	0	0	0	0	0
Nuovo Pignone	2	1	0	0	0	0
Nurel	0	0	0	2	0	0
Nyacol	0	0	0	0	0	1
Nybo Jensen	0	0	0	0	0	1
Nygard	0	0	0	0	3	0
Nylatex	0	0	0	0	0	1
Nylok	0	0	0	0	0	1
Nylstar	3	6	6	6	3	0
Nyltec	0	0	1	0	0	0
NZTops	1	0	1	0	0	0
Oakley	0	0	0	1	0	0
Oarsome	0	0	1	0	0	0
Oasis	1	1	0	0	0	0
OBEM	0	4	1	3	0	0
Oberrauch Group	0	0	0	0	0	1
OCCA	0	0	1	0	0	0
OCE	1	0	0	0	0	0
Ocean Lanka	0	0	0	0	0	1
Octatube	0	0	0	0	0	1
Octir	0	0	1	0	0	0
Odermark	0	0	0	0	0	1
Odevni Podnik	0	1	0	0	0	0
Odlo	0	0	0	0	1	0
Ofa	0	0	0	0	1	0
Officine Bisio	0	1	0	0	0	0
Officine Del Piano	0	0	1	0	0	0
Officine Gaudino	0	0	1	0	0	0
Officine Giovanelli	0	0	0	1	0	0
Officine Meccaniche di Lesmo	0	0	0	0	0	1
Offshore and Trawl Supply	0	0	0	0	0	1
Offshore Composite Structures	0	0	0	0	1	0
Ogallala Comfort Company	0	0	0	0	0	1
Ogretmen Corap	0	0	0	1	0	0
Ohara Paragium Chemical	0	0	0	0	0	1
Ohki Company	0	0	0	0	0	1
Oji Paper Company	0	0	0	0	1	0
Okaidi	0	0	0	1	0	0
Okamoto Industries	0	0	0	1	0	1
Oktavia	0	0	0	0	0	1
Olcese	0	0	0	1	0	0

	1999	2000	2001	2002	2003	2004
Oldcastle Precast	0	0	0	0	1	0
Olimpia Stefani	0	1	0	0	0	0
Oliver Group	1	0	1	0	0	0
Olymp Bezner and Company	0	0	0	0	0	2
Olympia	0	0	1	0	0	0
Olympic General	0	0	0	0	1	0
Omega	1	0	0	0	0	0
Omet	0	0	0	1	0	0
Omikenshi	1	2	1	0	0	0
Ommi	0	2	0	0	0	0
Omniglass	0	0	0	0	0	1
Omnitex	0	1	0	0	0	0
Omnium	0	0	0	0	1	0
OMP Racing	0	0	1	0	0	0
One Street Holding	0	0	0	0	0	1
Ontex	0	0	0	1	0	0
Onward Kashiyama	0	1	0	0	0	0
Orang Tiga	0	0	0	0	1	0
Orange County Choppers	0	0	0	0	0	1
Orange County Industrial Sewing Machine Company	0	0	0	0	1	0
Orbital Sciences	1	0	0	0	0	0
Orca Bodywear	0	1	0	0	0	0
Organ Needle Company	0	0	0	0	0	1
Orica Chemicals	1	0	0	0	0	0
Oriental Weavers	1	0	0	0	0	0
Origem	1	0	0	0	0	0
Orintex	1	0	0	0	0	2
Orizio Paolo	0	6	3	5	4	6
Orlandi	0	1	0	0	0	0
Orsallino Lane	0	0	1	0	0	0
Orteks	0	0	0	0	1	0
Osaka Gas Chemical Company	0	0	0	0	1	0
Osaka National Research Institute	0	1	0	0	0	0
Osborne and Little	1	1	0	0	0	0
Oshkosh B'Gosh	0	0	1	1	0	1
Osiris	0	0	0	0	1	0
Oskar Fisher	0	1	0	0	0	1
Osmonics	0	0	0	0	0	1
Ossur	0	0	0	0	0	1
Osta	0	0	1	0	0	0
OsteoBiologics	0	0	0	0	0	1
Osthoff Senge	0	0	0	0	1	1
Oswal	0	0	0	0	1	0
Otabo	0	0	0	0	0	1
Otis Elevator Company	0	0	0	0	0	1
Otsuka	0	0	0	0	1	0

WORLD TEXTILES Manufacturer Indexing

	1999	2000	2001	2002	2003	2004
Otto and Company	0	1	0	0	0	2
Otto Bock	0	0	0	0	0	1
Otto Golze	0	0	0	1	0	0
Otto International	0	0	0	0	1	0
Outlast Technologies	0	1	0	1	0	3
Ovattificio Resinatura Valpadana	1	0	0	0	0	1
Ovillos y Conos	0	0	0	0	1	0
Ovis	0	0	0	0	0	1
Owens Corning	6	6	3	2	8	16
Oxco	0	0	1	0	0	0
Oxeon	0	0	0	0	0	1
Oxford Industries	0	0	1	0	0	0
Oystar Textil Farben	0	0	0	0	0	1
Oystertec	0	0	0	0	0	1
Ozark	0	1	0	0	0	0
Ozeta	0	1	0	0	0	0
Paccar	0	0	0	0	0	1
Pace Setter Sportswear	1	0	0	0	0	0
Pacific Brands Group	0	1	0	0	0	1
Pacific Harish	0	0	1	0	0	0
Pacific Network	0	0	0	1	0	0
Pacific Safety Products	0	0	0	0	0	1
Pacific Sunwear	0	1	0	0	1	0
Pacific Textiles Company	0	0	0	0	0	1
Packy Poda	0	0	0	0	0	1
Pactiv	0	0	0	0	0	2
PAD System Technologies	1	1	1	0	0	0
Pagh Morup	0	0	0	0	0	1
Pai Lung	1	1	0	3	2	2
Pall	1	0	0	0	0	0
Palla Creativ Textiltechnik	1	0	0	0	0	0
Palliser	0	0	0	1	0	0
Palmers Textilkonzern	0	0	0	0	0	1
Palmetto Synthetics	0	1	0	0	0	1
Pangrim	0	1	0	0	0	0
Panolana Aps	0	0	0	0	0	1
Pantecnica	0	0	0	0	1	0
Panter	0	0	1	0	0	0
Pantex	0	1	0	1	1	0
Pantograms	1	0	0	0	0	0
Pantone	0	1	0	0	0	0
Paper Converting Machine Company	0	0	0	1	0	0
Paper Pak Products	0	1	0	0	0	0
Parabeam	0	0	1	0	0	0
Parade	0	0	0	1	0	0
Paragon Trade Brands	1	2	1	1	0	1
Parasuco	1	0	0	0	0	1

	1999	2000	2001	2002	2003	2004
Paratech	0	0	1	0	0	0
Parex-Mather	0	1	0	0	0	0
Park Technology	0	0	0	1	0	0
Parkdale Mills	0	3	2	0	0	0
Parker Knoll	0	1	0	0	0	0
Parkinson Machinery and Manufacturing	0	1	1	0	0	0
Paroc Group	0	0	0	0	0	1
Parras Cone	0	1	0	0	1	0
Pars Decor	0	0	0	1	0	0
Partners In Textiles Innovations	0	0	0	1	0	0
Pasco	0	0	0	1	0	0
Passport Knitwear	0	0	0	0	0	1
Patagonia Incorporated	1	0	0	0	0	1
Pathfinder	0	1	1	1	0	0
Paul Hartmann	0	0	3	0	0	2
Paulus Henri	0	1	0	0	0	0
Paxar	1	0	0	0	0	2
Paylana	0	0	0	1	0	0
PBT	0	0	0	0	0	1
Peach State Labs	0	0	0	0	1	0
Peacock Group	0	1	0	1	2	0
Pearl Academy of Fashion	0	0	0	0	0	1
Pearl Izumi	0	0	0	0	1	0
Peass	0	0	0	0	0	1
Peaudouce of Linselles	1	0	0	0	0	0
Pecci	1	1	1	0	0	0
Pedex and Company	0	0	0	0	0	1
Peek and Cloppenburg	0	0	0	0	0	1
Peerless Clothing	1	0	0	0	0	0
Peerless Garments	0	0	0	0	0	1
Pegas	0	1	1	1	0	0
Pegasus	0	3	0	1	2	4
Peguform	0	0	0	0	0	1
Pelikan Werke	0	0	0	0	0	1
Pellerin Milnor	0	0	0	0	0	1
Pematex	0	1	0	0	0	0
Pen Apparel	0	0	0	0	1	0
Pendleton Woolen Mills	0	0	1	0	0	0
Penn and Fletcher	0	0	0	0	0	1
Penn Nyla	0	0	2	2	1	0
Pentair Group	0	0	0	0	0	1
Penthouse Carpets	1	0	0	0	0	0
Pentwyn Splicer	1	0	0	0	0	0
PeopLoungers	0	0	0	0	1	0
Peratech	0	2	1	1	1	0
Perfect Equipments	0	0	1	0	0	0

WORLD TEXTILES Manufacturer Indexing

	1999	2000	2001	2002	2003	2004
Perfecta	0	1	0	0	0	0
Performance Air Products	0	0	0	0	0	1
Performance Group	0	1	0	0	0	0
Perry Ellis	0	0	1	0	0	0
Perseverance Mills	0	3	0	0	0	0
Perspectis	0	0	0	0	0	1
Petchkasem Weaving Company	0	0	1	0	0	0
Peter Law	0	0	0	1	0	0
Peter Schellenberg	0	1	0	0	0	0
Peter Scott and Company	0	0	0	0	1	2
Peter Wright	1	0	0	0	0	0
Petrie and McNaught	0	0	1	0	0	0
Petronas	0	0	0	0	0	1
Peugeot	0	0	1	0	2	0
Pfaff	2	3	1	2	6	1
Pfleiderer Dammstofftechnik	0	0	0	0	0	1
Phantom Industries	1	0	0	0	0	0
Pharetra	0	0	0	0	2	0
Pharmagg	0	0	0	0	0	3
Pheonix Tent and Awning Company	0	0	1	0	0	0
Pheonixx TPC	1	0	1	0	0	0
Phild Company	0	0	0	0	0	1
Philips Van Heusen	0	0	1	0	2	0
Philips	1	3	1	1	0	6
Phoenix Footwear Group	0	0	0	0	0	1
Phoenixx TPC	0	0	0	1	0	0
Phuong Dong Garment Company	0	0	0	0	0	1
Piacenza Wool Mill	0	0	0	0	1	0
Picadilly Fashions	1	0	0	0	0	0
Picanol	7	17	10	21	17	19
Pietrafesa	0	1	0	0	0	0
Pigeon Corporation	0	0	1	0	0	0
Pillowtex	0	2	1	1	0	0
Pilot Ink	0	1	0	0	0	1
Pilotelli	1	0	0	0	1	0
Pilsa	0	0	1	0	0	0
Pimlico Apparel	1	0	0	0	0	0
Pinault Printemps Redoute	0	0	0	0	0	1
Pincroft Dyeing and Printing Company	0	0	0	0	1	0
Pininfarina	0	0	0	1	0	0
Pinter	0	0	0	0	0	1
Pinzon	0	1	0	0	0	0
Pioneer	0	1	0	0	0	0
Pirelli	0	0	0	0	0	1
Pittards	0	0	1	0	0	0

	1999	2000	2001	2002	2003	2004
Pittsfield Weaving Company	0	0	0	0	0	3
Pittsfield Woolen Yarn Company	0	0	0	1	0	0
Pixalert	0	0	0	0	1	0
Plandeki	1	0	0	0	0	0
Plant Engineering	0	1	0	0	0	0
Plantex	1	0	0	0	0	0
PlasmaTreat	0	0	0	1	0	0
Plastform	0	0	0	0	0	1
Plasticisers	2	4	0	0	0	0
PlastiComp	0	0	0	0	1	0
Plasticos Industriales de Tampico	0	0	1	0	0	0
Platinum	0	0	0	0	0	1
Plauener Gardine	0	0	1	0	0	0
Plauener Spitze und Stickereien	0	0	1	0	0	0
Plauener Spitzenfabrikation Gustav Tegeler	0	0	1	0	0	0
Playtex Products	0	0	0	0	0	2
Pleva	0	0	0	0	2	0
Pliana	0	0	1	0	0	0
Pneumafil	0	1	0	0	0	0
Pneumatic Conveyors	0	0	0	1	0	0
Podium Distribution	0	0	0	0	0	1
Pointcarre	0	1	0	0	0	0
Polartec	0	1	0	1	1	0
Pole des Technologies Medicales	0	0	0	0	0	1
Polgat	0	0	0	0	1	1
Polo Ralph Lauren	0	0	1	0	0	1
Poly-Med	0	0	0	0	0	1
Polyamide High Performance	0	0	0	0	0	2
Polycoating	0	0	0	0	1	0
PolyEitan Composite	0	1	0	0	0	0
Polyfelt	0	3	0	0	0	0
Polymatech Company	0	0	0	0	0	1
Polymer Engineering	1	1	0	0	0	0
Polymer Group	2	1	3	6	3	24
Polymer Processing Research Institute	0	0	0	0	1	0
PolymerLatex	0	0	0	0	0	1
Polynorm	1	0	0	0	0	0
Polyplas	0	0	2	0	0	0
Polyplastics	0	0	0	0	0	2
Polysack Plastic	0	0	0	0	1	0
Polytropon	0	0	0	0	0	1
Polyvlies	0	1	0	0	0	0
Ponds	0	0	1	0	0	0

WORLD TEXTILES Manufacturer Indexing

	1999	2000	2001	2002	2003	2004
Pontetorto	0	0	1	0	0	0
Pony	0	1	0	0	0	0
Porcher	1	0	0	0	1	1
Porex	0	0	0	0	0	1
Porsche	0	0	0	0	1	0
Porterhouse	0	0	1	0	0	0
Portmeirion Potteries	1	0	0	0	0	0
Porvair	0	1	1	1	1	2
Positek	0	0	0	0	0	1
Power Paper	0	0	1	0	0	0
Power Plastics	0	1	0	0	0	1
Powertrusion International	0	0	0	0	0	1
PPG	1	0	3	3	0	3
Prada	0	1	0	0	0	0
Prado Rugs	0	1	0	0	0	0
Prairie Fiber Guild	0	0	0	0	1	0
Prana	0	0	0	0	1	0
Prato Manufacturers' Association	0	1	0	0	0	0
Precision Fabrics Group	0	0	0	0	0	4
Precision Fukuhara	0	0	0	0	0	1
Precision Loopers	0	1	0	0	0	0
Precision Processes Textiles	0	1	1	2	4	1
Precision Products	0	0	0	0	0	2
Precitex	0	0	0	0	1	0
Predilnica Litija	0	0	0	1	0	0
Premier Polytronics	0	0	0	2	2	3
Premium Private Label Manufacturers	0	0	0	0	1	0
Prestige Enterprise	0	0	0	0	1	0
Pretty Polly	0	2	0	0	0	0
Price Shepshed	1	0	0	0	0	0
Prima Design Systems	0	1	0	0	1	0
Prima Direct	0	0	1	0	0	0
Primaries	0	0	0	1	0	0
Primary Technologies	1	0	0	0	0	0
Primary Wool Cooperative	0	0	0	1	0	0
Primco	0	1	0	0	0	0
Primix	0	0	0	0	1	0
Prince	1	0	0	0	0	0
Principle Business Enterprises	0	1	0	0	0	0
Pringle	0	0	1	1	1	0
Prisma Fibers	0	0	0	0	0	5
Pro-Fit	0	0	0	0	0	1
Pro C Tech	0	0	0	0	0	1
Procher	0	0	0	1	0	0
Procotex	0	1	0	0	0	0
ProCtec	0	0	1	0	0	0

	1999	2000	2001	2002	2003	2004
Procter and Gamble	9	5	10	5	14	101
Productos del Sur	0	0	0	1	0	0
Products of Tomorrow	0	0	0	0	0	1
Produsa	0	1	0	0	0	0
Proel	0	0	0	0	1	0
Profiles Interweave	1	0	0	0	0	0
Proline	1	0	2	0	0	0
Proll and Lohmann	0	0	0	2	0	0
Promatech	0	0	7	7	2	7
ProStitch	0	0	0	0	1	0
Protech Armoured Products	1	0	0	0	0	0
Protechna	0	0	1	0	0	2
Protecnologica	0	1	0	0	0	0
Protti	0	1	0	1	1	0
Proulx	0	0	2	0	0	0
PSA Peugot Citroen	0	0	0	0	0	2
Puget Sound Rope	0	1	0	0	0	0
Pulse Furniture Design	0	0	0	0	1	0
Pulse	0	1	0	0	1	0
Pulsium	0	0	0	0	0	1
Pultron	0	0	0	1	0	0
Puma	0	0	0	1	0	0
Pyongyang Women's Trading and Garment Centre	0	0	0	0	1	0
Pyro Industries	0	0	0	0	0	1
Q-Panel	0	0	1	0	0	0
Qifa Textile Group	0	0	1	0	0	0
QinetiQ	0	0	0	0	1	2
Qiviuk	0	1	0	0	0	0
QMS	0	1	0	0	0	0
Quadrant Plastic Composites	0	0	0	0	1	1
Quadro di Montale	0	0	0	1	0	0
Quaker	1	1	0	2	1	0
Qualytel de Puebla	0	3	0	0	0	0
Quantum Group	0	0	0	1	1	0
Quantum Knitting Technology	0	0	0	0	0	1
Quartz and Silice	0	1	0	0	0	0
Quest	0	0	0	0	3	1
Quickstep Technologies	0	0	0	0	0	1
Quiksilver	0	1	0	1	1	0
Quill Medical	0	0	0	0	0	1
Quimidroga	0	1	0	0	0	0
R Dona	0	0	1	0	0	0
R P Associates	1	0	0	0	0	0
R Wimmers	0	0	0	1	0	0
Rabatex Group	0	0	0	1	0	0
Rabe Moden	0	0	1	0	0	0
Radici	1	2	2	6	3	5

WORLD TEXTILES Manufacturer Indexing

	1999	2000	2001	2002	2003	2004
Rainbow Masterbatch	0	0	1	0	0	0
Raisbeck	0	0	0	1	0	0
Raitech	0	0	1	0	1	0
Rajasthan Spinning and Weaving Mills	0	0	0	2	0	0
Rajasthan Textile Mills	0	1	0	0	0	0
Ralawise	0	0	1	0	0	0
Rally	0	0	0	0	0	1
Ralph Lauren	0	1	0	0	0	0
Ramatex	0	0	1	0	1	0
Ramsay McDonald	0	0	0	0	0	1
Ramtex	1	0	2	1	0	0
Ran Demo	0	0	1	0	0	0
Ranch Fiberglas	0	0	0	1	0	0
Ranck	0	0	1	0	0	0
Rantex	0	0	0	1	0	0
Rapra Technology	1	0	0	0	0	0
Ratti	0	0	0	1	1	1
Raymakers	0	1	0	0	0	0
Raymond Dale Wool Marketing	0	0	0	1	0	0
Raymond	0	1	1	0	0	0
Rayonier	2	0	0	1	0	2
Raytech	0	0	0	0	0	1
Raytheon	0	1	0	1	0	1
React	0	0	0	0	0	2
Reactofil	0	0	0	1	0	0
Real	0	0	0	1	0	0
Recaro	0	0	0	0	1	0
Recovery Room	1	0	0	0	0	0
Red Cross	0	0	0	0	0	2
Red Ray	0	0	1	0	0	0
Reebok	0	0	0	1	1	0
Reel	0	1	0	0	0	0
Reeves Brothers	0	0	1	0	0	0
Reflec	0	0	0	0	1	0
Reflective Technologies	1	0	0	0	0	0
Reflexite	0	0	0	0	1	0
Refnol	0	1	0	0	0	0
Regal	0	1	0	0	0	0
Regatta	1	0	0	0	0	1
Regeneracija dd Zabok	0	0	0	0	0	1
Reggiani	1	2	2	2	9	0
Regional Compact Car	0	0	0	0	0	1
Reid and Taylor	1	1	1	1	1	1
Reifenhauser	3	3	11	1	1	0
Reimotec	0	0	0	0	0	1
Reiner Diez	1	0	0	0	0	0
Relats	0	0	0	0	1	0

	1999	2000	2001	2002	2003	2004
Reliance	0	4	1	2	0	4
Reliant	0	0	0	1	0	0
Remei	0	1	0	0	0	0
Rena Lange	0	0	0	0	0	1
Renault	0	0	0	0	2	0
Renown	0	1	0	0	0	0
Replication Medical	0	1	0	0	0	0
Replin	0	1	1	0	0	0
Resch	1	0	1	0	0	0
Reshmi Industries	0	0	0	1	0	0
Resil	0	0	0	0	1	2
Resta	1	0	0	0	0	1
Retail Assist	0	0	0	0	0	1
Retail Brand Alliance	0	0	1	0	0	0
Retech	1	2	6	2	1	1
Reticle	0	0	0	0	0	1
Reusch	0	0	0	0	1	0
Reutter	0	0	0	1	0	0
Reva Medical	0	0	0	0	1	0
Revolution Enterprises	0	0	0	0	0	1
Reward	0	1	0	0	1	0
Rex Composites	0	0	0	0	0	1
RGS Group	0	0	1	0	0	0
Rhein Nadel	0	1	0	0	2	0
Rheinisch Westfalische Technische Hochschule	0	0	3	1	0	1
Rhodia Acetow	0	0	0	0	0	1
Rhodia Industrial Yarns	0	0	0	0	1	0
Rhodia Technical Fibers	1	0	0	0	1	0
Rhodia	2	5	6	2	1	5
Rhodius	0	0	0	0	0	1
Rhomtuft	0	1	0	0	0	0
Rhone Poulenc	0	1	1	0	0	0
Rhovyl	2	1	1	0	0	0
Riby	0	1	0	0	0	0
Richard Anderson	0	0	1	0	0	0
Richard Roberts	0	1	0	0	0	0
Richards	0	0	1	1	0	0
Ricoh	0	0	0	0	0	1
Riddhi Plastics	0	0	0	1	0	0
Riddle and Cockrell	0	1	0	0	0	0
Riedel and Tietz	0	0	1	1	0	0
Rieter ICBT	0	0	0	1	0	0
Rieter Perfojet	0	0	0	5	0	5
Rieter Scragg	0	0	0	3	0	0
Rieter	16	36	42	43	33	42
Rimach	0	0	0	0	2	0
Rimoldi	2	2	2	2	0	0
Rimslow	0	0	0	1	0	1

	1999	2000	2001	2002	2003	2004
Rins	1	0	0	0	0	0
Rip Curl	0	0	0	1	0	0
Ripe Maternity	0	0	0	0	1	0
Riso Kagaku	0	0	0	0	0	1
Ritas	0	0	1	0	0	0
Riverina Wool Combing	0	1	1	0	1	0
Riverwind	0	1	0	0	0	0
RKB	0	0	0	0	0	1
RKW	0	0	0	0	0	1
RMV	0	0	1	0	0	0
Roaches	0	1	0	0	0	0
Roadmaster	0	0	0	0	1	0
Robama	0	1	0	0	0	0
Robert Jowitt and Sons	0	1	1	0	1	0
Robert Noble	1	2	1	1	0	0
Robinson Healthcare	1	0	0	0	0	0
Roblon	0	1	0	0	0	0
Robo Products	0	0	0	1	0	0
Rocklea	1	2	2	2	0	0
Rockwear	0	0	1	0	0	0
Rockwell	0	1	0	0	0	1
Rockwool	0	0	2	0	0	4
RocTool	0	0	0	0	0	1
Roebuck	0	0	0	0	0	1
Roehm Pharma	0	0	1	0	0	0
Rofa Bekleidungswerk	0	1	0	0	0	0
Rogers	0	0	1	0	0	1
Rohm and Haas	0	0	1	0	0	2
Rohner	0	1	0	1	0	0
Roj Electrotex	1	0	0	0	0	0
Rokona	0	0	0	0	6	0
Rontex	0	0	0	0	0	1
Roots	0	0	0	0	0	1
Rose Hill Company	0	0	0	0	0	1
Rosebys	0	1	0	0	0	0
Rosink	0	0	0	1	0	0
Rositta	1	0	0	1	0	0
Ross and Glendining	0	0	0	0	0	1
Rossari	0	0	0	1	1	1
Rossignol	0	0	0	0	1	0
Rosslym Precision	0	0	0	0	1	0
Rosso	0	1	0	0	0	1
Rostaing	0	0	0	0	1	0
Rostoni	0	0	2	3	0	1
Rothmans Benson and Hedges	0	0	0	0	0	1
Rotondi	1	0	0	1	1	1
Rotorcraft	0	1	0	0	0	2
Rotta	0	3	1	0	0	0

	1999	2000	2001	2002	2003	2004
Roy Robson Fashion Company	0	0	0	0	0	1
Royal Park Uniform	0	1	0	0	0	0
Royal Ten Cate	0	0	0	0	4	1
Rtica	0	0	0	0	0	1
RTP Company	0	1	1	0	1	0
Rucoguard	0	0	0	0	0	1
Rudnik	0	0	0	1	0	0
Rudolf and Company	0	0	0	0	0	1
Rue Royale Diana	0	0	1	0	0	0
Ruen Machinari	0	0	0	0	0	1
Rug Doctor	0	0	0	0	0	1
Rugs by Robinson	0	0	1	0	0	0
Rumi	0	0	0	1	0	2
Running Bare	0	0	0	0	1	0
Russell Corporation	1	0	2	3	1	0
Ruther and Einenkel	0	0	0	0	0	1
Ruti	0	0	1	0	0	0
Ryalux	0	0	1	0	0	0
S A T E	1	0	0	0	0	0
S C Maremod	1	0	0	0	0	0
Sabanci	0	0	0	0	1	2
Sachsisches Textilforschungsinstitut	1	0	2	1	1	2
Sader Racineux	1	0	0	0	0	0
Saehan	1	1	0	1	0	1
Saertex	0	0	1	0	1	0
Safety Components	0	0	0	0	0	1
Sage	0	1	0	0	0	0
Sahbi Belarbi	1	0	0	0	0	0
Sahinler Holding	0	0	1	1	0	0
Sahm	1	0	1	0	0	0
SAI Automotive	0	0	0	0	1	0
Saint Gobain Vetrotex	0	0	0	1	0	0
Saint Gobain	0	1	0	2	4	12
Sakai Ovex	0	2	0	0	0	0
Saks	0	0	0	0	1	0
Salmoiraghi	1	3	3	1	1	0
Salomon	0	0	0	0	0	1
SaluMedica	0	0	1	0	0	0
Salvia Electronic	0	0	0	1	0	0
Salvona	0	0	0	0	1	0
Salzmann	0	0	0	1	0	1
Samil	0	0	1	0	0	0
Samit	0	1	0	0	0	0
Samsung Needle	0	0	0	0	1	0
Samsung	0	0	0	0	1	8
Samuelsohn	1	0	0	0	0	0
Samur	0	1	0	0	0	0

	1999	2000	2001	2002	2003	2004
San-Ted	0	0	0	1	0	0
San Fang	0	0	0	0	1	0
San Grato	0	0	1	0	1	0
San Leucio Silk Consortium	0	0	0	0	1	0
Sand and Sea	0	0	0	0	0	1
Sander	0	0	0	0	0	1
Sandoz	0	0	1	0	0	0
Sandra Textile Mills	0	0	0	1	0	0
Sanfangxiang	0	0	0	0	1	0
Sangam Group	0	1	0	0	0	0
Sanghi	0	1	0	0	1	0
Sangiacomo	0	3	0	5	0	6
Sankousha Company	0	0	0	0	0	1
Sans Fibres	0	0	1	0	0	0
Sant Andrea Novara	0	0	3	1	0	1
Sant Andres	0	0	1	0	0	0
Santaconstancia	0	0	0	0	1	0
Santex	3	2	7	0	3	2
Santexo	0	0	0	1	0	0
Santista	0	1	1	1	0	1
Santoni	1	5	2	8	4	6
Sanuk	0	0	0	0	0	1
Sanyo	0	0	0	0	0	7
Sapa Textil	0	1	0	1	0	0
Sapona	0	2	1	0	0	0
Sara Lee	0	1	1	0	2	10
SaraMax	0	0	0	0	1	0
Sarex	0	0	1	0	0	1
Sasson Jeans	0	0	0	0	0	1
Satere Oy	0	0	0	1	0	0
Sato Textile Company	0	0	0	0	0	1
SATRA	0	0	1	0	0	0
Satron	0	0	0	0	0	1
Sattler	0	0	0	0	0	1
Saudi Basic Industries	0	0	1	0	1	0
Saudi Carpets Factory	0	0	0	1	0	0
Saudi Yarn and Knitting Technology Factory	1	0	0	0	0	0
Saueressig	0	1	0	0	0	0
Saurer Allma	0	0	0	2	3	0
Saurer	6	5	12	5	6	26
Savio	3	9	4	2	3	6
Saxon Carpets	0	1	0	0	0	0
Saxon Screens	1	0	0	0	0	0
Saxonia	1	0	0	0	1	1
Saxony Textile Research	0	0	0	0	0	1
SC Johnson and Son	0	0	0	0	0	1
SCA	1	4	2	0	5	7
Scaglia	1	0	0	0	0	0

	1999	2000	2001	2002	2003	2004
Scapa Group	2	1	1	1	2	2
Schaefer	1	0	0	0	0	1
Schaetti	1	1	2	2	0	1
Scharer Schweiter Mettler	0	1	0	1	1	3
Scherf	1	0	0	0	0	0
Schild	0	0	0	0	0	1
Schilgen and Company	0	0	0	0	1	0
Schill and Seilacher	1	0	0	1	0	0
Schips	0	0	1	0	0	0
Schlafhorst	9	4	19	26	18	15
Schlemming	0	1	0	0	0	0
Schlumberger	2	6	9	1	2	5
Schmetz	0	2	0	1	1	0
Schmitz-Werke	0	0	0	0	0	1
Schneider Group	1	1	1	4	2	1
Schnier Elektrostatik	0	0	0	0	0	1
Schoeller and Hoesch	0	0	0	0	0	1
Schoeller	2	2	2	4	0	2
Schoenherr	0	0	2	0	0	1
Schoffel	0	0	0	0	0	1
Schofield and Smith	1	1	0	0	0	0
Schofield Cloth Finishers	1	0	1	0	0	0
Schofield Dyers and Finishers	0	0	0	1	0	0
Scholl Then	0	0	0	0	3	0
Scholl	2	4	0	0	2	0
School Trends	0	0	0	0	0	1
Schott and Meissner	1	1	0	0	0	0
Schroers	0	0	0	1	0	0
Schurr	0	1	0	0	0	0
Schwarza	0	1	0	0	0	0
Schwing	0	0	1	0	0	1
Scient'x	0	0	1	0	0	0
Scientific Molding	0	0	0	0	0	1
Scimed	0	0	0	0	2	3
Scitex Vision	0	0	0	0	1	0
Sclavos	0	1	0	2	0	0
Scola Teloni	1	0	0	0	0	0
Scomar	1	0	0	0	0	0
Scotch House	1	0	0	0	0	0
Scott and Fyfe	0	0	0	2	0	0
Scott Bader	1	0	0	0	0	1
Scottish Borders Enterprise	1	0	0	0	0	0
Scottish Cashmere Club	0	0	0	1	0	0
ScotWeave	1	0	0	1	0	0
Scratch Art	0	0	0	0	0	1
ScS Upholstery	0	0	1	1	1	0
Seafolly	0	0	2	0	0	0
Seal	0	0	1	1	0	0

WORLD TEXTILES Manufacturer Indexing

	1999	2000	2001	2002	2003	2004
Sealy	0	0	0	0	0	1
Seaman	0	0	0	1	0	1
Sears	0	1	0	0	0	1
SEB	0	0	0	0	0	3
Second Chance Body Armour	0	0	1	0	0	0
Second Nature	0	0	0	1	0	0
Security Qube System	0	0	0	0	0	1
Sederma	0	0	1	0	0	0
Sefar Group	0	0	0	0	1	1
Seft Development Laboratory Company	0	0	0	0	1	0
Seidensticker Group	0	0	0	0	0	1
Seiko Epson	0	0	0	0	0	1
Seilerei Bernhard Prutz	0	0	1	0	0	0
Seilerei und Flechterei Muffler	0	0	0	0	0	1
Seilerei Voigt	0	1	0	0	0	0
Seiren	1	2	0	0	0	0
Sellers	1	3	1	1	2	2
Sen Huang	0	0	0	0	1	0
Senior Experts Service	0	0	1	0	0	0
Sensura	0	0	1	0	0	0
Sequins International	0	1	0	0	0	0
Servitejo	0	0	0	0	0	1
Setec Maschinenbau	0	0	0	0	0	1
Seteria Bianchi	0	0	0	1	0	0
Setex	0	0	0	0	1	0
Setila	0	1	0	0	0	0
Seton Scholl Healthcare	1	0	0	0	0	0
Severin Daners	1	0	0	0	0	0
Sew Systems	0	1	0	0	0	0
Seydel	1	0	0	1	0	0
Sferra Brothers	0	0	0	0	0	1
SGF Chemie	0	0	0	0	0	1
SGL Carbon	1	1	1	1	0	4
SGS	0	0	0	0	1	0
SHAIRCO	0	0	0	1	0	0
Shakespeare Company	0	0	0	0	0	4
Shangdong Province Dongming Cotton Mill	0	1	0	0	0	0
Shanghai Longtou	1	0	0	0	0	0
Shanthi Gears	0	0	0	0	0	1
Shape 3	0	1	0	0	0	0
Sharon Rae	0	0	1	0	0	0
Shaw	1	1	1	0	1	1
Sheetal	0	0	1	0	0	0
Shei Chung Shin Industrial Company	0	0	0	0	0	1

	1999	2000	2001	2002	2003	2004
Shelby Trim Auto Upholstery	0	0	0	1	0	0
Shelf Mills	1	0	0	0	0	0
Shell	7	7	4	4	2	4
Shelter-Pro	0	0	0	0	1	0
Shelton	0	0	0	1	0	0
Sheridan	0	0	1	0	0	0
Sherif Clothing Company	0	0	0	1	0	0
Sherwood	1	1	0	0	0	0
Shibani Knitting Company	0	0	0	0	1	0
Shikibo	4	3	0	0	0	0
Shikokuizumisenni	0	0	0	0	0	1
Shilasdair	0	0	0	1	0	0
Shiloh	2	1	0	1	0	0
Shima Seiki	3	2	5	3	6	27
Shimadzu	0	0	0	2	0	0
Shimizu	1	0	0	0	0	0
Shin-Etsu Chemical	0	0	0	0	0	1
Shin Nihontex	1	0	0	0	0	0
Shirin	0	0	0	1	0	0
Shirley Developments	1	2	1	1	2	0
Shirmax	0	0	1	0	0	0
Shoou Shyng	0	0	0	1	0	0
Shore to Shore	0	0	0	1	0	0
Short and Robertson	0	0	1	0	0	0
Shorts	0	0	0	1	0	0
Showa Denko	0	0	0	0	1	3
Shri Lakshmi	1	0	0	0	0	0
Shriram Needle Bearing	0	1	0	0	0	0
Shuford Mills	0	1	0	0	0	0
Sider Arc	0	0	0	0	1	0
Siemens	1	0	1	1	1	2
Siemon Company	0	0	0	0	0	1
Sierem	0	0	0	1	0	0
Sierra Designs	0	1	0	0	0	0
Sierra Pacific	0	0	0	0	0	1
Sign Top	0	0	0	0	1	0
Sika	0	1	0	0	0	2
Sikoplast	0	0	1	0	0	0
Sikorsky	0	0	1	0	0	0
SIL Holdings	0	0	0	0	0	1
Silent Gliss	0	0	1	0	0	0
Silentnight	0	1	0	1	1	0
Silipos	0	0	0	0	0	1
Silk Mark Organization	0	0	0	0	0	1
Silver Jeans	1	0	0	0	0	0
Silverstar	0	0	0	0	1	0
Sim Tek	0	1	0	0	0	0
Sima	0	2	0	0	0	0

	1999	2000	2001	2002	2003	2004
Simba	1	1	0	0	1	0
Simco Holding	0	0	0	0	0	1
Simge	0	0	0	1	1	0
Simms	0	0	0	0	1	0
Simon de Winter	0	0	0	1	0	0
Simplex	0	1	0	0	0	0
Simply Oarsome	1	0	0	0	0	0
Simula	0	0	1	0	0	0
Sinfos	0	0	0	0	1	0
Sing Lun	0	0	0	0	0	1
Singer	0	0	0	0	0	1
Singham Fabrics	0	1	0	0	0	0
Sinoca	0	0	1	0	0	0
Sinochem	0	0	0	0	0	1
Sintelon	0	0	0	1	0	0
Sinterama	0	1	2	0	1	3
Sioen	1	0	1	0	1	0
Sirdar	0	0	3	0	0	0
Sirio	1	0	0	0	0	0
Siruba	0	0	0	0	1	0
Sistema Moda Italia	0	0	0	1	0	0
Sitip	0	0	0	1	0	0
Siyaram Silk Mills	0	1	0	0	0	0
SK Chemicals	0	1	0	0	0	1
Skloplast	0	0	0	1	0	0
Skulogix	0	0	1	0	0	0
Skyspan	0	0	3	2	0	0
Slack and Parr	0	1	1	0	1	1
Sleepcraft	0	0	1	0	0	0
Sleepmaster	0	0	0	0	1	0
Smart-Textil	0	1	0	0	0	0
Smart Fibres	2	0	0	0	0	0
Smartex	0	0	0	0	1	0
SmartMembrane	0	0	0	0	0	1
Smit Textile	0	0	0	0	1	2
Smith and Nephew	3	4	3	1	0	1
Smith Brothers of Berne	0	0	0	1	0	0
Smith Kline and French	0	1	0	0	0	0
SmithKline Beecham	0	1	1	0	0	0
SML Extrusion Technology	0	1	1	1	0	1
SMM	0	0	0	0	0	1
SMTS	0	0	0	1	0	0
Snaico	0	0	0	2	0	0
Snap Tite	0	1	1	0	0	0
Snow Lotus	0	0	0	0	0	1
Snowy Peak	0	0	0	0	0	1
Snyder	0	0	1	0	0	0
Societe Coton Textiles pour Materiaux Innovants	0	1	0	0	0	0

	1999	2000	2001	2002	2003	2004
Societe Nationale Industielle et Aerospatiale	0	0	0	0	1	0
Sodesa	0	0	0	0	0	1
Soex	0	0	1	0	0	0
Sofradim	0	0	0	1	2	1
Soft Body	0	0	0	0	0	1
Softmakers Solutions	0	0	0	1	0	0
SOFTswitch	0	0	0	1	0	0
Sohler Neuenhauser	0	0	0	1	0	0
Solar Fashion	1	0	1	0	0	0
Solaris	0	0	0	0	1	0
Solarveil America	1	1	0	0	0	0
Soldon Systems	0	0	0	0	0	1
Sole Technology	0	0	0	0	0	1
Solipat	0	0	0	1	0	0
Solis	0	1	0	0	0	0
Solutia	1	6	1	2	3	2
Solvay	0	1	0	0	1	1
Somcc	0	0	1	0	0	0
Somelos	0	0	0	1	0	0
Somet	1	13	4	0	2	0
Somic	0	0	0	0	1	0
Sonobond	0	0	0	1	1	0
Sonoco	1	2	0	1	0	2
Sony	0	0	1	0	0	1
Sophis	1	1	0	0	1	0
Sotira	0	0	0	0	0	1
Sotratex	1	0	0	0	0	0
South African Fine Worsteds	0	0	0	2	0	0
South African Polypropylene Yarns	0	2	0	0	0	0
South India Textile Research Association	0	1	2	5	2	3
South Korean Dyeing Technology Centre	0	0	0	2	0	0
Southern Illinois University	0	0	0	0	1	0
Southern Mills	1	2	0	0	0	2
Southern Vectis	0	1	0	0	0	0
Southern Webbing Mills	0	0	0	0	0	1
Southpac Trust	0	0	0	0	0	1
Southwest Recreational Industries	0	0	0	0	0	1
Southwest Screens and Filters	0	0	1	0	0	0
Southwest Technologies	0	0	0	0	0	1
Southwestern Sales	0	0	0	0	1	0
Southwestern Wire Cloth	0	0	0	1	0	0
SP Aerospace and Vehicle Systems	0	0	0	0	0	1

WORLD TEXTILES Manufacturer Indexing

	1999	2000	2001	2002	2003	2004
SP Systems	0	4	1	1	1	0
Spar Knitwear	1	0	0	0	0	0
Sparco	0	0	0	0	1	0
Speciality Processors	0	0	2	1	1	0
Speciality Textile Products	0	0	1	0	0	0
Specialty Textiles Interiors	0	1	0	0	0	0
Spectrum	0	1	1	0	0	1
Speedo	0	5	3	0	1	2
Spencer Composites	0	1	0	0	0	0
Spencer Wright	0	0	1	0	0	2
Sperotto Rimar	3	11	1	8	4	4
Spheric Products	0	0	0	0	0	2
Spiegel	0	0	0	0	1	0
Spiller Lane	0	1	0	0	0	0
Spinelli	1	0	0	0	0	0
Spinnbau	0	1	2	1	1	2
Spinnweberei Uhingen	0	0	0	1	0	0
Spirowind	0	0	1	0	0	0
Splendour.com	0	0	0	0	1	0
Spoerry	0	0	0	1	1	0
Sport Fashions	0	1	0	0	0	0
Sportler	0	0	0	0	0	1
Sports Authority	0	0	0	1	0	0
Sportscraft	0	1	0	0	0	0
Sportsgirl/Sportscraft Group	0	1	0	0	0	0
Sportsman Midwest	0	0	1	0	0	0
Sportsphere	0	0	0	1	0	0
Sportswear Company	0	0	0	0	1	0
Spotless Plastics	0	0	0	0	0	2
Springs	0	3	0	0	0	0
Sprint Metal	0	0	1	0	0	0
Spungold	0	0	0	0	0	1
SRA	1	1	0	1	0	0
SRF	0	0	1	0	0	0
Ssangyong Machine Industry Company	0	0	0	0	1	0
SSL	0	0	0	1	0	1
SSP and Technology	0	1	0	0	1	0
St John Knits	0	1	0	0	0	0
Stabio	0	0	0	0	1	0
Stackmann	0	0	0	0	0	1
Staedtler	0	1	0	0	0	2
Stafford Higgins	0	0	2	0	1	0
Stahle Eltex	0	0	1	0	0	0
Stahlecker Grill	0	1	0	0	0	0
Stahls	0	0	0	0	0	1
Stalam	0	0	0	0	0	1
StaMax	0	0	0	0	1	0

	1999	2000	2001	2002	2003	2004
Stamperia di Fenegro	0	0	1	0	0	0
Standard Wool	0	2	1	0	1	1
Standeven	0	0	0	0	0	1
Stanelco	0	0	1	0	0	0
Stanford	0	1	1	0	0	0
Star Binding and Trimming	0	0	1	0	0	0
Starfire Systems	0	0	0	0	1	0
Starlab	0	1	0	0	0	0
Starlinger	0	1	3	2	0	1
Staroup	0	0	0	0	1	0
Startes	0	1	0	0	0	0
State University of New York	0	0	0	0	0	1
Statex Engineering	1	0	0	1	1	1
Staubli	1	4	3	2	2	6
Staveley Fine Worsteds	0	0	0	1	0	0
Stedfast	0	0	0	0	1	0
Steel Heddle	0	0	0	2	0	0
Steelcase	1	0	0	0	0	0
Stefalex	1	0	0	0	0	0
Steiger Zamark	1	0	0	0	3	2
Steiger	0	1	7	3	5	1
Steilmann Group	1	2	0	0	0	0
Stephenson Thompson	0	1	1	1	0	1
Sterling Group	0	1	0	0	1	0
Stern	0	0	0	1	0	0
Stillwell	0	1	1	0	0	0
Stilon	0	1	0	0	0	0
Sting	0	0	0	0	1	0
Stitches	1	0	0	0	0	0
Stockhausen	0	0	0	3	0	1
Stoddard	1	0	0	3	0	0
Stoll	6	3	10	12	6	4
Stork Aerospace	0	0	0	0	0	1
Stork Textile Printing	3	5	6	4	2	3
Strahm	0	0	0	0	1	2
Stratasys	0	0	0	0	0	1
Stratus	0	1	0	0	0	0
Streats	1	1	1	0	0	0
Strellson	0	0	1	0	0	1
Stretch-Tex	0	0	1	0	1	1
Stretchline	0	0	0	1	0	0
Strines Textiles	0	0	0	1	0	0
Strongwell	1	2	0	0	2	0
Strudex	0	0	1	0	0	0
Stucken Group	0	1	1	0	0	0
Studibaker	0	0	0	1	0	0
Stummer	1	0	1	0	0	0
Styl	1	0	0	0	0	0

	1999	2000	2001	2002	2003	2004
Style-Line	0	0	0	1	0	0
Style Holdings	1	0	0	0	0	0
Styleknit	0	0	0	0	0	1
Stylus	0	0	0	0	1	0
Subrenat Expansion	0	2	0	0	0	0
Successori Reda	0	0	1	0	1	1
Suchy	0	0	0	1	1	0
Sucker Muller Hacoba	3	5	0	2	0	0
Suda Architects	0	1	0	0	0	0
Sudaglass	0	0	0	0	1	0
Sue Seiki	0	0	0	1	0	0
Suessen	2	22	13	18	7	13
Sugoi Performance Apparel	1	0	0	0	0	0
Sultex	0	0	0	1	11	13
Sulzer Tessile	0	0	0	0	1	0
Sulzer Textil	2	5	16	29	13	2
Sulzer	18	29	16	3	3	6
Sumitomo	1	0	3	3	1	4
Summit	1	0	0	0	1	1
Sumo	0	0	1	0	0	0
Sun Capital Partners	0	0	0	0	0	1
Sun Isle Casual Furniture	0	0	0	0	0	1
Sun Textiles	0	0	0	0	1	0
Sunbeam Products	0	0	0	0	0	1
Sunburst Chemicals	0	0	0	0	0	1
Sunbury Textile Mills	0	0	0	1	0	1
Sunil	0	1	0	0	0	0
Sunlight	0	1	0	0	0	0
Sunskins	0	0	1	0	0	0
Sunstar	0	0	0	0	0	2
Suominen	0	0	0	1	0	1
Super Sack Manufacturing	0	0	1	0	1	0
Super Spinning Mills	0	0	0	0	1	0
Superba	3	0	0	0	0	1
Superior Fabrics	0	1	0	0	0	0
Superior Uniform Group	0	1	0	0	0	0
Supertex Industries	1	0	0	0	0	1
Supertex Sarex	0	0	1	1	0	1
Supreme Elastic	0	0	1	0	1	0
Surface Transforms	0	0	0	0	0	1
Susa	0	0	1	0	0	0
Suss Design	1	0	0	0	0	0
Suvi	0	0	1	0	0	0
Suzhou Nove	0	1	0	0	0	0
Suzhou University	0	1	0	0	0	0
Suzuki Warper	0	0	0	0	0	1
Svenska Rayon	0	2	0	0	0	0
Svitap	0	0	1	0	0	0

	1999	2000	2001	2002	2003	2004
SWA	0	1	0	0	0	0
Swanndri	0	0	0	0	0	1
Swarovski Crystal	0	1	0	0	0	0
Swastik	0	0	1	0	0	0
Swindell Dressler	0	0	0	0	1	0
Swiss Federal Institute of Technology	0	0	2	2	0	1
Swiss Shield	0	0	1	0	0	0
Swisstex	0	1	0	0	0	0
Symalit	0	0	1	0	0	0
Symcad	0	1	0	0	0	0
Sympatex	0	1	0	0	1	2
Syntech Fibres	0	0	0	1	0	0
Synthetic and Art Silk Mills Research Association	0	1	0	7	0	1
Synthetic and Rayon Textile Export Promotion Council	0	0	0	0	1	0
Synthetic Fibre Development and Application Centre	0	1	0	0	0	0
Synthetic Industries	0	4	0	0	0	1
Synthetic Tubular Socks	0	0	0	0	1	0
Sys Tec	0	0	0	1	0	1
System Breitenbach	0	0	0	1	0	0
T Shirt and Son	0	0	0	0	0	1
Tachi-S Company	0	0	0	0	0	2
Tacoma Tent and Awning	0	0	0	1	0	0
Taconic	0	0	1	0	0	0
Taicang Litai	0	0	0	0	1	0
Taikyu Shoten	0	1	0	0	0	0
Tait and Style	0	0	0	1	0	0
Taiwan Textile Federation	0	0	0	0	1	0
Taiyo Membrane	0	0	0	0	0	1
Tajima	2	1	0	0	1	1
Takasago	1	0	0	0	0	0
Takata Pharmaceutical	1	0	0	0	0	0
Take Two	0	0	0	1	0	0
TAL Group	0	0	0	0	1	0
Talbots	0	1	0	0	0	1
Talon Technology	0	0	0	0	1	0
Tama Plastic Industry	1	0	0	0	2	0
Tame Valley Tufters	0	0	1	0	0	0
Tamfelt	0	0	0	0	0	7
Tamil Nadu Plastics Manufacturers Association	0	0	0	1	0	0
Tanaka	0	1	0	0	0	0
Tao	0	1	0	0	0	0
Tapex	0	0	0	0	1	0
Tapistron	2	0	1	0	0	0

	1999	2000	2001	2002	2003	2004
Target	1	0	0	0	0	1
Tarkett Sommer	0	0	0	0	1	0
Tarrant Apparel	1	0	1	0	0	0
Tata	0	0	0	1	0	0
Tate and Lyle	0	1	0	0	0	0
Tavolo della Moda	0	0	0	0	1	0
Taylor and Lodge	0	1	0	1	0	1
Taylor King Furniture Company	0	0	0	0	1	0
TC2	0	0	0	0	1	0
TCFL Centre	0	1	0	0	0	0
TDA Research	0	0	0	0	0	1
Te Strake	0	0	0	0	0	1
Teamdress Stein	0	0	0	1	0	0
teamX	0	0	0	1	0	0
TechFab	0	0	0	0	1	1
Techni Lin	0	0	0	0	0	3
Techni Tex	0	0	0	0	0	1
Technical Fibre Products	0	1	1	1	0	0
Technical Textiles and Nonwoven Association	0	0	0	0	1	1
Technical University of Munich	0	0	1	0	0	0
TechniLin	0	0	0	0	1	0
TechniTex Faraday Partnership	0	0	1	1	0	0
Techno-Plast	0	0	0	0	0	1
Technofilati	0	0	0	0	1	0
Technorama	0	0	0	0	0	1
TechnoTeam	0	0	1	0	1	1
Tecmath	0	1	0	1	0	0
Tecnica	0	0	0	0	0	1
Tecnologie Biologiche	0	0	1	0	0	0
Tecnomeccanica Biellese	2	2	0	1	2	1
Tecnorama	1	1	0	4	0	3
Tecnosoft	0	0	0	0	0	1
Tecnotessile	0	0	0	1	1	1
Ted Baker	1	1	0	1	1	2
Tefron	0	0	0	0	1	0
TEI Biosciences	0	0	0	1	0	0
Teijin Shoji	0	1	0	0	0	0
Teijin Twaron	1	2	2	2	1	2
Teijin	13	20	9	1	4	13
Teinturerie	1	0	0	0	0	0
Teitex	0	0	0	0	1	0
Tej	0	0	0	1	0	0
Tejicondor	0	0	0	0	1	0
Tekig Invest	0	0	0	1	0	0
Tekni-Quest	0	0	1	0	0	0

	1999	2000	2001	2002	2003	2004
Tekstina	0	0	0	1	0	0
Telair	0	0	0	2	0	1
Telenor	1	0	0	0	0	0
Temafa	0	2	0	0	0	2
Temco	2	1	5	1	1	4
Templeton Carpet Mills	0	0	1	0	0	0
Tenax	0	0	1	0	0	0
Tencel Limited	0	1	0	1	1	4
Tenotex	1	0	0	0	0	0
Tensar	0	0	1	0	0	0
Tergal	0	1	2	0	0	0
Termo	0	0	0	0	0	1
Termoelettronica	0	0	0	3	0	0
Terra Nova Equipment	0	0	0	1	0	0
Terraoka Seisakusho Company	0	0	0	0	0	1
Terrot	2	0	1	2	1	6
Tesco	0	1	0	0	0	0
Tessitura Enrico Sironi	0	0	1	0	0	0
Tessitura Luigi Santi	1	0	0	0	0	0
Tessiture Pietro Radici	0	0	0	1	0	0
Tessival	0	1	0	0	0	0
Testa	0	1	0	0	0	2
Tetra Laval	0	0	0	0	0	1
Tex-a-tec	0	0	0	0	0	1
Tex Electronic	0	0	1	0	0	0
Tex Tech	0	0	1	0	0	1
Texaa	0	0	0	1	0	0
Texacloth	0	0	0	1	0	0
TEXBIMA	0	0	0	0	1	0
Texcom	0	0	0	0	1	0
Texel	0	1	1	1	0	0
Texinpro	0	0	0	1	0	0
TexKimp	0	0	1	0	0	0
Texollini	1	0	0	0	0	0
Texoprint	0	0	0	0	1	0
Texpa	0	1	0	0	0	0
TEXparts	0	1	1	0	0	0
Texpinal	0	0	1	0	0	0
Texport	0	0	1	0	0	0
Texprojekt	0	0	0	0	1	0
Texskill	0	0	1	0	0	0
Textechno	0	1	2	0	3	0
Textest	1	0	0	0	0	0
Textil Bezerra de Menezes	0	0	0	1	0	0
Textil Cano y Segura	2	0	0	0	0	0
Textil Manuel Goncalves	0	0	1	0	0	0
Textile and Health Association	0	0	0	0	0	1

WORLD TEXTILES Manufacturer Indexing

	1999	2000	2001	2002	2003	2004
Textile and Industrial Sales	0	3	0	0	0	0
Textile Association of India	0	0	0	0	0	1
Textile Automation Sales	0	0	0	1	0	0
Textile Enhancements	0	0	1	0	1	2
Textile Machinery Manufacturers Association	0	0	0	0	0	1
Textile Machinery Society of Japan	0	0	0	1	0	0
Textile Partners	0	0	0	0	0	1
Textile Technology Centre	0	0	0	1	0	0
Textiles et Plastiques Chomarat	0	0	0	0	0	1
Textilforschunginstitut Thuringen Vogtland	0	0	1	1	1	2
Textilma	0	0	0	0	2	3
Textilwerk Schauenstein	0	0	0	0	1	0
Textilwerke Deggendorf	0	0	0	1	0	0
Textor	0	0	0	1	1	0
Textured Jersey	0	0	0	1	0	1
Texturing Services	0	1	0	0	0	0
Thai Acrylic Fibre	0	0	0	1	0	0
Thai Baroda	0	0	3	0	0	0
Thai Composites Company	0	0	0	0	0	1
Thai Garment Manufacturers Association	0	0	0	1	0	0
Thai Polyester Company	0	0	0	1	0	0
Thai Silk Company	0	0	0	0	1	0
Thales Nederland	0	0	0	0	1	0
The North Face	1	2	1	0	3	0
Then	2	2	2	0	2	0
Theory LLC	0	0	0	0	0	1
Therma-Tru Doors	0	0	0	0	1	1
ThermoView	0	0	0	1	0	0
Thewissen	0	1	0	0	0	0
Thibeau	1	4	4	2	1	1
Thies	1	2	4	4	1	1
Thistle Mills	0	0	1	0	0	0
Thom	0	0	0	0	1	1
Thomann Hanry	0	0	0	0	1	0
Thomas Bradford	1	0	0	0	0	0
Thomas Chadwick and Sons	0	0	0	1	0	0
Thomas Smith	1	1	0	1	0	0
Thomson Research Associates	0	1	0	0	1	1
Thor	1	1	2	0	1	0
Thrace Plastics	1	0	0	0	0	0
Three Rings	0	0	0	0	1	0
Three Rivers Knit Printers	0	0	0	1	0	0
Threshold Flooring	0	1	0	0	0	0

	1999	2000	2001	2002	2003	2004
Thuringian Institute for Textile and Plastics Research	0	1	0	0	0	2
Thyssen	0	0	0	0	0	1
Tiax	0	0	0	0	0	2
Ticona	0	0	0	2	5	3
Tidland	0	0	0	1	0	0
Tiesse	0	1	0	0	0	0
Tietex	0	0	1	1	3	1
Tiger Sales	0	0	0	0	0	1
Tilessen	1	0	0	0	0	0
Tilsa Technical	0	0	0	0	0	1
Tim Hoyle Digital Print and Design	0	0	1	1	0	0
Timber Intent	0	0	0	1	1	0
Timberland	0	0	1	0	2	1
Timtex	0	2	0	0	0	0
Tinarelli	0	0	0	1	0	0
Tintoria Comense	0	1	0	0	0	0
Tintoria di Quaregna	0	0	0	0	0	1
Tintoria Giovanni Crespi	0	0	0	1	0	0
Tintoria Zerbi	0	1	0	0	0	0
Tipoiti	1	0	0	0	0	0
Tissage et Enduction Serge Ferrari	0	0	0	0	0	3
Tissages de l'Aigle	0	0	1	0	0	0
Tissu Rothrist	0	1	0	0	0	0
TJX	0	0	0	1	0	0
TLM Linea	0	0	0	0	1	0
TNT	0	0	0	0	0	1
Toa Wool	0	1	0	0	0	0
Toagosei	1	0	0	0	0	1
Tobias Embroidery	0	0	1	0	0	0
Todd and Duncan	0	0	0	0	0	1
Todor	0	0	0	1	0	0
Toho Rayon	1	1	1	0	0	0
Toho Tenax	0	0	0	0	1	2
Tokai Senko	1	1	0	0	0	0
Tokusen	0	0	0	0	0	1
Tomassini Spinnerettes	0	1	0	0	0	0
Tomkinsons	1	0	0	0	0	0
Tommy Hilfiger	0	0	1	0	0	0
Tompkins	0	0	0	0	0	1
Tonello	0	0	0	1	0	0
Tontine Industries	1	0	0	0	0	1
Too	0	0	0	0	1	0
Top Form	0	0	0	1	0	0
Top Glass	1	0	0	0	0	0
Topkapi	0	0	0	1	1	0
Tops Fray Marcos	0	0	0	1	0	0

	1999	2000	2001	2002	2003	2004
Toray	25	10	10	4	7	19
Torcitura B and B	0	0	0	1	0	0
Torcitura	0	0	0	1	0	0
Torres	0	1	0	0	0	0
Tosama	0	0	0	1	0	0
Tosco	0	1	0	0	0	0
Toshiba	0	0	0	0	1	5
Toshin Kogyo Ichinose	0	1	1	0	0	0
Total Filter Technology	0	0	0	0	0	1
Touma Shirts	0	0	0	0	1	0
Townsend Design	0	0	0	0	0	1
Toyada	1	0	0	0	0	0
Toyo Boseki	0	0	0	0	1	6
Toyo Electric Company	0	0	0	0	0	1
Toyo Promotion	1	0	0	0	0	0
Toyo Tanso Company	0	0	0	0	0	1
Toyoba	1	0	1	0	0	0
Toyobo	8	6	7	3	2	1
Toyoda	4	3	0	0	0	1
Toyoshima	1	1	0	0	0	0
Toyota	1	1	3	1	0	5
TPC	0	0	0	0	0	1
Trade Circle Technical Industries	1	0	0	0	0	0
Trans-Textil	0	0	0	0	1	0
Transfertex	0	0	0	0	1	0
Transhield Technology	0	0	0	0	0	1
TransMIT	0	0	0	0	0	1
Transprints	0	1	0	0	0	0
Traptek	0	0	0	0	0	1
Treasure	0	1	0	0	0	0
Tredegar	0	0	0	2	1	2
Treepoint	0	0	0	1	0	0
Tregambe	0	0	0	1	0	0
Treklite	0	0	1	0	0	0
Trelleborg	0	0	0	1	0	0
Trespaphan	0	0	0	0	0	1
Tretorn	0	0	1	0	0	0
Trevira	2	3	9	7	5	6
Tria	0	0	1	0	0	0
Triad Disposables	0	0	0	1	0	0
Trigema	1	0	0	2	0	0
Trimble	0	0	1	0	0	0
TrimTex	0	0	1	0	0	0
Trina Turk	0	1	0	0	0	0
Trinca	0	0	0	0	0	1
Trinity College Dublin	0	0	0	0	1	0
Trion	0	0	0	0	1	0
Tritex	1	1	0	1	0	0

	1999	2000	2001	2002	2003	2004
Triumph	0	1	3	0	0	2
Triveneta Grandi Impainti	0	0	0	0	0	1
Tropical Sportswear	2	1	0	0	0	0
Troy Mills	1	0	0	0	0	0
True Temper	0	1	0	0	0	0
Trustfin Monarch	0	0	1	0	0	0
Trustfin	0	1	0	1	1	1
Trutzschler	4	4	6	2	5	14
Tsudakoma	2	1	2	0	1	27
Tsuruga Asahi Doken	0	0	0	1	0	0
TT Tents	0	0	1	0	0	0
Tubingen	0	0	0	0	1	0
Tufnol	1	0	0	0	0	0
Tuftco	1	0	0	1	1	2
Tuftex	2	0	0	0	0	0
Tukatech	0	0	1	0	1	3
Tukaweb	0	0	0	0	1	0
Tuntex Group	0	0	1	0	0	0
Tuohy Furniture	0	0	0	0	1	0
Turf Stabilization Technologies	0	0	1	0	0	0
Turtle Plastics	2	0	0	0	0	0
Turtle Wax Incorporated	0	0	0	0	0	1
Tutto Garndrukerei	0	1	0	0	0	0
Tuva Looms	0	0	1	0	0	0
Tweave	0	1	0	0	0	0
Twin Upholstery	0	0	0	0	0	1
Twistechnology	0	0	0	0	1	0
Tybor	0	0	0	0	1	0
Tyco	0	0	2	2	1	1
Tyr Sport	0	1	0	0	0	1
U-Freight	0	0	0	0	1	0
U Engineering	0	0	0	1	1	0
Ube	0	0	0	0	0	3
UCAR Carbon Company	0	0	0	0	0	2
Ulster Carpet Mills	0	0	0	1	0	1
Ulster Weavers	0	0	1	1	0	0
Ultimair	0	1	0	0	0	0
Ultimate Textile	1	0	0	0	0	0
Ultra Tex Knitting	0	1	0	0	0	0
Under Armour	0	0	1	0	0	0
UnderCoverWear	0	0	0	0	0	1
Unicharm	1	1	1	1	2	12
Uniek	0	0	0	0	0	1
Unifi	4	8	10	4	3	3
Unilan	0	0	0	1	0	0
Unilever	0	0	2	0	3	27
Unimin	0	1	0	0	0	0
Union Special	0	2	0	1	0	0

WORLD TEXTILES Manufacturer Indexing

	1999	2000	2001	2002	2003	2004
Union	1	0	1	0	0	0
Uniplet Trustfin	0	1	0	0	0	0
Uniqema	0	1	1	2	0	0
Unitech Texmech	0	0	0	0	0	1
United Bonded Fabrics	0	0	0	0	0	1
United East India Company	0	0	0	0	0	1
United EFP	0	0	0	1	0	0
United States Air Force	0	0	1	0	0	0
United States Biomaterials	0	0	0	1	0	0
United States Surgical Corporation	0	0	1	2	0	0
United Wire	0	0	0	0	0	1
Unitex	1	0	0	1	0	1
Unitika	5	4	1	0	0	2
Universal Fabric Structures	0	0	0	0	0	1
Universal Knitting	1	0	0	0	0	1
Universal Master Products	0	0	0	0	0	1
Universal Pharmacy Bonded Laboratories	0	1	1	0	1	0
Unix	0	0	0	0	0	1
Urban Outfitters	0	0	0	0	1	0
Uruguayan Society of Corriedale Breeders	0	0	0	0	1	0
Uruguayan Wool Secretariat	0	0	0	0	1	0
US Army Research Office	0	0	1	0	0	0
US Army Soldier and Biological Chemical Command	0	0	1	0	0	0
US Global Aerospace	0	0	0	0	1	0
US Global Nanospace	0	0	0	0	1	0
US Guilford Mills	1	0	0	0	0	0
US Laminating	0	0	0	1	0	0
US PAcific Group	0	0	0	1	0	0
Uster Technologies	0	1	0	0	0	3
UTIT	0	0	1	0	2	0
Vaahto Group	0	0	0	0	1	0
Valblanc Chemie	0	0	0	0	0	1
Valco	0	0	0	0	0	2
Valdese Weavers	0	1	1	0	0	0
Valeo	0	0	0	0	0	1
ValLesina	0	0	0	1	0	0
Valmet	1	0	0	0	1	0
Valtion Teknillinen Tutkimuskeskus	0	0	0	1	0	0
Vamatex	4	2	8	0	1	1
Van de Wiele	1	4	1	4	4	17
Van Tieghem	0	1	1	0	0	0
Van Wees	0	0	0	0	0	1
Van Went	1	0	0	0	0	0

	1999	2000	2001	2002	2003	2004
Vanden Berghe	0	0	0	0	0	1
Vanguard	0	1	0	0	0	0
Vanners	0	0	1	0	0	0
Vanwyk Group	0	0	0	0	1	0
Vardhman	0	1	0	0	1	2
Varemac	0	0	0	1	0	0
Vateks	1	0	0	0	0	0
Vaughans	0	0	1	1	0	0
Vaupel	0	0	1	0	0	0
Veba	0	0	0	1	0	0
VEC Technology	0	0	0	0	1	0
Vecotex	0	0	0	1	0	0
Veejay Lakshmi	0	0	0	3	0	1
Veejay Savio Lakshmi	0	0	1	0	0	0
Veit	0	2	1	2	0	1
Veith Systems	0	0	0	0	0	1
Velcro Industries	0	0	0	1	0	1
Velutex	0	0	0	1	0	0
Ventex	0	0	0	0	1	0
Venti Oelde	0	0	0	0	0	1
Venture Tape	0	0	0	0	0	1
Venus Textiles	0	0	0	0	0	1
Vepetex	0	0	0	1	0	0
Verband Schweizerischer Bandfabrikanten	0	0	0	1	0	0
Verbrugge	0	0	0	2	0	0
Verdant Technologies	0	0	0	0	0	1
Vernaro	1	0	0	0	0	0
Veronica Textiles	1	0	0	0	0	0
Verpackungszentrum Graz	0	0	0	0	1	0
VersaCore	0	2	0	0	0	0
Vertigo	0	0	0	0	1	1
Vestas	0	0	0	0	0	1
VF	0	2	3	0	2	4
Vi be Mac	0	1	0	0	0	0
Viba Siat	1	1	0	0	0	0
Vibram	0	0	0	0	0	1
Vicki James Sleepwear	0	0	0	0	1	0
Victor Innovatex	0	0	0	1	0	0
Victoria Carpets	2	2	0	1	2	0
Victoria Wool Processors	0	0	1	2	1	0
Victorias Secret	0	1	1	0	0	0
Victrex	0	0	0	1	1	0
Viehoff	0	0	1	0	0	0
Viero	1	0	0	0	1	0
Vietnam National Textile and Garment Corporation	0	0	0	0	1	1
Vignoni	0	0	1	1	1	2
Villiani	0	0	0	0	0	1

WORLD TEXTILES Manufacturer Indexing

	1999	2000	2001	2002	2003	2004
Vinamul Polymers	0	0	0	1	0	0
VINATEX	0	2	0	0	0	0
Virgin	1	0	0	1	0	0
Virkler	0	1	0	0	0	0
Virudhunagar Textile Mills	0	1	0	0	0	0
Vistagy	0	0	0	1	0	1
Visteon	0	0	0	0	0	1
Vita	0	0	0	1	0	0
Vitale Barberis Canonico	0	0	0	1	0	0
Vitaweb	0	1	0	0	0	0
Viva Sistems	0	0	2	0	0	0
VivoMetrics	0	0	0	2	3	1
Vlifotex	0	1	0	0	0	0
Vlisco	0	0	1	0	0	0
Voith	0	0	2	0	3	10
Volga Wool	0	0	1	0	0	0
Volkmann	0	1	1	0	1	5
Volkswagen	0	0	0	0	1	1
Vollenweider	0	1	0	0	0	0
Volvo	0	0	0	1	0	1
Vomatex	0	0	1	0	0	0
Vorwek	0	0	0	0	1	0
Voswinkel and Company	0	0	0	0	0	1
Vought Aircraft Industries	0	0	0	0	0	1
Vouk	2	1	0	0	0	0
Vowalon	0	0	0	0	0	2
VSM Group	0	0	0	0	1	1
VUTEk	0	0	0	0	0	1
VXL Systems	0	0	0	0	1	1
VyaTek Sports	0	0	0	0	0	1
Vyncolit	0	0	1	0	0	1
W and J Whitehead	1	1	0	0	0	0
W L Gore	4	1	4	2	4	7
W R Grace and Company	0	0	0	0	0	2
W Schillig	0	0	0	0	1	0
Waagmeester	0	0	0	0	1	1
Wac Data Service	0	0	0	0	0	2
Wacker Chemie	1	0	0	1	0	1
Wacker Polymer Systems	0	0	0	1	0	0
Wacoal	0	0	0	0	0	1
Wagenfelder Group	0	0	0	0	1	0
Wakayama Senko	1	0	0	0	0	0
Wal Mart	0	2	0	0	0	2
Walk Off Mats	0	0	0	0	2	1
Walker Greenbank	0	1	1	2	0	0
Walkisoft	1	1	0	0	0	0
Walter Group	0	0	1	0	0	0
Walter Textil	0	1	0	0	0	0

	1999	2000	2001	2002	2003	2004
Wangner Finckh and Company	0	0	0	0	0	1
Warehouse One	0	0	1	0	0	0
Waring Products	0	1	0	0	0	0
Warmkraft	0	0	0	0	0	2
Warnaco	0	0	1	0	0	0
Warwick Mills	0	0	0	0	0	1
Waterworks	0	0	0	0	0	1
Wattex	1	0	0	0	0	0
Wax Converters Textiles	0	0	0	0	1	0
Weavexx	0	0	0	0	1	3
Web Converting	0	0	0	1	0	0
Web Dynamics	2	4	4	1	0	0
Web Processing	2	1	1	2	2	0
Webasto	0	0	1	0	0	0
Webbing and Trimming	0	0	0	1	0	0
WebCore	0	0	0	0	1	0
Weber	0	1	0	0	0	0
Weberei Appenzell	0	0	0	1	0	1
Wefatex	0	0	0	0	2	0
WEIMA	0	0	0	1	0	0
Weiman	0	0	0	1	0	0
Weinmann	0	0	0	0	0	1
Weisbrod Zurrer	0	1	0	1	0	0
Weisters	0	0	1	0	0	0
Weko	0	0	1	0	0	0
Welbeck	1	1	1	1	0	0
Welco	0	0	0	1	0	0
Wellman	0	1	10	1	2	2
Welspun Syntex	0	0	0	0	1	0
Weserland	0	0	1	0	1	0
Wesfarmers Dalgety	0	1	0	0	0	0
Wesfarmers Landmark	0	0	0	1	0	0
West Point Foundry and Machine Company	0	0	1	0	0	0
West Point Stevens	1	1	0	0	0	0
West Virginia University	0	0	0	0	0	1
Westaim	0	1	1	0	0	0
Westbond	1	0	0	0	0	0
Westdale Dyeing and Finishing	1	2	0	0	0	0
Westerman	0	1	0	0	0	0
Western Nonwovens	0	2	0	0	0	0
Westfarmers Landmark	0	0	0	0	1	0
Westinghouse Savannah River Company	0	0	0	2	0	0
WestPoint Stevens	1	2	2	1	0	0
Westwood	0	0	1	0	0	0
Westwools	0	1	0	0	0	0

	1999	2000	2001	2002	2003	2004
Wet Seal	1	0	0	0	0	0
WET Thermplate	0	0	0	0	2	0
Wetzel	0	2	0	2	0	0
Weyerhaeuser	1	0	0	3	6	8
WGF Colcoton	1	0	0	0	0	0
Whirlpool	0	0	0	0	0	9
White Knight	0	0	0	0	2	0
Whitehouse Institute of Design	0	0	0	1	0	0
Whiteley and Green	0	0	1	0	0	0
Whitestone Weavers	0	1	0	0	0	1
Whitford Worldwide	0	0	0	0	1	0
Whizard Protective Wear	0	0	1	0	0	0
Wicks and Wilson	0	0	1	0	0	0
Wide	0	0	1	0	0	0
Widemex Geldrop	0	0	0	0	0	1
Wieland Lufttechnik	0	1	0	0	0	0
Wiener Werkstatte	0	0	0	0	0	1
Wiessner	0	0	0	0	0	1
Wilcom	0	0	1	0	1	0
Wilhelm	1	0	0	0	0	0
Willcox and Gibbs	0	1	0	0	0	0
William Baird	0	0	1	0	0	0
William Hall and Company	0	0	0	1	0	0
William Lockie	0	0	0	1	0	0
William Pownall and Sons	0	0	0	0	0	1
William Sinclair	0	1	0	0	0	0
William Tatham	1	1	2	1	0	0
Williamson Dickie Manufacturing Company	0	0	0	1	1	0
Wilson Composite Group	0	1	0	0	0	0
Wilson Fabrics and Wallpapers	0	0	0	1	0	0
Wilson Sporting Goods	0	0	2	0	0	0
Windel	0	2	0	0	1	0
Windsor Apparel	0	0	0	0	0	1
Winfield Consumer Products	0	0	0	0	0	1
Wings Systems	0	0	0	0	1	0
Winn	1	0	0	0	0	0
Winrock Grass Farm	0	0	1	0	0	0
Wise	0	1	0	0	0	0
Wistex	0	0	0	0	1	0
Witco	0	1	0	0	0	0
Witte	0	0	0	0	2	0
Witton Mills	0	1	0	0	0	0
WM Fibres	0	0	1	2	0	0
Woco Franz Josef Wolf	0	0	1	0	0	0
Wohrl	0	0	0	1	0	0
Wolfangel	1	0	0	0	0	0

	1999	2000	2001	2002	2003	2004
Wolford	1	0	1	0	1	1
Wollux	0	1	0	0	0	0
Woodshed Technologies	0	0	0	0	0	1
Wool Equities	0	0	0	0	0	1
Wool Research Organization of New Zealand	3	6	7	7	3	1
Wool Services International	1	1	1	1	0	0
Wool Strategy Group	1	0	0	0	0	0
Wool Testing Authority UK	0	0	0	0	0	1
Wool Tex Stock Company	0	0	0	1	0	0
Woolmark Company	6	8	4	6	8	5
Wools of New Zealand	3	2	1	1	0	1
Woolworth	0	0	0	0	0	1
World Company	0	0	1	0	0	0
World Fibers	0	0	0	0	0	3
World Trade Organization	0	0	7	8	15	1
Worthington	1	1	0	0	0	0
Woven Legends	0	1	0	0	0	0
Wrace	0	0	0	2	0	0
Wrightson	1	1	0	0	0	0
WRONZ Euralab	0	0	0	1	0	0
Wumag	0	1	1	0	0	0
Wuxi	1	0	0	0	0	0
Wykes	0	1	0	1	0	0
Wyoming Woolens	1	0	0	0	0	0
X-Rite	1	1	0	0	0	0
Xerox	0	0	0	0	0	3
Xetma	0	0	0	1	0	1
Xiameter	0	0	0	0	1	0
Xinjiang Bazhou Taichang Industrial	1	0	0	0	0	0
Xinjiang Green Health Luobuma	0	0	0	0	1	0
Xiros	0	0	0	0	0	2
Xorella	1	1	1	0	2	1
XterraGear	0	1	0	0	0	0
Xtreme Fibers	0	0	0	1	0	0
Xyleco	0	0	0	0	0	1
Xymid	0	0	0	0	0	1
Yakka	0	1	0	0	0	0
Yale	0	0	0	0	0	1
Yamaha	0	0	1	0	0	0
Yamato	0	2	0	1	0	1
Yamauchi	0	0	0	0	0	1
Yarn Mavens	0	1	0	0	0	0
Yarra Falls	2	0	0	0	0	0
Yarra Trail	0	0	1	0	0	0
Yatlantico	0	1	0	0	0	0
Yazaki	0	0	0	0	0	1

WORLD TEXTILES Manufacturer Indexing

	1999	2000	2001	2002	2003	2004
Yeo Paull	1	0	0	0	0	0
Yesim	0	0	0	1	0	0
Yetek Company	0	0	0	0	0	1
Yeu Ming Tai	0	0	0	0	0	1
YKK	1	0	0	0	1	3
Yokohama Rubber Company	0	0	0	0	0	1
Yonca Onuk	0	0	0	0	0	1
Yorklyde	0	1	0	1	0	0
Yorkshire Group	1	4	3	1	0	1
Yoshida	0	0	0	1	0	0
Youghal Carpets	0	2	0	0	0	0
Yoz-Ami	0	0	0	0	1	0
Yugen Kaisha Ichiryu	0	0	0	0	1	0
Yumco	0	0	0	1	0	0
Yves Saint Laurent	0	1	0	1	0	0
Yxendis	0	0	0	0	0	2
Zaitex	0	0	0	0	0	1
Zanco	0	1	0	0	0	0
Zanotti Elasticos	0	0	0	0	1	0
Zegna Baruffa	1	3	1	2	1	0
Zellmeger	1	0	0	0	0	0
Zellulosefaser Beteiligungsgesellschaft	0	0	1	0	0	0
Zellweger Luwa	0	0	0	0	0	1
Zellweger Uster	8	14	6	3	2	4
Zenit	0	0	0	0	1	0
Zenith Pumps	0	1	1	0	1	0
Zentex	0	0	0	0	0	1
Zeon	0	0	0	1	0	0
Zhelaniye	1	0	0	0	0	0
Zian Hi Tech	0	0	1	0	1	0
Zig Sheng	0	0	0	1	1	0
Zignago Tessile	0	0	0	0	0	1
Zima	0	1	0	1	0	0
Zimmer	1	5	10	4	18	6
Zimmermann	0	1	0	0	0	1
Zinser	5	3	16	5	5	3
Zivtex	0	0	0	0	1	0
Zollanvari	0	0	2	0	0	0
Zoltek	2	2	3	1	2	4
Zoysian	0	0	1	0	0	0
Zschimmer and Schwarz	0	2	2	4	1	3
ZSK	2	1	0	1	2	1
Zucchi-Bassetti	1	1	0	0	0	0
Zweigle	0	1	2	0	0	0
Zwickauer Kammgarn	0	1	0	0	0	0
Zydex	0	0	0	2	0	0
Zyex	0	0	4	0	0	0
Zyvax	0	0	0	0	0	1

USE OF TRADE NAME INDEXING PER YEAR

	1999	2000	2001	2002	2003	2004
3Weave	0	0	2	1	0	0
A M Y	0	0	2	1	0	0
Accepta	0	0	2	0	0	0
Accu Sorb	0	0	0	0	1	0
Acrilan	0	0	0	0	0	1
Acticoat	0	1	0	0	0	0
Actigard	0	0	1	0	0	0
ActionBac	0	0	1	0	0	0
Actisorb	0	0	0	0	1	0
AdBac	0	0	2	0	0	0
Aerelle	1	0	0	0	0	0
Aerorig	0	1	0	0	0	0
Air-Com-Tex 700	0	0	3	1	0	0
Airex	0	0	1	0	0	0
Airlock	0	0	1	0	0	0
Airmint	0	0	0	1	0	0
Airo Quattro	0	1	0	0	0	0
Airtex	0	0	0	1	0	0
Airtubes	0	0	1	0	0	0
AirWeb	0	0	2	1	0	0
Akulon Ultraflow	0	0	0	0	0	1
Akulon	0	0	0	0	0	1
Alceru	0	0	0	0	2	0
Alcoprint	0	1	0	0	0	0
ALforfix	0	0	0	1	0	0
Alizarin	0	1	0	1	0	0
Alomable	1	0	0	0	0	0
AlphaSan	0	0	0	2	0	2
Always	0	1	1	0	0	0
Amicor Pure	0	1	1	1	0	0
Amicor	2	2	3	1	2	1
Amilan	0	0	1	0	0	0
Anecot	0	0	1	0	0	0
Anox	0	0	0	0	2	0
Antron	0	0	2	0	2	1
Aqua Lok	0	0	1	0	0	0
AquaJet	1	3	6	1	4	2
Aqua Lok	0	0	1	0	0	0
Aqualon	0	0	1	0	0	0
Aquamarble	0	0	0	0	1	0
AquaMax	0	0	1	0	0	0
Aquator	0	0	0	1	1	0

WORLD TEXTILES Trade Name Indexing

	1999	2000	2001	2002	2003	2004
Aracon	0	0	0	0	1	0
Araldite	0	0	0	0	0	1
Arathane	0	0	0	0	0	1
Arc-Man	0	0	1	0	1	0
Arkofix	0	0	1	0	0	0
Armos	1	0	1	0	0	0
Arnitel	0	0	1	1	0	0
Arsura	2	0	0	0	0	0
ArtFlock	0	0	0	1	0	0
Artistri	0	0	3	0	0	0
Asahiguard	0	0	1	0	0	0
ASPUN	0	0	2	1	0	0
Assist	0	0	1	0	0	0
Astramol	0	0	0	0	1	0
Astrazon	0	0	0	0	1	0
Astroturf	0	0	0	0	2	0
Athtoma	0	1	0	0	0	0
Atmos	0	0	0	0	0	1
Atryl	0	0	0	0	0	1
Atvantage	0	1	0	0	0	0
Augusta	0	0	0	1	0	0
Aurora	0	0	0	0	1	0
Auspreg UVO	0	0	0	0	0	1
Autoconer	0	1	6	2	1	1
Autocoro	0	1	11	17	12	6
Autodose	0	0	0	1	0	0
Autowarp	0	0	1	1	0	0
Averde	1	0	0	0	0	0
Avery	0	1	0	0	0	0
Avicel	0	0	0	1	0	0
Avora	0	0	1	0	0	0
Awnmax	0	0	0	1	0	0
Awntex	0	0	0	2	0	0
Azdel	0	0	0	0	0	1
Back Vitalizer	0	0	0	0	0	1
Bactosol	0	1	0	0	0	0
Band Aid	0	0	0	1	0	0
Barricade	0	0	1	0	0	0
Basofil	1	1	4	1	0	0
Basojet	0	0	0	0	1	0
Bayflex	0	0	0	0	1	0
Baylase	0	0	2	2	2	1
Baypreg	0	0	0	0	1	0

	1999	2000	2001	2002	2003	2004
Bayscent	0	0	0	0	1	0
Baytec	0	0	0	0	1	0
Belcoro	0	0	1	7	3	2
belCoTex	0	0	3	0	0	0
Bellneat	1	0	0	0	0	0
Berkol	0	0	3	0	0	0
Berol Spin	0	1	0	0	0	0
Berol Visco	0	2	0	0	0	0
Bicoflex	0	0	0	0	1	1
Bilorex	0	0	0	0	1	0
BioClip	0	0	0	1	0	0
BiodivYsio	0	0	0	1	0	0
Bioefect	0	0	0	0	1	0
Bioform	0	0	1	0	0	0
Biofresh	0	0	0	1	0	0
Biokryl	1	0	1	0	0	0
Biomesh	0	0	0	0	0	1
Bionolle	0	0	0	0	0	1
Bionyl	0	0	1	0	0	0
Biopol	0	0	2	0	0	0
BioPrep	0	0	0	1	0	0
BioSteel	0	0	0	0	2	1
Blankit	0	0	0	1	0	0
Blankophor	0	0	0	0	1	1
Brani Waistware	0	0	0	0	1	0
Calcobond	0	0	0	1	0	0
Caliweb	0	1	0	1	0	0
Caprosyn	0	0	0	1	0	0
CarboCrete	0	0	0	1	0	0
Carbofol	0	0	0	0	1	0
Carboform	0	0	0	0	0	1
Carilon	1	1	2	0	0	0
Celanna	0	1	1	0	0	0
Celcon	0	0	0	1	0	0
Cellusoft	0	0	1	2	0	0
Celstran	0	0	0	0	1	1
Cerafil	0	0	0	1	0	0
Charcoalan	0	0	0	1	0	0
Chimassorb	0	0	1	0	0	0
Chinon	0	0	0	0	2	0
Chitopoly	0	0	0	0	0	1
Chloroban	0	0	1	0	0	0
Chromophtal	0	0	1	0	0	0

WORLD TEXTILES Trade Name Indexing

	1999	2000	2001	2002	2003	2004
Chromotex	0	0	0	1	1	0
Chromspun	0	1	0	0	0	0
Cibacron	0	2	1	3	0	3
Cirrasol	0	0	0	1	0	0
Cleanburn	0	0	1	0	0	0
ClimaBalance	0	0	0	0	1	1
Climatex Lifeguard	0	0	0	1	0	0
Climatex LifeguardFR	0	0	0	1	0	0
Cliptronic	0	3	1	1	0	0
Clovertec	0	0	0	1	0	0
Cloxacillin	0	0	0	1	0	0
Colana	0	1	0	0	0	0
Colback DuoLock	0	0	1	0	0	0
Colback EasyTuft	0	0	1	1	0	0
Colback	0	0	1	1	0	0
Colbond	0	1	1	0	0	0
ColdGear	0	0	1	0	0	0
Colorite	0	0	0	0	1	0
ColorStrip	0	0	1	0	0	0
ColorTools	0	0	1	0	0	0
ColourBrain	0	0	1	1	0	0
Colpoca	0	0	1	0	0	0
COM4	0	0	8	4	4	1
Comforel	0	0	0	0	1	0
ComforSpin	0	0	6	5	6	0
Comfort Bath	0	1	0	0	0	0
ComfortBack	0	0	1	0	0	0
ComforTemp	1	0	1	1	0	1
Comfortlace	0	0	0	0	0	1
Conex	0	0	0	0	1	0
Contexxor	0	0	1	0	0	0
Cool Pad	0	0	1	1	0	0
Cool Wool	0	0	1	0	1	0
CoolMax	3	2	5	2	4	2
Cordenka	0	0	0	1	0	0
Cordura	0	0	3	3	2	2
Corian	0	0	0	0	0	1
Corn Fiber	0	0	0	1	0	0
Cornplast	0	0	0	0	1	0
Corolab	0	0	0	0	1	0
Corterra	5	3	3	7	1	1
Cosmocil	0	0	1	0	0	0
Courtelle Definition	0	0	1	0	0	0

	1999	2000	2001	2002	2003	2004
Courtelle Matis	0	0	0	2	1	0
Courtelle Micro	0	0	2	0	0	0
Courtelle Recoil	0	1	0	0	0	0
Courtelle Sun	0	0	1	1	0	0
Courtelle	1	0	1	1	2	0
Crabyon	0	1	0	2	0	0
CreelManager	0	0	0	0	4	0
Cretex	0	0	1	0	0	0
Crypton	0	0	1	0	0	1
Curadis	0	0	1	0	0	0
Curaleg	0	0	1	0	0	0
Curv	0	0	0	1	0	0
Cyclanon	1	0	2	3	0	0
Cycolac	0	0	0	0	1	0
Cycom	0	0	0	0	0	1
D-Core	0	0	1	0	0	0
Dacron Hexsa	0	0	0	1	0	0
Dacron	1	0	0	0	0	1
Dancing Billboards	0	0	1	0	0	0
Danufil	1	0	0	0	0	0
Deci-Tex	0	0	0	0	1	0
Delnet	0	0	1	0	0	0
DelPore	0	0	3	0	0	0
Delrin	0	0	0	0	0	1
Demique	0	0	0	0	1	0
Demron	0	0	0	0	0	1
DeniLite	0	0	1	0	1	0
DeniMax	0	0	0	0	1	1
Denimwool	0	0	0	0	1	0
DeniPrime	0	0	0	0	1	0
Depend	0	1	0	0	0	0
Detect	0	0	0	1	0	0
Diadavin	0	0	0	2	1	0
Dianix	0	0	0	2	3	0
Diaserver	0	0	0	1	0	0
Diolen Care	0	0	0	1	0	0
Diolen	0	0	0	1	1	0
Diresul	0	0	0	1	0	0
Disperstat	0	0	1	0	0	1
Divinycell	0	0	1	0	0	0
Dorlastan	4	3	4	3	3	1
Dos and Dye	0	0	0	2	0	0
Dow XLA	0	0	0	3	6	0

WORLD TEXTILES Trade Name Indexing

	1999	2000	2001	2002	2003	2004
Dralon	0	1	1	0	0	0
Dri-release	0	0	0	0	0	1
Drimarene	0	0	2	0	1	1
Dritex	0	0	0	1	0	0
Dry 'n Lock	0	0	1	0	0	0
Dry Tech	0	0	0	1	0	0
Dry Weave	0	0	1	0	0	0
Dryarn	0	0	0	2	0	0
Dryline	1	0	0	0	0	0
Drytex	0	0	1	0	0	0
Drywear	0	0	1	0	0	0
DryWick	0	0	1	0	0	0
DuoMod	0	0	0	1	0	0
Dur-o-set	0	0	0	1	0	0
Duralcan	0	0	1	0	0	0
Durapex	0	0	2	0	0	0
Durethan	0	0	0	0	1	0
Duroslip	0	0	1	0	0	1
Dylar	0	0	0	1	0	0
Dymetrol	0	0	0	1	0	0
Dynafil	0	0	3	0	3	0
Dynamar	0	0	0	1	0	0
Dyneema	0	4	4	5	2	4
Dyneon	1	0	0	0	0	0
E-control	0	0	1	0	1	0
Eastar Bio	0	0	0	1	0	1
EasyLeno	0	0	0	0	0	1
ECObleach	0	0	0	1	0	0
EcoFix	0	0	4	0	1	0
Econase	0	0	0	0	1	0
Ecopure	0	0	0	1	0	0
Ecosafe	0	0	0	0	1	0
EcoSinge	0	0	1	0	0	0
Ecostone	0	0	0	0	1	0
ElastoTwist	0	0	0	0	1	0
Elektex	0	0	1	0	0	0
Eliotex	0	0	0	0	1	0
Elite	0	1	2	0	2	1
EliTe	0	0	2	8	5	3
Elitwist	0	0	0	0	2	4
Elk	0	0	1	0	0	0
Elvanol	0	0	0	1	0	0
Emblazia	0	0	1	0	0	0

	1999	2000	2001	2002	2003	2004
Engage	0	0	0	0	1	0
enVix	0	0	0	0	0	1
Enzoguard	0	0	0	0	1	0
Epic	0	0	1	0	0	0
Eriofast	0	0	0	0	1	2
Eslex	0	0	0	1	0	0
Evalon	0	1	0	0	0	0
Evolon	0	2	9	2	0	1
Evospeed	0	0	1	1	0	0
Exceval	0	0	1	2	0	0
ExoGrid	0	0	0	0	0	1
Fabrasorb	0	0	0	1	0	0
FabricEye	0	0	0	0	1	0
Faradex	0	0	0	0	0	1
Fashiontronic	0	0	1	0	0	0
Fastskin	0	1	2	0	0	1
Favimat	0	0	0	0	1	0
Felosan	0	0	0	1	0	0
Fiberdri	0	1	0	0	0	0
Fibersorb	0	1	0	0	0	0
Fiberstab	0	0	1	0	0	0
Fibrafinn	0	0	0	3	1	0
Filifine	0	0	2	0	0	0
Fillwell	0	0	1	0	1	0
Firesorb	0	0	0	1	0	0
Fixapret	0	0	0	1	0	0
Flamestab	0	0	2	0	0	0
Flotex	0	1	0	1	0	0
Fluor	0	0	1	0	0	0
Fluotop T	0	2	0	0	0	0
Flushaway	0	0	2	0	0	0
Fly Guys	0	0	1	0	0	0
Foron	0	0	0	0	0	1
Fortrel	0	0	1	0	0	0
Fortron	0	0	0	1	2	0
Fossfibre	1	0	0	0	0	0
Fosshape	0	0	0	1	0	0
Fosshield	0	0	1	0	0	1
Freshloft	0	0	1	0	0	0
Futura	0	0	0	1	0	0
FyreRoc	0	0	0	0	1	0
Gamex	0	0	0	0	1	0
Gebecon	0	0	0	0	0	1

	1999	2000	2001	2002	2003	2004
Gelanots	0	0	1	0	0	0
Genesis	0	0	0	0	1	0
Glacial	0	0	0	0	2	0
Glare	0	0	2	0	1	0
GoodNites	0	0	0	0	0	1
Gore-Tex	4	1	5	3	4	5
Gorix	0	1	0	1	0	0
GrassGro	0	0	0	1	0	0
GreenTex	0	0	0	1	0	0
Grid-X	0	0	0	0	1	0
Grilene	0	0	1	0	0	0
Grilon	0	0	1	0	0	0
Halar	0	0	0	1	0	0
HaloShield	0	0	0	0	0	1
HarvestForm	0	0	0	1	0	0
Health Shield	0	0	0	1	0	0
HeatGear	0	0	1	0	0	0
Helizarin	0	0	0	3	2	0
HemaJet	0	0	0	0	1	1
Herex	0	0	1	0	0	0
HexMC	0	0	1	0	0	0
HexPly	0	0	0	0	0	1
Hexweb	0	0	1	0	0	0
Hi-Nicalon	0	0	1	1	1	0
HighDistance	0	0	0	0	0	5
HiPer	0	0	0	0	1	0
Hisorb	0	0	1	0	0	0
Hollofil	1	0	0	0	0	0
Holofiber	1	0	0	0	0	0
Huggies	0	0	0	1	0	1
Husky	0	0	0	0	0	1
Huytexx	0	0	1	0	0	0
Hycoknit	0	0	0	0	0	1
Hydrocel	0	1	0	0	0	0
Hydrodrain	0	1	0	0	0	0
Hydrofiber	0	0	0	2	0	0
HydroFormer	0	0	1	0	0	0
HydroLock	0	0	0	0	0	1
Hydrolon	0	0	1	0	0	0
Hydroplus	0	0	1	0	0	0
Hydroseal	1	0	0	0	0	0
Hydroweave	0	2	0	0	0	0
HyFill	0	0	0	1	0	0

	1999	2000	2001	2002	2003	2004
Hyflon	0	0	0	1	0	0
Hypox	0	0	0	0	0	1
Hytrel	0	0	0	2	0	0
I-Tex	0	0	1	0	0	0
Imbue	0	0	2	1	0	0
Imperon	0	0	0	1	0	1
Impreglon	1	0	0	0	0	0
Incognito	0	0	1	0	0	0
Indanthren	0	0	4	0	1	2
IndiAge	0	0	0	0	1	0
Infinity	0	0	1	0	0	0
Ingeo	0	0	0	0	15	4
Innodye	0	0	0	0	1	0
Innova	0	1	0	0	0	0
Instachem	0	1	0	0	0	0
Insulyte	0	0	1	0	0	0
Intelliwinder	0	0	0	1	0	0
Intera	0	0	1	0	0	0
InterSpun	0	1	1	2	0	0
IQClean	0	0	1	0	0	0
Irgaguard	0	0	2	0	0	0
Irgaphor	0	0	2	3	0	0
Irgastat	0	0	1	0	0	0
Irgatec	0	0	0	0	0	1
Isolan	0	0	0	0	2	6
ISOray	0	0	0	0	0	1
Jacquardtronic	0	1	1	1	0	6
Jetskins	0	0	0	1	0	0
Jettex	0	0	0	0	1	0
Kalrez	0	0	0	0	1	0
Kanecaron	0	0	2	0	0	0
Kapex	0	0	1	0	0	0
Kapton	0	1	0	0	0	0
Kensel	0	0	1	0	0	0
Kerlix	0	0	1	0	0	0
Kermel	1	0	1	2	4	1
Kevlar	12	23	22	23	10	14
Kiara	0	0	1	0	0	0
Kieralon	0	0	0	1	0	0
Klearspan	0	0	1	0	0	0
Kotex	0	0	0	0	1	0
Kraton	0	0	0	1	0	0
Kristall	0	0	0	1	0	0

	1999	2000	2001	2002	2003	2004
Kuralon K-II	0	0	1	0	0	0
Kuralon	1	0	2	0	0	0
Kynol	0	1	0	0	0	1
Lactron	2	1	0	1	0	0
Lanaset	0	0	2	0	2	1
Lanasol	0	0	2	0	0	0
Lanasyn	0	0	0	0	0	1
Lanseal	0	1	0	0	0	0
Leacril	0	1	0	1	1	0
Leak Lock	0	0	0	0	1	0
Lecture	1	0	0	0	0	0
Ledra	0	0	0	1	0	0
Lenpur	0	0	0	1	1	0
Lenzing Lyocell	2	0	4	1	3	0
Leophen	0	0	0	1	0	0
Leucophor	0	0	0	0	0	1
Levafix	0	2	1	1	2	1
Lexan	0	0	0	0	1	0
Lextra 3D	0	0	0	1	0	0
Lextra	0	1	1	2	0	0
Lifepath AAA	0	0	1	0	0	0
LifeShirt	0	0	0	1	3	0
Lightron	1	0	0	0	0	0
lincLITE	0	0	3	1	0	0
Linel Comfort	0	1	0	0	0	0
Linel	0	2	0	0	0	0
Lite Loft	0	0	0	0	1	0
Liteflex	0	1	0	0	0	0
Livefresh	0	0	1	0	0	0
Loftguard	0	0	1	0	0	0
Loosegear	0	0	1	0	0	0
Lorica	0	0	0	1	0	0
Lubrifol	0	0	0	1	0	0
Lubrol	0	0	0	1	0	0
Lufnen	0	0	2	0	1	0
Lumaxx	0	0	1	2	1	0
Luminex	0	0	0	0	2	0
Luprimol	0	0	0	0	1	0
Luprintol	0	0	0	1	2	0
Lurapret	0	0	0	0	0	1
Lurex	0	1	0	0	0	0
Lusynton	0	0	0	2	0	0
Lutexal	0	0	0	0	1	0

	1999	2000	2001	2002	2003	2004
Lycra Soft	0	0	1	0	0	0
Lycra XA	0	0	0	1	0	0
Lycra	9	16	21	30	27	7
Lysol	0	0	0	0	0	1
Lysorb	0	0	0	1	0	0
M5	0	0	0	1	0	0
Macrosorb	0	0	1	0	0	0
Makspec	1	0	1	0	0	0
Mater-Bi	0	0	0	0	2	2
Matrimid	0	0	0	0	0	1
Maxflex	0	1	0	1	0	0
Maxilon	0	0	0	1	0	0
MediSoft	0	0	0	0	0	1
Mepilex	0	0	1	0	0	0
Meryl	1	2	3	6	4	1
Met5	0	0	1	0	0	0
Metocene	0	0	0	1	0	0
MetPore	0	0	0	1	0	0
MiCralon	0	0	0	1	0	0
MiCrell	0	0	0	1	0	0
Micro-rebar	0	0	0	0	1	0
Microban	0	0	0	1	0	1
Microbe Shield	0	0	0	1	2	1
Microbisan	0	0	0	1	0	0
Microcare	0	0	0	1	0	0
MicroFresh	0	0	0	0	0	1
Microft Light	0	0	1	0	0	0
Microlattice	0	0	0	0	1	0
Microlisse	0	0	1	0	0	0
Micromass	0	0	0	1	0	0
Micromattique	0	0	1	0	0	1
Micropake	0	1	1	1	0	0
MicroSafe	1	1	0	0	0	0
Microtex	0	0	0	0	1	0
MiCrovacuum	0	0	0	1	0	0
Microwetting	0	0	0	1	0	0
Migra-Cap	0	0	0	0	1	0
MigraJet	0	0	0	0	0	2
Minlon	0	0	0	0	1	0
Miraflex	1	0	0	0	0	0
Miratec	1	0	0	0	1	0
Mirhon	0	0	1	0	1	0
Modal Fresh	0	1	0	1	0	0

WORLD TEXTILES Trade Name Indexing

	1999	2000	2001	2002	2003	2004
Moistegy	0	0	1	0	0	0
MPDiol Glycol	0	0	0	0	1	0
Multiflow	0	0	1	1	1	0
Multipor	0	0	0	0	0	1
Mynx	0	0	1	1	1	0
Nafion	0	0	0	3	1	0
Naizil	0	0	1	0	0	0
Nano-Care	0	0	0	0	1	0
Nano-Pel	0	0	0	0	1	0
Nanoweb	0	0	0	0	0	1
NatureWorks	0	6	9	13	3	0
Naturflow	0	0	0	1	0	0
Nega-Stat	0	0	0	0	1	0
Neogel	0	0	0	0	0	1
Neoprene	1	0	6	3	3	3
Neotis	0	0	0	1	0	0
NewCell	3	0	0	0	0	1
Newtex	0	0	1	0	0	0
Nextel	0	0	1	0	0	0
NextStitch	0	0	1	0	0	0
NForz	0	0	0	1	1	0
Nicalon	1	2	3	0	2	0
Nidaplast	0	0	0	0	0	1
Nitron	0	1	0	0	1	0
No Wind	0	0	1	0	0	0
Nodax	0	0	0	0	1	0
Nomex	0	5	6	11	5	4
Nostatex	0	0	0	0	1	0
NovaThin	1	0	0	1	0	0
Novolan	0	1	2	0	1	0
Nu Trak	0	0	1	0	0	0
Nuva	0	0	1	0	0	0
Nylosan	0	0	0	0	1	1
Nylostab	0	1	1	0	0	0
Oasis	0	1	0	1	0	0
OceanWeave	0	0	0	1	0	0
OMNIplus	0	2	3	2	0	1
Omniwind	0	0	1	0	0	0
Optidye	0	0	0	0	0	2
Optim	0	1	0	0	2	0
Orex	1	0	0	0	0	0
ORMOCERe	0	0	0	0	1	0
Ostanthren	0	0	0	0	1	0

	1999	2000	2001	2002	2003	2004
Ostazin	0	0	0	0	1	0
OutDry	0	0	0	2	0	0
Outlast	0	0	3	0	1	1
P84	0	0	0	0	1	0
PadKit	0	1	1	0	0	0
Pampers	0	1	0	0	0	1
Panex	0	0	0	1	0	0
Pantone	0	0	0	1	0	0
ParaGlass	0	0	1	0	0	0
Paragon	0	1	0	0	0	0
Pasmo	1	0	0	0	0	0
Pecoback	0	0	1	0	0	0
PEEK-Optima	0	0	1	0	0	0
Pellicle	0	1	0	1	0	0
Pentex	0	0	1	0	0	0
Perlon	0	0	0	0	1	0
Perma Core	0	0	1	0	0	0
Permafresh	4	0	0	0	0	0
Permalon	0	0	1	0	0	0
Permasep	0	0	1	0	0	0
Permavel	0	0	1	0	0	0
Persoftal	0	0	1	0	0	0
Pertex	0	1	0	0	0	0
Petra	0	1	0	0	0	0
Planova	0	0	1	0	0	0
PlasmaTreat	0	0	0	1	0	0
Plastisol	0	0	0	2	0	0
PLX	0	0	0	0	0	1
Poise	0	1	0	0	0	0
Polarguard	0	1	0	0	0	0
Polartec Power Dry	1	0	0	1	0	0
Polartec Wind Pro	0	1	0	0	0	0
Polartec	1	2	2	0	2	1
Polybac	0	0	1	0	0	0
Polycolon	0	0	1	0	0	0
PolyJet	0	0	0	0	0	2
Polyon II	0	0	0	0	0	1
Poro-Tex	0	1	0	0	0	0
Power-Tex	0	0	2	0	0	0
Power Stretch	0	0	1	0	0	0
Powerfil	0	0	0	0	0	1
PowerRip	0	3	0	0	0	0
Powerskin Xtreme	0	0	0	0	0	1

WORLD TEXTILES Trade Name Indexing

	1999	2000	2001	2002	2003	2004
Powerskin	0	1	0	0	0	1
Powerstretch	0	0	1	0	0	0
Precontraint	0	0	0	1	0	0
PressVu	0	0	0	0	0	1
Prestogen	0	0	0	2	0	0
PRIFORM	0	0	0	0	0	1
Primapor	0	1	0	0	0	0
Primasol	0	0	0	1	0	0
Proact	0	0	0	0	0	1
Proban	0	0	0	0	0	1
Procion	0	2	3	6	4	2
ProDyn	0	1	6	0	0	0
ProFil	0	0	0	0	0	2
Protechtor	0	1	0	0	0	0
PTP	0	0	0	0	0	1
Purecape	0	0	0	1	0	0
Purista	0	1	0	0	1	0
Pyralux	0	0	0	0	0	1
Pyron	0	0	0	0	2	0
Pyrotone	0	0	1	0	0	0
Pyrovatim	0	0	0	0	1	0
Quallofil	1	0	0	0	0	0
Quickview	0	0	1	0	0	0
Quickwash Plus	0	0	2	0	0	0
RadYarn	0	0	0	1	0	0
Rascheltronic	0	1	1	3	1	0
Rayolan	0	0	0	0	0	1
Rayosan	0	0	0	0	1	0
Realan	0	0	0	1	2	0
Reemay	0	0	1	0	1	0
Reflexx	0	0	0	1	0	0
Reicofil	0	0	4	2	1	3
Remazol	0	1	4	1	4	0
Renfil	0	0	1	1	0	0
Reozon	0	0	0	0	1	0
Repellan	0	0	0	2	0	0
Replay	0	0	0	0	0	1
Resistat	0	0	0	0	1	1
Resistex	0	0	0	0	2	0
Rewin	0	0	0	0	1	0
Rhovyl	1	0	0	1	0	0
Riteflex	0	0	0	0	0	1
Rodrun	0	0	0	0	1	1

	1999	2000	2001	2002	2003	2004
Rohacell	0	0	1	0	0	0
Roica	1	1	0	1	0	0
Rotona	0	0	0	0	0	2
Runderwear	0	0	0	0	0	1
Rynite	0	0	0	0	1	0
Ryton	1	0	0	0	0	0
Ryvalon	0	0	0	0	1	0
S-2 Glass	0	0	0	0	0	1
Safesorb	0	1	0	0	0	0
Safetac	0	0	1	0	0	0
Saffil	0	0	0	0	1	0
Sandofix	0	1	0	0	0	0
Sanitized	0	0	1	0	0	0
SanProCare	0	0	0	0	1	0
Saran	0	0	1	0	0	0
SAVANT	0	1	1	1	0	0
Sawascreen	0	1	0	1	0	0
Sawasorb	0	0	0	0	1	0
Scotchgard	0	0	1	0	1	0
Scotchlite	0	0	0	0	2	0
Scotdic	0	0	0	1	0	0
Scrub Free	0	0	0	0	0	1
Scrubbing Bubbles	0	0	0	0	0	1
Scuorozyme	0	0	0	0	0	1
SeaCell	0	0	4	4	5	2
Secugrid	0	0	0	0	1	0
Securus	0	0	4	1	0	0
Sensil	0	0	0	1	0	0
Sensory Perception Technologies	0	0	0	0	3	1
Sensura	0	0	4	0	0	0
SewFree	0	0	0	1	1	0
Shamir	0	0	0	1	0	0
Shrape	0	0	0	0	1	0
Sikawrap	0	1	0	0	0	0
Sillook	1	0	0	0	0	0
Silpalon	0	0	0	0	1	0
SilvaSorb	0	0	0	0	1	1
Simtex	0	0	0	0	1	0
Sirolan Yarnspec	0	0	1	0	0	0
Sirolan CF	0	0	1	0	0	0
Sirolan Fleecescan	0	0	1	3	0	0
Sirolan Laserscan	0	0	0	3	0	0

	1999	2000	2001	2002	2003	2004
Sirospun	0	1	1	0	0	2
SkinTech	0	0	0	0	0	1
Skintex	0	0	0	0	0	2
Snowskins	0	0	0	0	1	0
Soft Rolling Skin	0	0	0	1	0	1
SoftSeal	0	0	0	1	0	0
SOFTswitch	0	0	1	1	0	0
Solacryl	0	0	0	1	0	0
SoleFresh	0	0	0	0	0	1
Solopol	0	0	1	0	0	0
SoloSpun	2	2	2	1	2	2
Solscour	0	0	0	1	0	0
Solupor	0	0	1	0	0	0
Sonora	0	0	1	0	0	0
Sontara	0	0	1	0	0	1
Sorbfresh	0	0	0	1	0	0
Sorbtek	0	0	0	1	0	0
Sorona	0	2	2	2	0	1
Spectra	2	1	6	4	1	0
Sportwool	1	0	1	1	1	0
SpotShield	0	0	0	0	0	2
Sprint	0	0	0	0	1	0
SpunJet	0	0	1	1	0	0
Spyra	0	0	1	2	1	0
SpyroDerm	0	0	0	1	0	0
Stainmaster	0	0	0	1	1	0
StaMax	0	0	0	1	1	0
Stantex	0	1	1	0	0	0
Stanyl	0	0	1	0	0	0
Stanylenka	1	0	0	1	0	0
Starlight	0	0	0	1	0	0
StarWeb	0	0	1	0	0	0
Stay Clean	0	0	0	0	0	1
Stedshield	0	0	0	0	1	0
Stoll multi gauges	0	0	0	1	0	0
Stoll racking effects	0	0	0	1	0	0
Stratex	0	0	1	0	0	0
Stretch-aire	0	0	2	0	0	0
Strucore	0	0	0	0	0	1
Strudon	0	0	1	0	0	0
Suberis	0	1	0	0	0	0
Sultra	0	0	1	0	0	0
Sumifix Supra	0	0	0	1	0	0

	1999	2000	2001	2002	2003	2004
Sumifix	0	0	2	4	0	0
Sunbrella Firesist	0	0	1	0	0	0
Sunbrella	0	0	0	6	0	3
Sunpaque	0	0	0	0	1	0
Sunsure	0	0	1	0	0	0
Super Miracosmo	0	0	2	0	0	0
SuperCore	0	0	0	1	0	0
SuperFabric	0	0	0	0	0	1
Supermicroft	0	0	1	0	0	0
Supima	0	0	0	1	0	0
Supplex	0	1	2	1	2	0
Supralan	0	0	0	0	4	0
Suprel	0	0	0	0	1	0
Surface Master	0	0	0	0	0	1
Surgicel	0	0	0	0	1	0
SweatX	0	0	0	1	0	0
Swellcoat	0	0	0	0	0	1
Swiffer	0	0	1	0	0	0
Sylkharesse	0	0	0	1	0	0
Sympatex	0	0	1	1	0	0
Tactel Colorsafe	0	1	0	0	0	0
Tactel	1	13	5	7	6	1
Tactesse	0	0	0	1	0	0
Tampax	0	1	0	0	0	0
Tan Thru	0	0	1	0	0	0
Tapistron	0	0	1	0	0	0
Taslan	1	0	0	0	0	0
Technora	0	0	3	3	1	0
Technylstar	0	0	1	0	0	0
Tecnostan	0	0	1	0	0	0
Teflon	4	3	7	12	5	2
TekFlex	0	0	0	0	0	1
Telon	0	0	0	0	2	2
TempTrak	0	0	0	0	0	1
Tenacord	0	0	1	0	0	0
Tenara	1	0	0	0	0	1
Tencel	13	16	15	14	12	9
Tensylon	0	1	0	1	0	0
Terasil	0	1	3	1	3	1
Teratop	0	0	0	0	0	1
Terital	0	0	3	2	1	0
Terminox	0	0	0	1	0	0
Terrafix	0	0	0	1	0	0

WORLD TEXTILES Trade Name Indexing

	1999	2000	2001	2002	2003	2004
Terramesh	0	0	0	0	0	1
Textilene	0	0	0	1	0	0
TeXtreme	0	0	0	0	0	1
Textronic	0	1	0	1	0	0
Therapatch	0	0	0	1	0	0
Therma-Float	1	0	0	0	0	0
Therma-Tru	0	0	0	0	0	1
ThermaCare	0	0	0	1	1	0
ThermaCells	0	0	0	1	0	0
Thermastat	0	0	0	0	1	0
Thermo-Composite	0	0	0	1	0	0
Thermocule	0	0	0	1	0	0
Thermolite	0	0	1	1	2	0
Thermosol	0	0	0	1	0	0
Thermovyl	0	1	0	0	0	0
Thinsulate	0	0	0	0	1	0
Thiobac	0	0	0	2	1	0
Thiolon	0	0	0	0	1	0
Tilex	0	0	0	0	0	1
Timbrelle Care	0	0	0	1	0	0
TiMesh	0	0	0	0	3	1
Tinofast	0	0	3	3	0	0
Tinosan	0	5	2	2	0	0
TissueMend	0	0	0	1	0	0
Tissupor	0	0	0	0	0	2
TopAir	0	0	0	0	0	1
Torlon	0	0	0	0	0	1
Totalbath	0	0	0	1	0	0
TowFlex	0	1	0	0	0	0
Trevira	1	6	16	14	8	5
Triactor	0	0	0	1	0	0
Tribo	0	1	0	0	0	0
Triclosan	0	1	1	0	0	1
Trilon	0	0	0	1	0	0
Triton	0	0	1	0	0	0
Trol	0	0	0	0	0	1
Tubiguard	0	0	0	0	1	1
Tubisoft	0	0	0	0	0	2
Tufcel	0	2	0	0	0	0
Tuffshell	0	0	0	1	0	0
Tverlana	0	0	1	0	0	0
Twaron	0	5	4	7	3	0
TwiNjeans	0	0	0	1	0	0

	1999	2000	2001	2002	2003	2004
Twintex	0	0	0	1	0	0
TwistJet	0	0	0	1	0	0
Tychem	0	0	2	0	0	0
TYCOR	0	0	0	0	1	0
Tyfo	0	0	0	0	0	1
Tylose	0	0	0	0	0	1
Tyvek	1	3	4	1	0	1
Tyzor	0	0	0	1	0	0
Ultima	0	0	1	0	0	0
Ultra-Fresh	0	2	1	0	1	1
Ultra Mat	0	0	0	1	0	0
Ultralux	0	0	0	0	1	0
Ultramid	0	1	1	3	3	0
Ultraphor	0	0	0	1	1	0
Ultrason	0	0	0	0	1	0
Ultratect	0	0	0	0	1	0
Uniplex	0	0	0	0	1	1
Unisperse	0	1	0	0	0	0
Unithox	0	0	0	0	1	0
Uster	0	1	2	0	1	0
Vamac	0	1	0	0	0	0
VAP	0	0	0	0	0	1
VarioFil	0	0	2	0	0	0
VarioProtect	0	0	0	0	1	0
Vectra	0	0	0	0	1	1
Vectran	1	0	1	2	1	0
Velcro	1	8	2	4	4	4
VenTron	0	0	0	0	0	1
Versashield	0	0	0	1	0	0
Versiva	0	0	0	2	0	0
Vertitubes	0	0	1	0	0	0
Verton	0	0	0	0	3	0
Vespel	0	0	0	0	1	0
Vestamid	0	0	1	0	0	0
VeTron	0	0	0	1	0	0
Viafil	0	0	1	0	0	0
Viero	0	0	0	1	0	0
Vignoni	0	0	1	0	0	0
Viledon	0	0	0	1	0	0
Vilene	0	0	0	2	0	0
Viloft	1	0	0	1	0	0
Vinnex	0	0	0	1	0	0
Virtual Woolclasser	0	0	0	1	0	0

WORLD TEXTILES Trade Name Indexing

	1999	2000	2001	2002	2003	2004
Viscolight	0	0	0	1	0	0
Viscozyme	0	0	0	1	0	0
Vitech	0	2	0	0	0	0
Vivanet	0	0	0	1	0	0
Vivasorb	0	0	0	1	0	0
Vonnel	0	1	0	0	0	0
Vortex	0	0	0	1	0	0
Vulcan	0	0	1	0	0	0
Vyrene	0	1	0	2	0	0
Walotex	0	0	0	0	1	0
Warmsensor	0	0	1	0	0	0
Waternity	0	0	1	0	0	0
Weatherwize	0	0	0	0	0	1
WebFlex AX	0	2	1	0	0	0
Wellbond	0	0	1	0	0	0
WHOLEGARMENT	0	0	0	1	0	0
WikMax	0	0	1	0	0	0
Windstopper	0	0	0	0	1	1
Wolkiss	1	0	1	1	0	0
Wonderflex	0	0	0	1	0	0
Woolscience	0	0	0	0	2	0
X-Cor	0	0	0	0	0	1
X-Static	1	0	0	1	1	4
Xiameter	0	0	0	1	0	0
Xmod	0	0	0	0	2	0
Xorel	0	0	1	0	0	0
XTEND	0	0	0	1	0	0
Z-Fiber	0	0	0	0	0	1
Zeftron	0	0	1	0	0	1
Zelan	0	0	0	1	0	0
ZenTron	0	0	0	0	0	1
Zeovate	0	0	1	0	0	0
Zerostat	0	0	1	0	0	0
Zorflex	0	0	0	0	1	0
ZPREG	0	0	0	0	0	1
Zyflex	0	0	1	0	0	0
Zylon	0	1	3	1	3	0
Zytel	0	0	0	0	1	2

LIST OF JOURNALS INDEXED

AATCC Review
ISSN: 1532-8813
American Association of Textile Chemists and Colorists, 1 Davis Drive, Box 12215, Research Triangle Park, NC 27709-2215, USA Tel: +1-919-5498141 Fax: +1-919-5498933 ISSUES/YR: 12

Abstracts of UK Patent Applications
ISSN: not available
The Patent Office Concept House, Cardiff Road, Newport, Gwent, NP9 1RH, UK Tel: +44-1633-813930 Fax: +44-1633-813600 ISSUES/YR: 1

Advanced Composites Bulletin
ISSN: 0951-953X
International Newsletters Ltd, 9A Victoria Square, Droitwich Spa, Worcs, WR9 8DE, UK Tel: +44-1993-824130 Fax: +44-1993-824150
ISSUES/YR: 12

Advances in Textiles Technology
ISSN: 1472-0256
International Newsletters Ltd, 9A Victoria Square, Droitwich Spa, Worcs, WR9 8DE, UK Tel: +44-1993-824130 Fax: +44-1993-824150
ISSUES/YR: 12

African Textiles
ISSN: 0144-7521
Alain Charles Publishing Ltd, Alain Charles House, 27 Wilfred Street, London, SW1E 6PR,UKTel: +44-20-78347676 Fax: +44-20-79730076
ISSUES/YR: 4

Alpacas Australia
ISSN: 1328-8318
Australian Alpaca Association Inc., P. O. Box 1076, Mitcham North, VIC 3132, Australia Tel: +61-3-98737700 Fax: +61-3-98737711 ISSUES/YR: 4

Annual Conference Report - International Textile Manufacturers Federation
ISSN: 1727-4761
International Textile Manufacturers Federation, Am Schanzengraben 29, Postfach, Zurich, 8039, Switzerland Tel: +41-1-2017080 Fax: +41-1-2017134 ISSUES/YR: 1

Apparel
ISSN: 1543-2009
P.O. Box 1273, Skokie, IL 60076-8273, Tel: +1-610-6827495 Fax: +1-610-6827351 ISSUES/YR:12

Apparel Strategist
ISSN: 1085-6781
P.O. Box 406, Fleetwood, PA 19522, USA
ISSUES/YR: 11

Archaeological Textiles Newsletter
ISSN: 01697331
Norwegian University of Science and Technology, Vitenskapsmuseet, Trondheim, 7034, Norway
ISSUES/YR: 2

Artes de Mexico
ISSN: 0300-4953
Artes de Mexico y del Mundo S.A. de C.V., Cordoba 69, Col Roma, Mexico City, DF, 06700, Mexico Tel: +52-55-55255273 Fax: +52-55-55255925 ISSUES/YR: 6

Asian Textile Journal
ISSN: 0971-3425
G P S Kwatra, 201 Sonal Link Industrial Estate, Building 2, Link Road, Malad (W), Mumbai, 400 064, India Tel: +91-22-28812709 Fax: +49-22-2889873 ISSUES/YR: 12

ATA Journal
ISSN: 1015-8138
Adsale Publishing Company, Unit 1101-06, 11/F, Island Place Tower, 520 King's Road, North Point, Hong Kong, China Tel: +852-25118897 Fax: +852-25163530 ISSUES/YR: 6

ATIRA Communications on Textiles
ISSN: 0971-0833
Ahmedabad Textile Industry's Research Association, Ambawadi Vistar P.O., Ahmedabad, 380 015, India Tel: +49-79-6304671 Fax: +49-79-6302874 ISSUES/YR: 4

Australasian Textiles and Fashion
ISSN: 1327-4414
Australasian Textiles Publishers, P.O. Box 286, Belmont, VIC 3216, Australia Tel: +61-3-52613966 Fax: +61-3-52616950 ISSUES/YR: 6

Autex Research Journal
ISSN: not available
Technical University of Lodz, Faculty of Textile Engineering and Marketing, ul. Zeromskiego 116, Lodz, 90-543, Poland Tel: +48-42-6313350 Fax: +48-42-6313343 ISSUES/YR: 4

AVR Allgemeiner Vliesstoff-Report
ISSN: 0170-4060
dpw-Verlagsgesellschaft MbH, Postfach 1353, Heusenstam, D-63131, Germany Tel: +49-6104-6060 Fax: +49-6104-606317 ISSUES/YR: 6

Band- und Flechtindustrie
ISSN: 1432-3745
Melliand Textilberichte GmbH, Mainzer Landstrasse 251, Frankfurt am Main, D-60326, Germany Tel: +49-69-75951652 Fax: +49-69-75951650 ISSUES/YR: 4

WORLD TEXTILES List of Journals Indexed

Biomedical Materials
ISSN: 0955-7717
International Newsletters Ltd, 9A Victoria Square, Droitwich Spa, Worcs, WR9 8DE, UK Tel: +44-1993-824130 Fax: +44-1993-824150
ISSUES/YR: 12

Boletin Intexter del Instituto de Investigacion Textil y de Cooperacion Industrial
ISSN: 1131-6756
Instituto de Investigacion Textil y de Cooperacion Industrial, Colon 15, Terrassa, 08222, Spain Tel: +34-93-7398277 Fax: +34-93-7398272
ISSUES/YR: 2

Bragantia
ISSN: 0006-8705
Instituto Agronomico, Servico de Divulgacao Tecnico-Cientifica, Avenida Barao de Itapura 1481, P.O. Box 28, Campinas, 13020-902, Brazil Tel: +55-19-32315422 Fax: +55-19-32315422
ISSUES/YR: 2

Brand Strategy
ISSN: 0965-9390
Centaur Publishing Ltd., St. Giles House, 49-50 Poland Street, London, W1V 4AX, 7 Tel: +44-20-9704000 Fax: +44-20-79704191 ISSUES/YR: 12

BTRA Scan
ISSN: 0972-8341
Bombay Textile Research Association, Lal Bahadur Shastri Marg, Ghatkopar, Mumbai, 400 086, India
ISSUES/YR: 4

Bulletin - OIDFA
ISSN: 0771-7571
OIDFA, 29 Avenues Georges Mandel, Paris, F-75116, France ISSUES/YR: 4

Canadian Apparel
ISSN: 1484-3684
Canadian Apparel Federation, 504-124 O'Connor Street, Ottawa, Ontario, K1P 5M9, Tel: +1-613-2313220 Fax: +1-613-2312305 ISSUES/YR: 6

Canadian Textile Journal
ISSN: 0008-5170
Groupe CTT Group, 3000 Boulle, St-Hyacinthe, QC, J2S 1H9, Canada Tel: +1-450-7781870 Fax: +1-450-7789016 ISSUES/YR: 6

Cellulose
ISSN: 0969-0239
Kluwer Academic Publishers, Van Godewijckstraat 30, Dordrecht, 3311 GZ, Netherlands Tel: +31-78-6576000 Fax: +31-78-6576254 ISSUES/YR: 4

Chemical Fibers International
ISSN: 0340-3343
Deutscher Fachverlag GmbH, Mainzer Landstrasse 251, Frankfurt a. M., 60326, Germany Tel: +49-69-75951721 Fax: +49-69-75951720 ISSUES/YR: 6

China Textile and Apparel
ISSN: 1021-5824
Adsale Publishing Company, Unit 1101-06, 11/F, Island Place Tower, 520 King's Road, North Point , Hong Kong, China Tel: +852-25118897 Fax: +852-25163530 ISSUES/YR: 6

Clothing and Textiles Research Journal
ISSN: 0887-302X
International Textile and Apparal Association, P.O. Box 1360, Monument, CO 80132, USA Tel: +1-719-4883716 ISSUES/YR: 4

Coloration Technology
ISSN: 1472-3581
Society of Dyers and Colourists, Box 244, Perkin House, 82 Grattan Road, Bradford, Yorkshire, BD1 2JB, UK Tel: +44-1274-725138 Fax: +44-1274-392888 ISSUES/YR: 6

Colourage
ISSN: 0010-1826
Colour Publications Pvt. Ltd, 126-A Dhuruwadi, A.V. Nagwekar Marg, Prabhadevi, Mumbai, 400 025, India Tel: +49-22-4309318/4306319/4309610 Fax: +49-22-4300601 ISSUES/YR: 12

Composite Structures
ISSN: 0263-8223
Elsevier Ltd, The Boulevard, Langford Lane, Kidlington, Oxford, OX5 1GB, UK Tel: +44-1865-843000 Fax: +44-1865-843010 ISSUES/YR: 12

Composites Science and Technology
ISSN: 0266-3538
Elsevier Ltd, The Boulevard, Langford Lane, Kidlington, Oxford, OX5 1GB, UK Tel: +44-1865-843000 Fax: +44-1865-843010 ISSUES/YR: 12

Composites Technology
ISSN: 1083-4117
Ray Publishing Inc., 4891 Independence Street, Suite 270, Wheat Ridge, CO 80033, USA Tel: +1-303-4671776 Fax: +1-303-4671777 ISSUES/YR: 6

Costume
ISSN: 0590-8876
Costume Society, 56 Wareham Road, Lytchett Matravers, Poole, Dorset, BH16 6DS, UK
ISSUES/YR: 1

Crafts
ISSN: 0306-610X
Crafts Council, 44a Pentonville Road, London, N1 9BY, UK Tel: +44-20-72787700 Fax: +44-20-78370858 ISSUES/YR: 6

Design Issues
ISSN: 0747-9360
MIT Press, Five Cambridge Center, Cambridge, MA 02142-1493, USA Tel: +1-617-2535646 Fax: +1-617-5771545 ISSUES/YR: 4

Design Journal
ISSN: 1460-6925
Ashgate Publishing Ltd., Gower House, Croft Road, Aldershot, Hampshire, GU11 3HR, UK Tel: +44-1252-331551 Fax: +44-1252-344405
ISSUES/YR: 3

Digital Textile
ISSN: 1742-1128
World Textile Publications Ltd, Perkin House, 1 Longlands Street, Bradford, W. Yorks., BD1 2TP, UK Tel: +44-1274-378800 Fax: +44-1274-378811
ISSUES/YR: 6

Disposables and Nonwovens
ISSN: 0012-3811
Chandler Publications Ltd., 5 Evans Road, Eynsham, Oxon, OX8 1QT, UK Tel: +44-+44-1803-864668 Fax: ISSUES/YR: 6

Drapers Record
ISSN: 0967-3776
EMAP Fashion Ltd, Angel House, 338-346 Goswell Road, London, EC1V 7QP, UK Tel: +44-20-75201500 Fax: +44-20-75201501 ISSUES/YR: 51

DuPont Magazine
ISSN: 0095-8808
E.I. du Pont de Nemours and Co. Inc., Dupont Building, 1007 Market Street, Wilmington, DE 19898, USA Tel: +1-302-7741000 ISSUES/YR: 4

DWI Reports
ISSN: 0942-301X
Deutsches Wollforschungsinstitut, R.W.T.H., Veltmanplatz 8, Aachen, 52062, Germany Tel: +49-241-44690 Fax: +49-241-4469100 ISSUES/YR: 1

Dyes and Pigments
ISSN: 0143-7208
Elsevier Ltd, The Boulevard, Langford Lane, Kidlington, Oxford, OX5 1GB, UK Tel: +44-1865-843000 Fax: +44-1865-843010 ISSUES/YR: 12

Embroidery
ISSN: 1477-3724
Embroiderers' Guild, P.O. Box 42B, East Molesey, Surrey, KT8 9BB, UK Tel: +44-20-89431229 Fax: +44-20-89779882 ISSUES/YR: 6

Euratex Bulletin
ISSN: 1028-5954
EURATEX, Rue Montoyer 24, Box N-10 Brussels, 1000, Belgium Tel: +32-2-2854881 Fax: +32-2-2306054 ISSUES/YR: 5

Euroseil
ISSN: 0012-0758
Verlag Ernst Bauer, Breite Gasse 2, Ulm, 89073, Germany Tel: +49-731-64051 Fax: +49-731-6021276 ISSUES/YR: 4

Eurostitch Magazine
ISSN: 1387-0076
Eurostitch Magazine, P.O. Box 50084, Almere, 1305 AB, Netherlands Tel: +31-36-5314505 Fax: +31-36-5349003 ISSUES/YR: 6

Extracts from European Patent Applications, Part 1B: Primary Industry, Fixed Constructions, Mining
ISSN: 0943-1268
Thomson Scientific GmbH, Landbergerstrasse 191a, Munchen, 80687, Germany Tel: +49-89-547560 Fax: +49-89-54756309 ISSUES/YR: 52

Fabric Architecture
ISSN: 1045-0483
Industrial Fabrics Association International, 1801 County Road B W., Roseville, MN 55113-4061, USA Tel: +1-651-2222508 Fax: +1-651-6319334
ISSUES/YR: 6

Fangzhi Gaoxiao Jichukexue Xuebao
ISSN: 1006-8341
Basic Sciences Journal of Textile Universities, 19 South Jinhua Road, Xi'an, 710048, China
ISSUES/YR: 4

Fashion Business International
ISSN: 1473-0391
World Textile Publications Ltd, Perkin House, 1 Longlands Street, Bradford, W. Yorks., BD1 2TP, UK Tel: +44-1274-378800 Fax: +44-1274-378811
ISSUES/YR: 12

Fiber Organon
ISSN: 0040-5132
Fiber Economics Bureau Inc., 1530 Wilson Blvd, Ste.690, Arlington, VA 22209, USA Tel: +1-201-2281107 Fax: +1-201-2287598 ISSUES/YR: 12

Fiberarts
ISSN: 0164-324X
Fiberarts Magazine, 201 E. Fourth Street, Loveland, CO 80537, USA Tel: +1-970 6134679 Fax: +1-970 6134656 ISSUES/YR: 5

Fibres and Textiles in Eastern Europe
ISSN: 1230-3666
Institute of Chemical Fibres, Ul. Sklodowskiej-Curie 19/27, Lodz, 90-570, Poland ISSUES/YR: 4

Filiere Maille
ISSN: 1146-0733
Editions de l'Industrie Textile, 16 rue Ballu, Paris Cedex 09, 75311, Tel: +33-1-48741596 Fax: +33-1-48740189 ISSUES/YR: 4

Flock
ISSN: 0933-2316
Joachim Muller, Postfach 1370, Buedingen, 63643 G, Germany Tel: +49-6042-2938 Fax: +49-6042-7871 ISSUES/YR: 4

WORLD TEXTILES List of Journals Indexed

Geosynthetics International
Thomas Telford Services Ltd, Thomas Telford House, 1 Heron Quay, London, E14 4JD, 7 Tel: +44-20-9876999 Fax: +44-20-75389620 ISSUES/YR: 6

Geotechnical Fabrics Report
ISSN: 0882-4983
Industrial Fabrics Association International, 1801 County Road B W., Roseville, MN 55113-4061, USA Tel: +1-651-2222508 Fax: +1-651-6319334 ISSUES/YR: 9

Geotechnik
ISSN: 0172-6145
Verlag Gluckauf GmbH, P.O. Box 185620, Essen, 45206Germany Tel: +49-2054-92412023 Fax: +49-2054-924129 ISSUES/YR: 4

Geotextiles and Geomembranes
ISSN: 0266-1144
Elsevier Ltd, The Boulevard, Langford Lane, Kidlington, Oxford, OX5 1GB, UK Tel: +44-1865-843000 Fax: +44-1865-843010 ISSUES/YR: 6

Hali
ISSN: 0142-0798
Hali Publications Ltd, Hali Subscriptions Department, Tower House, Lathkill Street, Market Harborough, Leicestershire, LE16 9EF, UK Tel: +44-185-8438818 Fax: +44-185-8468969 ISSUES/YR: 6

Handwoven
ISSN: 0198-8212
Interweave Press Inc., 201 E. Fourth Street, Loveland, CO 80537-5655, USA Tel: +1-970-6697672 ISSUES/YR: 5

Heimtex
ISSN: 0017-9876
Panorama Verlags- und Werbegesellschaft mbH, Sudbrackstrasse 14-18, Bielefeld, 33611, Germany Tel: +49-521-5850 Fax: +49-521-585480 ISSUES/YR: 12

High-Performance Composites
ISSN: 1081-9223
Ray Publishing Inc., 4891 Independence Street, Suite 270, Wheat Ridge, CO 80033, USA Tel: +1-303-4671776 Fax: +1-303-4671777 ISSUES/YR: 6

IFAI's Marine Fabricator
ISSN: 1079-8250
Industrial Fabrics Association International, 1801 County Road B W., Roseville, MN 55113-4061, USA Tel: +1-651-2222508 Fax: +1-651-6319334 ISSUES/YR: 4

Indian Journal of Fibre and Textile Research
ISSN: 0971-0426
National Institute of Science Communication and Information Resources (NISCAIR), Dr. K. S. Krishnan Marg (Near Pusa Gate), New Delhi, 110-012, India Tel: +49-11-5786301-07 Fax: +49-11-5787062 ISSUES/YR: 4

Indian Journal of Labour Economics
ISSN: 0971-7927
University of Lucknow, Indian Society of Labour Economics, Badshaw Bagh, Lucknow, Uttar Pradesh, India ISSUES/YR: 4

Indian Journal of Sericulture
ISSN: 0445-7722
Central Sericultural Research and Training Institute, Srirampuram, Manandavadi Road, Mysore, 570 008, India Tel: +49-821-531406 Fax: +49-821-520845 ISSUES/YR: 2

Indian Silk
ISSN: 0019-6355
Central Silk Board, Ministry of Textiles, CSB Complex, BTM Layout, Madiwala, P.B. 6825, Bangalore, Karnataka, 560 068 , India Tel: +49-80-6689152 Fax: +49-80-6681511 ISSUES/YR: 12

Industria Laniera Tessile Abbigliamento
ISSN: 1593-2338
Faenza Editrice S.p.A., Via Pier de Crescenzi 44, Faenza, 48018, Italy Tel: +39-546-670411 Fax: +39-545-660440 ISSUES/YR: 6

Industria Tessile
ISSN: 0019-7491
Mediatex International Srl, Via Fabio Filzi 27, Milan, 20124, Italy Tel: +39-2-6773261 Fax: +39-2-67493714 ISSUES/YR: 6

Industria Textila
ISSN: 1222-5347
Inst. Nat. Cercetare-Dezvoltare Text., Pielarie Str. Lucretiu Patrascanu 16, Bucharest, Sect. 3, 74674, Romania Tel: +40-21-6434402 Fax: +40-21-3405515 ISSUES/YR: 4

Industrial Crops and Products
ISSN: 0926-6690
Elsevier, P.O. Box 211, Amsterdam, 1000 AE, Netherlands Tel: +31-20-4853757 Fax: +31-20-4853432 ISSUES/YR: 6

Industrial Fabric Products Review
ISSN: 0019-8307
Industrial Fabrics Association International,1801 County Road B W., Roseville, MN 55113-4061,USA Tel: +1-651-2222508 Fax: +1-651-6319334 ISSUES/YR: 12

Industrie Textile
ISSN: 0019-9176
Editions de l'Industrie Textile, 16 rue Ballu, Paris Cedex 09, 75311, France Tel: +33-1-48741596 Fax: +33-1-48740189 ISSUES/YR: 11

InTents
ISSN: 1090-8366
Industrial Fabrics Association International, 1801 County Road B W., Roseville, MN 55113-4061, USA Tel: +1-651-2222508 Fax: +1-651-6319334
ISSUES/YR: 4

International Carpet Bulletin
ISSN: 0268-2966
World Textile Publications Ltd, Perkin House, 1 Longlands Street, Bradford, W. Yorks., BD1 2TP, UK Tel: +44-1274-378800 Fax: +44-1274-378811
ISSUES/YR: 4

International Carpet Yearbook
ISSN: 1357-5201
World Textile Publications Ltd, Perkin House, 1 Longlands Street, Bradford, W. Yorks., BD1 2TP, UK Tel: +44-1274-378800 Fax: +44-1274-378811
ISSUES/YR: 1

International Dyer
ISSN: 0020-658X
World Textile Publications Ltd, Perkin House, 1 Longlands Street, Bradford, W. Yorks., BD1 2TP, UK Tel: +44-1274-378800 Fax: +44-1274-378811
ISSUES/YR: 12

International Fiber Journal
ISSN: 1049-801X
International Media Group Inc., 7401 Carmel Executive Park Drive, Suite 202, Charlotte, NC 28226, USA Tel: +1-704-5441969 Fax: +1-704-5446559 ISSUES/YR: 6

International Journal of Clothing Science and Technology
ISSN: 0955-6222
Emerald, 60/62 Toller Lane, Bradford, West Yorkshire, BD8 9BY, UK Tel: +44-1274-777700 Fax: +44-1274-785200 ISSUES/YR: 6

International Journal of Sheep and Wool Science
ISSN: not available
University of New England, Wool Technology and Sheep Breeding Journal, Sheep CRC, CJ Hawkins Homestead, Armidale, NSW 2351, Australia
ISSUES/YR: 4

International Textile Bulletin
ISSN: 1029-8525
ITS Publishing, International Textile Service, Kesslerstrasse 9, Schlieren-Zurich, CH-8952, Switzerland Tel: +41-1-7384800 Fax: +41-1-7384832 ISSUES/YR: 6

International Textile Bulletin: Nonwovens, Industrial Textiles
ISSN: 1024-6592
ITS Publishing, International Textile Service, Kesslerstrasse 9, Schlieren-Zurich, CH-8952, Switzerland Tel: +41-1-7384800 Fax: +41-1-7384832 ISSUES/YR: 4

Izvestiya Vysshikh Uchebnykh Zavedenii, Seriya Teknologiya Tekstil'noi Promyshlennosti
ISSN: 0021-3497
Izvestiya Vysshikh Uchebnykh Zavedenii, Prosp. F. Engelska 21, Ivanova, 153000, Russia
ISSUES/YR: 6

JEC Composites Magazine
ISSN: 0754-0876
19 Boulevard De Courcelles, Paris, 75008, France Tel: +33-1-58361509 Fax: +33-1-58361513
ISSUES/YR: 8

Journal du Textile
ISSN: 0293-0757
61, rue du Malte, Paris, Cedex 11, 75541, France Tel: +33-1-43572189 Fax: +33-1-47000835
ISSUES/YR: 42

Journal for Weavers, Spinners and Dyers
ISSN: 0267-7806
Association of Guilds of Weavers, Spinners and Dyers, 4 St. Jude's Close, Queen's Grove, Southsea, Hampshire, PO5 3HQ, UK ISSUES/YR: 4

Journal of Applied Polymer Science
ISSN: 0021-8995
John Wiley and Sons Inc., P.O.Box 18667, Newark, NJ 07191-8667, USA Tel: +1-212-8506000 Fax: +1-212-8506012 ISSUES/YR: 24

Journal of Beijing Institute of Clothing Technology (Natural Science Edition)
ISSN: 1001-0564
Beijing Institute of Clothing Technology, North Heping Street, Beijing, 010-64210952, China
ISSUES/YR: 4

Journal of Biotechnology and Sericology
ISSN: 1346-8073
National Institute of Sericultural and Entomological Science, 1-2 Owashi, Tsukuba-shi, Ibaraki-ken, 305, Japan Tel: +81-298-386056 Fax: +81-298-386028 ISSUES/YR: 6

Journal of China Textile University
ISSN: 1000-1476
China Textile University, 1882 Xa'an Xilu, Shanghai, 200051, China Tel: +86-21-62193057 Fax: +86-21-62089144 ISSUES/YR: 6

Journal of Cotton Science
ISSN: not available
Cotton Foundation, National Cotton Council of America, P.O. Box 820285, Memphis, TN 38182-0258, USA ISSUES/YR: 4

Journal of Fashion Marketing and Management
ISSN: 1361-2026
Emerald, 60/62 Toller Lane, Bradford, West Yorkshire, BD8 9BY, UK Tel: +44-1274-777700 Fax: +44-1274-785200 ISSUES/YR: 4

WORLD TEXTILES List of Journals Indexed

Journal of Geotechnical and Geoenvironmental Engineering
ISSN: 1090-0241
American Society of Civil Engineers, 1801 Alexander Graham Bell Drive, Reston, VA 20191-4400, USA Tel: +1-703-2956300 Fax: +1-703-2956211 ISSUES/YR: 12

Journal of Industrial Hemp
ISSN: 1537-7881
Haworth Press Inc., 10 Alice Street, Binghamton, NY 13904-1580, USA Tel: +1-607-7225857 Fax: +1-607-7710012 ISSUES/YR: 2

Journal of Industrial Textiles
ISSN: 1528-0837
SAGE Publications Ltd, 1 Olivers Yard, 55 City Road, London, EC1Y 1SP, UK Tel: +44-20-73248500 Fax: +44-20-73248600 ISSUES/YR: 4

Journal of Natural Fibers
ISSN: 1544-0478
Haworth Press Inc., 10 Alice Street, Binghamton, NY 13904-1580, USA Tel: +1-607-7225857 Fax: +1-607-7710012 ISSUES/YR: 4

Journal of Soochow University Engineering Science Edition
ISSN: 1673-047X
Soochow University Press, School of Material Engineering, No.1 Shizi Street, Suzhou, Jiangsu, 215021, China Tel: +86-512-65112799 Fax: +86-512-65221028 ISSUES/YR: 6

Journal of Sustainable Product Design
ISSN: 1367-6679
Kluwer Academic Publishers, Van Godewijckstraat 30, P.O. Box 17, Dordrecht, 3300 AA, Netherlands Tel: +31-78-6392392 Fax: +31-78-6542254 ISSUES/YR: 4

Journal of Textile and Apparel, Technology and Management
ISSN: not available
North Carolina State University, College of Textiles, P.O. Box 8301, Raleigh, NC 27695, USA Tel: +1-919-5156646 Fax: +1-919-5153733 ISSUES/YR: 4

Journal of the China Textile Institute
ISSN: 1019-0473
China Textile Institute, 6 Chen-Tian Road, Tu-Chen City, Taipei, Taiwan Tel: +886-2-2644720 Fax: +886-2-2675110 ISSUES/YR: 4

Journal of the Japan Research Association for Textile End-Uses
ISSN: 0037-2072
Japan Research Association for Textile End-Uses, Yoshin Ogimachi City Heights, Room 201, 11-5, Doshin 2-chome, Kita-ku, Osaka, 530, Japan Tel: +81-6-3581441 Fax: +81-6-3581442 ISSUES/YR: 12

Journal of the Suzhou Institute of Silk Textile Technology
ISSN: 1000-1999
Suzhou Institute of Silk Textile Technology, Suzhou University, Ganjiang Donglu 200, North Campus of Suzhou University, P.O.Box 156, Suzhou, Jiangsu, 215021, China Tel: +86-512-7480098 Fax: +86-512-7258875 ISSUES/YR: 6

Journal of the Textile Association
ISSN: 0368-4636
Textile Association (India), 72-A Santosh, Dr. M.B. Raut Road, Shivaji Park, Dadar, Mumbai, 400 028, India Tel: +49-22-4449286 Fax: +49-22-4447082 ISSUES/YR: 6

Journal of the Textile Institute
ISSN: 0040-5000
Textile Institute International, St James's Buildings, 4th Floor, Oxford Street, Manchester, M1 6FQ, UK Tel: +44-161-2371188 Fax: +44-161-2361991 ISSUES/YR: 7

Journal of the Textile Machinery Society of Japan
ISSN: 0371-0580
Textile Machinery Society of Japan, Osaka Science and Technology Center Building, Utsubo Koen, 8-4-1 Utsubo-hon-machi, Nishi-ku, Osaka, 550, Japan Tel: +81-6-4434691 Fax: +81-6-4434694 ISSUES/YR: 12

Journal of Tianjin Institute of Textile Science and Technology
ISSN: 1000-1557
Tianjin Institute of Textile Science and Technology, 63 Chenglinzhuang Road, Hedong District, Tianjin, China ISSUES/YR: 6

Journal of Tianjin Polytechnic University
ISSN: 1671-024X
Tianjin Polytechnic University, 63 Chenglinzhuang Road, Hedong District Tianjin, 300160, China ISSUES/YR: 4

Kettenwirk-Praxis
ISSN: 0170-401X
Karl Mayer GmbH, P.O. Box 1120, Obertshausen, 63166, Germany Tel: +49-6104-4020 Fax: +49-6104-43574 ISSUES/YR: 4

Knitting International
ISSN: 0266-8394
World Textile Publications Ltd, Perkin House, 1 Longlands Street, Bradford, W. Yorks., BD1 2TP, UK Tel: +44-1274-378800 Fax: +44-1274-378811 ISSUES/YR: 1

Kobunshi Ronbunshu
ISSN: 0386-2186
Society of Polymer Science, Nagaoka Building, 2-4-2, Tsukiji, Chuo-ku, Tokyo, 104, Japan Tel: +81-3-35433771 ISSUES/YR: 12

Kunsthandwerk und Design
ISSN: 0454-6539
Ritterbach Verlag GmbH, Rudolf-Diesel-Strasse 5-7, Frechen, 50226, Germany Tel: +49-22-3418996 Fax: +49-22-34186690 ISSUES/YR: 6

Lace
ISSN: 0308-3039
Lace Guild, The Hollies, 53 Audnam, Stourbridge, West Midlands, DY8 4AE, UK Tel: +44-1384-390739 Fax: +44-1384-444415 ISSUES/YR: 4

LDB Interior Textiles
ISSN: 0892-743X
E.W. Williams Publications Co., 2125 Center Avenue, Fort Lee, NJ 07024, USA Tel: +1-201-5927007 Fax: +1-201-5927171 ISSUES/YR: 12

Maille et Technique
ISSN: 1146-5735
Editions Vauclair, 41 Boulevard du General Martial Valin, Paris, 75015, France Tel: +33-1-45576060 ISSUES/YR: 12

Man-Made Textiles in India
ISSN: 0377-7537
The Synthetic and Art Silk Mills Research Association, Sasmira, Sasmira Marg, Worli, Mumbai, 400 025, India Tel: +49-22-4935351 Fax: +49-22-4930225 ISSUES/YR: 12

Materials World
ISSN: 0967-8638
Institute of Materials, 1 Carlton House Terrace, London, SW1Y 5DB, UK Tel: +44-20-78394071 Fax: +44-20-78392078 ISSUES/YR: 12

Medical Textiles
ISSN: 0266-2078
International Newsletters Ltd, 9A Victoria Square, Droitwich Spa, Worcs, WR9 8DE, UK Tel: +44-1993-824130 Fax: +44-1993-824150 ISSUES/YR: 12

Mekcmua
ISSN: 1311-9095
9/G, Kl. Ohridski Blvd., Sofia, 1756, Bulgaria Tel: +359-2-759219 Fax: +359-2-9873378 ISSUES/YR: 6

Melliand International
ISSN: 0947-9163
IBP - International Business Press Publishers, Deutsche Fachverlag, Mainzer Landstrasse 251, Frankfurt am Main, 60326, Germany Tel: +49-69-759501 Fax: +49-69-75951390 ISSUES/YR: 4

Melliand Textilberichte
ISSN: 0931-9735
Deutscher Fachverlag GmbH, Mainzer Landstrasse 251, Frankfurt a. M., 60326, Germany Tel: +49-69-75951721 Fax: +49-69-75951720 ISSUES/YR: 12

Men's Fashion International
ISSN: 1439-7080
Deutscher Fachverlag GmbH, Mainzer Landstrasse 251, Frankfurt a. M., 60326, Germany Tel: +49-69-75951721 Fax: +49-69-75951720 ISSUES/YR: 2

Nonwovens Industry
ISSN: 0163-4429
Rodman Publications Inc., 17 S. Franklin Turnpike, Box 555, Ramsey, NJ 07446 , USA Tel: +1-201-8252552 Fax: +1-201-8250553 ISSUES/YR: 12

Nonwovens Report International
ISSN: 0953-1092
World Textile Publications Ltd, Perkin House, 1 Longlands Street, Bradford, W. Yorks., BD1 2TP, UK Tel: +44-1274-378800 Fax: +44-1274-378811 ISSUES/YR: 6

Nonwovens World
ISSN: 0888-1979
MTS Publications, 4100 South 7th Street, Kalamazoo, Michigan 49009, USA Tel: +1-269-3751236 Fax: +1-269-3756710 ISSUES/YR: 6

Ornamente
ISSN: 0931-0657
Verlag M. und H. Schaper GmbH, Borsigstrasse 5, Alfeld, 31061, Germany Tel: +49-5181-80090 Fax: +49-5181-800933 ISSUES/YR: 5

Pakistan Textile Journal
ISSN: 0048-2757
Mazhar Yusuf, 304 Shaheen Centre, Main Clifton Road, Karachi, Kehkashan, Pakistan Tel: 92-21-431167475 Fax: 92-21-4533911 ISSUES/YR: 12

Performance Apparel Markets
ISSN: 1477-6456
Textiles Intelligence Ltd., 10 Beech Lane, Wilmslow, SK9 5ER, UK Tel: +44-1625-536 136 Fax: +44-1625-536 137 ISSUES/YR: 4

Performance Textiles
ISSN: 0968-0861
Impact Company Publications, Media House, 55 Old Road, Leighton Buzard, Beds., LU7 7RB, UK Tel: +44-1525-370013 Fax: +44-1525-382487 ISSUES/YR: 4

Polymer
ISSN: 0032-3861
Elsevier Ltd, The Boulevard, Langford Lane, Kidlington, Oxford, OX5 1GB, UK Tel: +44-1865-843000 Fax: +44-1865-843010 ISSUES/YR: 26

Polymer - Plastics Technology and Engineering
ISSN: 0360-2559
Taylor and Francis Inc., 325 Chestnut St, Suite 800, Philadelphia, PA 19106, USA Tel: +1-800-3541420 Fax: +1-215-6258914 ISSUES/YR: 5

WORLD TEXTILES List of Journals Indexed

Polymer International
ISSN: 0959-8103
John Wiley and Sons Ltd, The Atrium, Southern Gate, Chichester, West Sussex, PO19 8SQ, UK Tel: +44-1243-779777 Fax: +44-1243-775878 ISSUES/YR: 12

Polymers for Advanced Technologies
ISSN: 1042-7147
John Wiley and Sons Ltd, The Atrium, Southern Gate, Chichester, West Sussex, PO19 8SQ, UK Tel: +44-1243-779777 Fax: +44-1243-775878 ISSUES/YR: 12

Przeglad Wlokienniczy
ISSN: 1731-8645
Wydawnictwo SIGMA, Ul. Mazowiecka 12, Warsaw, 00-950, Poland Tel: +48-22-8268016 Fax: +48-22-8268016 ISSUES/YR: 12

Reinforced Plastics
ISSN: 0034-3617
Elsevier Ltd, The Boulevard, Langford Lane, Kidlington, Oxford, OX5 1GB, UK Tel: +44-1865-843000 Fax: +44-1865-843010 ISSUES/YR: 12

Research Disclosure
ISSN: 0374-4353
Kenneth Mason Publication Ltd., The Book Barn, Westbourne, Hampshire, PO10 8RS, UK Tel: +44-1243-377977 Fax: +44-1243-379136 ISSUES/YR: 12

Review of Progress in Coloration and Related Topics
ISSN: 0557-9325
Society of Dyers and Colourists, Box 244, Perkin House, 82 Grattan Road, Bradford, Yorkshire, BD1 2JB, UK Tel: +44-1274-725138 Fax: +44-1274-392888 ISSUES/YR: 1

Revista de la Industria Textil
ISSN: 0210-0800
Revitextil S.L., Cerdena 269, entlo. 2a, Barcelona, 08013, Spain Tel: +34-93-4571220 Fax: +34-93-2075568 ISSUES/YR: 10

Revista de Quimica Textil
ISSN: 0300-3418
Asociacion Espanola de Quimicos y Coloristas Textiles, Gran Via de Corts Catalanes 670, 6, Barcelona, 08010, Spain ISSUES/YR: 6

Rivista delle Tecnologie Tessili
ISSN: 0394-5413
Editoriale Elsevier S.p.a, Via Vittoria Colonna 4, Milano, 1-20149, Italy Tel: +39-2-485918264 Fax: +39-2-485918280 ISSUES/YR: 8

Selezione Tessile
ISSN: 1125-5579
Tecniche Nuove, Via C. Menotti 14, Milan, 20129, Italy Tel:+39-2-7570329 Fax: +39-2-7570364 ISSUES/YR: 6

Sen'i Gakkaishi
ISSN: 0037-9875
Society of Fiber Science and Technology, c/o Japan Publications Trading Co., Box 5030, Tokyo, 100-31, Japan ISSUES/YR: 12

Separation Science and Technology
ISSN: 0149-6395
Taylor and Francis Inc., 325 Chestnut St, Suite 800, Philadelphia, PA 19106, USA Tel: +1-800-3541420 Fax: +1-215-6258914 ISSUES/YR: 16

Spectrum Textile
ISSN: not available
Siemens Aktiengesellschaft, Automation and Drives Group, Frauenauracher Str. 80, Erlangen, D-91056, Germany ISSUES/YR: 4

Spinnovation
ISSN: not available
Spindelfabrik Suessen, Postfach 1320, Suessen, 73075, Germany ISSUES/YR: 2

Sportswear International
ISSN: 1021-0989
Deutscher Fachverlag GmbH, Mainzer Landstrasse 251, Frankfurt a. M., 60326, Germany Tel: +49-69-75951721 Fax: +49-69-75951720 ISSUES/YR: 6

Stitches Magazine
ISSN: 0899-5893
Intertec Publishing Corp., 9800 Metcalf Avenue, Overland Park, KS 66212-2215, USA Tel: +1-913-3411300 Fax: +1-913-9671898 ISSUES/YR: 12

Surface Design Journal
ISSN: 0197-4483
Surface Design Association, P.O. Box 360, Sebastopol, CA 95473-0360, USA Tel: +1-816-8360913 Fax: +1-816-2548521 ISSUES/YR: 4

Synthetic Fibres
ISSN: not available
Association of Synthetic Fibre Industry, Raj Mahal 1st Floor, 84 Veer Nariman, Mumbai, 400 020, India ISSUES/YR: 4

Technical Textiles International
ISSN: 0964-5993
International Newsletters Ltd, 9A Victoria Square, Droitwich Spa, Worcs, WR9 8DE, UK Tel: +44-1993-824130 Fax: +44-1993-824150 ISSUES/YR: 8

Technische Textilien
ISSN: 0323-3243
Deutscher Fachverlag GmbH, Mainzer Landstrasse 251, Frankfurt a. M., 60326, Germany Tel: +49-69-75951721 Fax: +49-69-75951720 ISSUES/YR: 4

Tecnica Textil Internacional
ISSN: 0040-1900
Editiones Tecnicas Especializades, Traversera de Gracia 15, 4, Barcelona, 08021, Tel: +34-93-2097933 Fax: +34-93-2096918 ISSUES/YR: 6

Tekstil
ISSN: 0492-5882
Savez Inzenjera i Technicara Tekstilaca Hrvatske, Novakova 8, P.O.Box 829, Zagreb, HR-10001, Croatia Tel: +385-1-4818252 Fax: +385-1-4818252 ISSUES/YR: 12

Tekstil i Obleklo
ISSN: 1310-912X
Tekstil i Obleklo, 108 Rakovski Street, P.O. Box 379, Sofia, 1000, Tel: +359-2-9803045 Fax: +359-2-801230 ISSUES/YR: 10

Tekstilec
ISSN: 0351-3386
Urednistvo Tekstilec, Snezniska 5 - p.p. 311, Ljubljana, 61000, Tel: +386-1-224417 ISSUES/YR: 12

Tekstilna Industrija
ISSN: 0040-2389
Savez Inzenjera i Technicara Tekstilaca SR Srbije, Novakova 8, P.O.Box 829, Zagreb, HR-10001, Croatia Tel: +385-1-4818252 Fax: +385-1-4818252 ISSUES/YR: 12

Tessili per Impieghi Tecnici
ISSN: 1125-9183
Ascontex Editoriale Srl, V. Filzi 27, Milano (MI), 20124, Italy Tel: +39-2-67100731 ISSUES/YR: 4

Textile Asia
ISSN: 0049-3554
Business Press Ltd., California Tower, 11-F, 30-32 d'Aguilar Street, P.O. Box 185, Hong Kong, China Tel: +852-25233744 Fax: +852-28106966 ISSUES/YR: 12

Textile Forum
ISSN: 1431-3529
Textil-Forum-Service/ B. Sterk, Friedenstrasse 5, Postfach 5944, Hannover, D-30175, Germany Tel: +49-511-817007 Fax: +49-511-813108 ISSUES/YR: 4

Textile History
ISSN: 0040-4969
Maney Publishing, Hudson Road, Leeds, LS9 7DL, UK Tel: +44-113-2497481 Fax: +44-113-2486983 ISSUES/YR: 2

Textile Horizons
ISSN: 1353-6184
World Textile Publications Ltd, Perkin House, 1 Longlands Street, Bradford, W. Yorks., BD1 2TP, UK Tel: +44-1274-378800 Fax: +44-1274-378811 ISSUES/YR: 6

Textile Journal
ISSN: 0008-5170
Groupe CTT Group, 3000 Boulle, St-Hyacinthe, QC, J2S 1H9, Canada Tel: +1-450-7781870 Fax: +1-450-7789016 ISSUES/YR: 6

Textile Magazine
ISSN: 0040-5078
Gopali and Co., Quanta Zen Apartments, Ground Floor, No.2. Thomas Road, T. Nagar, Chennai, 600 017, India Tel: +49-44-24330979 Fax: +49-44-24332413 ISSUES/YR: 12

Textile Month
ISSN: 0040-5116
World Textile Publications Ltd, Perkin House, 1 Longlands Street, Bradford, W. Yorks., BD1 2TP, UK Tel: +44-1274-378800 Fax: +44-1274-378811 ISSUES/YR: 12

Textile Network
ISSN: 1612-5096
Meisenbach GmbH, Postfach 2069, Bamberg, D-96011, Germany Tel: +49-951-861135 Fax: +49-951-861158 ISSUES/YR: 12

Textile Outlook International
ISSN: 0268-4764
Textiles Intelligence Ltd., 10 Beech Lane, Wilmslow, SK9 5ER, UK Tel: +44-1625-536 136 Fax: +44-1625-536 137 ISSUES/YR: 6

Textile Rental
ISSN: 0195-0118
Textile Rental Association of America, 1800 Diagonal Road, Suite 200, Alexandria, VA 22314, USA Tel: +1-703-5190029 ISSUES/YR: 12

Textile Research Journal
ISSN: 0040-5175
Textile Research Institute, 601 Prospect Avenue, Box 625, Princeton, NJ 08542, USA Tel: +1-609-9243150 Fax: +1-609-6837836 ISSUES/YR: 12

Textile Technology
ISSN: not available
W. Schlafhorst AG und Co., Blumenberger Strasse 143-145, Monchengladbach, D-41004, Germany Tel: +49-21-61280 Fax: +49-21-61282645 ISSUES/YR: 2

Textile Topics
ISSN: 1043-5239
International Textile Center, P.O. Box 45019, Lubbock, TX 79409-5019, USA Tel: +1-806-7473790 Fax: +1-806-7473796 ISSUES/YR: 1

Textile Trends
ISSN: 0040-5205
Eastland Publications Private Ltd., 44 Chittaranjan Avenue, Calcutta, 700 012, India Tel: +49-33-273096 Fax: +49-33-2251096 ISSUES/YR: 12

Textile View Magazine
ISSN: 1384-5306
Metropolitan Publishing bv, Parkwijk, Waldeck Pyrmontlaan 20, Amsterdam, 1075 BX, Netherlands Tel: +31-20-6177624 Fax: +31-20-6179357 ISSUES/YR: 4

Textile World
ISSN: 0040-5213
Billian Publishing Inc., 2100 Powers Ferry Road, Atlanta, GA 30339, USA Tel: +1-770-9555656 Fax: +1-770-9520669 ISSUES/YR: 12

Textiles and Clothing International
ISSN: 1470-6296
Textile Institute International, St James's Buildings, 4th Floor, Oxford Street, Manchester, M1 6FQ, UK Tel: +44-161-2371188 Fax: +44-161-2361991 ISSUES/YR: 1

Textiles Eastern Europe
ISSN: 1354-5981
Textile Media Services Ltd, 2A Bridge Street, Silsden, Keighley, BD20 9NB, UK Tel: +44-1535 656489 Fax: +44-8700 940868 ISSUES/YR: 12

Textiles Magazine
ISSN: 1367-1308
Textile Institute International, St James's Buildings, 4th Floor, Oxford Street, Manchester, M1 6FQ, UK Tel: +44-161-2371188 Fax: +44-161-2361991 ISSUES/YR: 4

Textiles Panamericanos
ISSN: 0049-3570
Billian Publishing Inc., 2100 Powers Ferry Road, Atlanta, GA 30339, USA Tel: +1-770-9555656 Fax: +1-770-9520669 ISSUES/YR: 6

Textiles South East Asia
ISSN: 1743-3231
Textile Media Services Ltd, 2A Bridge Street, Silsden, Keighley, BD20 9NB, UK Tel: +44-1535 656489 Fax: +44-8700 940868 ISSUES/YR: 12

Textilkunst International
ISSN: 0934-3342
Verlag M. und H. Schaper GmbH, Borsigstrasse 5, Alfeld, 31061, Germany Tel: +49-5181-80090 Fax: +49-5181-800933 ISSUES/YR: 4

Textil-Revue
ISSN: 0040-4861
St.Galler Tagblatt AG, Fuerstenlandstrasse 122, Postfach 6362, St. Gallen, 9001, Switzerland Tel: +41-71-2727431 Fax: +41-71-2727449 ISSUES/YR: 41

Textilveredlung
ISSN: 0040-5310
Swiss Society Textile and Chemistry, Kesslerstrasse 9, Schlieren/Zurich, CH-8952, Switzerland Tel: +41-1-7384828 Fax: +41-1-7384829 ISSUES/YR: 12

TextilWirtschaft
ISSN: 0040-487X
Deutscher Fachverlag GmbH, Mainzer Landstrasse 251, Frankfurt a. M., 60326, Germany Tel: +49-69-75951721 Fax: +49-69-75951720 ISSUES/YR: 52

Tinctoria
ISSN: 0040-7984
Progetto Nobilitazione Tessile SRL, V.le Sarca, 223, Milano, 20126, Italy Tel: +39-2-66103404 Fax: +39-2-66103444 ISSUES/YR: 12

TUT Textiles a Usages Techniques
ISSN: 1161-9317
Editions de l'Industrie Textile, 16 rue Ballu, Paris Cedex 09, 75311, France Tel: +33-1-48741596 Fax: +33-1-48740189 ISSUES/YR: 4

U.S. Patent and Trademark Office
ISSN: not available
U.S. Patent and Trademark Office, General Information Services Division, Crystal City Plaza 3, rm. 2C02, Washington, DC 20231, USA Tel: +1-703-3084357 Fax: +1-703-3057786 ISSUES/YR: 52

Unitex
ISSN: not available
Unitex a.s.b.l., Rivierstraat 19/102, Beervelde, 9080, Belgium Tel: 9-3552388 Fax: 9-3567880 ISSUES/YR: 6

Upholstery Design and Management
ISSN: 1531-7234
Chartwell Communications Inc., 380 E. Northwest Highway, Des Plaines, IL 60016, USA Tel: +1-847-3906700 Fax: +1-847-3907100 ISSUES/YR: 12

Vlakna a Textil
ISSN: 1335-0617
Research Institute for Man-Made Fibers, Svit, Okr. Poprad, 059-21, Slovakia Tel: +421-92-55688 Fax: +421-92-55663 ISSUES/YR: 4

Weaver's Digest
ISSN: 1424-5752
Sultex Limited, Head Office, Marketing/Public Relations, Ruti ZH, CH-8630, Switzerland Tel: +41-55-2502923 Fax: +41-55-2502171 ISSUES/YR: 2

Wool and Woollens of India
ISSN: 0043-7808
Indian Woollen Mills' Federation, Churchgate Chambers 5, 7th Floor, 5 New Marine Lines, Mumbai, 400 020, India Tel: +49-22-2624372 Fax: +49-22-2624675 ISSUES/YR: 4

Wool Press
ISSN: 1445-1123
CRC for Premium Quality Wool, Suite 6, 25 Argyle Street, Fitzroy, VIC 3065, Australia Tel: +61-3-94165180 Fax: +61-3-94165185 ISSUES/YR: 1

Wool Record
ISSN: 0263-6131
World Textile Publications Ltd, Perkin House, 1 Longlands Street, Bradford, W. Yorks., BD1 2TP, UK Tel: +44-1274-378800 Fax: +44-1274-378811 ISSUES/YR: 12

Wool Research Organisation of New Zealand Reports
ISSN: 0112-2851
Wool Research Organisation of New Zealand Inc., Private Bag 4749, Christchurch, New Zealand Tel: +64-3-3252421 Fax: +64-3-3242717 ISSUES/YR: 1

Wool Technology and Sheep Breeding
ISSN: 0043-7875
University of New England, Wool Technology and Sheep Breeding Journal, Sheep CRC, CJ Hawkins Homestead, Armidale, NSW 2351, Australia ISSUES/YR: 4

WSA
ISSN: 1356-644X
World Trades Publishing, P.O. Box 6, Liverpool, L22 0QN, UK Tel: +44-151-9289288 Fax: +44-151-9284190 ISSUES/YR: 6

Yalkut Israel Textile and Fashion Journal
ISSN: 0372-7777
Israel Textile and Fashion Association, 12 Anna Frank Street, P.O. Box 243, Ramat Gon, 52526, Israel Tel: +972-3-7521133 Fax: +972-3-7521141 ISSUES/YR: 4

MAJOR TEXTILE ORGANISATIONS

AATCC
American Association of Textile Chemists and Colorists,
1 Davis Drive, Box 12215,
Research Triangle Park,
NC 27709-2215, USA
Tel: +1-919-5498141
Fax: +1-919-5498933
www.aatcc.org

ACIMIT
Italian Association of Textile Machinery Manufacturers,
Via Tevere 1
20123 Milano
Italy
Tel: +39 02 4693611
Fax: +39 02 48008342
www.acimit.it

SASMIRA
Synthetics and Art Silks Mills Research
Sasmira Marg
Worli
Mumbai – 40025
India
Tel: 022 24932016
Fax: 022 24930225
www.sasmira.org

ATIRA
Ahmedabad Textile Industry's Research Association
P.O. – Ambawadi Vistar
Ahmedabad – 380015
India
Tel: (079) 26307921, 26307922, 26307923, 26305131, 26305132, 26305133
Fax: (079) 26302874, 26301969
www.atira-tex-rnd.org

NITRA
North Indian Textile Research Association
Sector 23
Raj Nagar
Ghazibad – 201002
India
Tel: 914 4783335
Fax: 914 783638
www.nitratextile.org

BTRA
The Bombay Textile Research Association
Lal Bahadur Shastri Marg
Ghatkopar (W)
Mumbai - 400 086
India
Tel: 2500 3651 / 2652 / 2117 / 1119
Fax: 91 - 22 - 2500 0459
www.btraindia.com

SITRA
South Indian Textile Research Association
P.B.No.: 3205,
Coimbatore Aerodrome Post,
Coimbatore - 641 014,
Tamil nadu,
India
Tel: 91 - 422 - 2574367-9,
Fax: 91 - 422 – 2571896
www.sitra.org.in

MANTRA
Man-Made Textiles Research Organisation
Near Market Telephone Exchange
Ring Road
Surat – 395002
Tel: 0261 – 623211
Fax: 0261- 622500

EATP
European Association for Textile Polyolefin's
Avenue E Van Nieuwenhyss 4,
B-1160 Brussels,
Belgium
Tel: +32-2-676-7472
Fax: +32-2-676-7474
www.eatp.org

ETH Zurich
Eidgenössische Technische Hochschule Zürich
ETH Zentrum, HG
Rämistrasse 101,
CH-8092 Zürich,
Switzerland
Tel: +41 44 632 11 11
Fax: +41 44 632 10 77
www.ethz.ch

EDANA
The European Disposables and Nonwovens Association
Avenue Eugène Plasky, 157,
B-1030 Brussels,
Belgium
Tel: +32 2 734 93 10
Fax: +32 2 733 35 18
www.edana.org

IFAI
Industrial Fabrics Association International
1801 County Road B W,
Roseville, MN 55113,
USA
Tel: 651/222-2508, 800/225-4324
Fax: 651/631-9334
www.ifai.com

INDA
Association of the Nonwoven Fabrics Industry
PO Box 1288,
Cary, NC 27512
USA
Tel: +1-919-233-1210 x123
Fax: +1-919-233-1282
www.inda.org

Japan Sewing Machinery Manufacturers Association
Fuji Building,
1-14-12 Nishi-Shinbashi
Minato-ku
Tokyo 105-0003
Japan
Tel:+81-3-3597-0470
Fax:+81-3-3597-0477
www.fair.or.jp/jiam

CIRCOT
Central Institute for Research on Cotton Technology
(Indian Council of Agricultural Research)
Adenwala Road,
Matunga, Mumbai 400 019
India
Tel: 91-022-24127273,76
Fax: 91-022-24130835/24157239
www.circot.res.in

SEAMS
The National Association for the Sewn Products Industry
4921-C Broad River Road
Columbia
South Carolina
USA
www.seams.org

Cotton Incorporated
World Headquarters
6399 Weston Parkway
Cary, North Carolina 27513
USA
Tel: (919) 678-2220
Fax: (919)678-2230
www.cottoninc.com

DWI
Deutsches Wollforschungsinstitut an der RWTH Aachen e.V.
Pauwelsstraße 8
52056 Aachen
Germany
Tel. ++49 241 80-233-00
Fax ++49 241 80-233-01
www.dwi.rwth-aachen.de

National Cotton Council of America
1918 N Parkway
Memphis TN 38112-5000
USA
Tel: (901) 274-9030
Fax: (901) 725-0510
www.cotton.org

IAF
International Apparel Federation
5 Portland Place
London W1B 1PW
UK
Tel: +44(0)20-7636 7788
Fax: +44(0)20-7436 8261
www.iafnet.org

CIRFS
International Rayon and Synthetic Fibres Committee
Avenue E. Van Nieuwenhuyse 4,
B-1160 Brussels,
Belgium
Tel: +32 (2) 676 74 55
Fax: +32 (2) 676 74 54
www.cirfs.org

The Silk Association of Great Britain
5 Portland Place,
London W1N 3AA
UK
Tel: 0171 636 7788
Fax: 0171 636 7515

SAMPE
Society for the Advancement of Material and Process Engineering
1161 Park View Drive,
Covina, CA 91724-3751,
USA
Tel: +1 626-331-0616
Fax: 626-332-8929
sampeibo@sampe.org
www.sampe.org

University of Leeds
Department of Colour and Polymer Chemistry,
University of Leeds,
Leeds LS2 9JT
UK
Tel: 0044 (0)113 343 2776
Fax: 0044 (0)113 343 2947
www.colour.leeds.ac.uk

SDC
Society of Dyers and Colourists
PO Box 244,
Perkin House, 82 Grattan Road,
Bradford, BD1 2JB
UK
www.sdc.org.uk

Major Organisations

ITMF
International Textile Manufacturers Federation
Am Schanzengraben 29,
Postfach,
CH-8039 Zürich
Switzerland
Tel: +41 -1 283 63 80
Fax: +41 -1 283 63 89
www.itmf.org

Textile Institute
1st Floor, St. James's Buildings,
Oxford Street,
Manchester, M1 6FQ
UK
Tel: +44(0)161 237 1188.
Fax: +44(0)161 236 1991
www.texi.org

The University of Bolton,
Deane Road,
Bolton, BL3 5AB
UK
www.bolton.ac.uk

TANDEC
Textiles and Nonwovens Development Center
1321 White Ave.,
Knoxville, TN 37996-1950,
USA
Tel: 865-974-6298
Fax: 865-974-3580
tandec@utk.edu

ITAA
Information Technology Association of America
ITAA Offices,
1401 Wilson Boulevard,
Suite 1100,
Arlington, VA 22209
USA
Tel:(703)522-5055
www.itaa.org

Associazione Tessile e Salute
Corso G Pella 2
13900 Biella
Italy
Tel: 015 8488348
Fax: 015 8408064
www.tessilesalute.it

Canadian Textiles Institute
222 Somerset Street West
Ottawa
Ontario
K2P 2G3
Canada
Tel: 613 232 7195
Fax: 613 232 8722
www.textiles.ca

CSTRI
Central Silk Technological Research Institute
B.T.M. Layout
Madivala
Bangalore – 560068
India
Tel: 080 6687808
Fax: 080 6680435

Wool Research Association
Kolshet Road
P.O. Sandoz Baug
Thane – 400607
India
Tel: 022 25346453
Fax: 022 25343629

IJIRA
Indian Jute Industries Research Association
17 Taratola Road
Calcutta – 700088
India
www.ijira.org

Alberta Research Council
250 Karl Clark Road
Edmonton, Alberta
Canada T6N 1E4
Tel: (780) 450-5111
Fax: (780) 450-5333
www.arc.ab.ca

BTSR
Best Technologies Study and Research
21057 Olgiate Olona
VA Italy
Tel: (39) 0331/323202
Fax: (39) 0331/323282
www.btsr.com

CSIRO
Commonwealth Scientific and Industrial Research Association
Bag 10
Clayton South
Victoria
Australia 3169
Tel: +61 3 9545 2176
Fax: +61 3 9545 2175
www.csiro.au

CRS-ACS
Cooperative Research Centre for Advanced Composite Structures
506 Lorimer Street
Fishermans Bend
Victoria 3207
Australia
Phone: +61 3 9646 6544
Fax: +61 3 9646 8352
www.crs-acs.com.au

ITV Denkendorf
German Institute for Textile and Fiber Research
Koerschtalstrasse 26
73770 Denkendorf
Germany
Tel: +49 (0) 711 / 93 40 – 0
www.itv-denkendorf.de

DTNW
Deutsches Textilforschungszentrum Nord West
Institut an der Universität Duisburg - Essen
Adlerstrasse 1
D-47798 Krefeld
Germany
Tel: 0049 - 21 51 - 84 3 0
Fax: 0049 - 21 51 - 84 3 143
www.dtnw.de

DLR Institut
German Aerospace Center
Linder Höhe
51147 Köln
Germany
Tel: +49 2203 601-0
Fax: +49 2203 67310
www.dlr.de

IFI
International Fabricare Institute
14700 Sweitzer Lane,
Laurel, MD 20707
USA
Tel: 301-622-1900
Fax: 240-295-0685
www.ifi.org

FAPRI
Food and Agricultural Policy Researxh Institute
Iowa State University
578 Heady Hall
Ames, Iowa 50011-1070
Tel: (515) 294-7519
Fax: (515) 294-6336
www.fapri.org

INTEXTER
Colom, 15,
08222 Terrassa,
Barcelona
Spain
Tel: +34 93 739 82 70
Fax: +34 93 739 82 72

TEXTRANET
Textile Technology Transfer Network
14, rue des Reculettes
75013 PARIS
Tel.: (33) 1 44 0819 00
Fax.: (33) 1 44 08 19 39

BTTG
British Textile Technology Group
Wira House
West Park Rinq Road
LEEDS LS16 6QL - ENGLAND
Tel.: (44) 113 259 19 99
Fax: (44) 113 278 03 06
www.bttg.co.uk

Norsk Tekoinstitutt
The Norwegian Textile and Clothing Institute
Thormøhlensgate 55
5008 Bergen, NORWAY
Tel. +47 55 54 38 00
Fax +47 55 31 06 09

CITEVE
Centro Tecnológico das Indústrias
Quinta da Maia
P - 4760 Vila Nova
De Famalicao
Portugal
Tel.: (351) 52 300 300
Fax: (351) 52 76 563 or 748

CANESIS
Canesis Network Ltd
Cnr Springs Road and Gerald Street
Lincoln, Canterbury
Private Bag 4749
Christchurch
New Zealand
Tel: +64 (3) 325 2421
Fax: +64 (3) 325 2717
www.canesis.com

MAJOR CONFERENCES

PREMIERE VISION
Head Office,
20, boulevard Eugène Deruelle,
69432 Lyon, Cedex 3,
France
Tel: 33(0)4 72 60 65 00
Fax: 33(0)4 72 60 65 09
info@premierevision.fr
www.premierevision.fr

DOMOTEX
Organiser:
Deutsche Messe AG
Press and Public Relations,
Messegelände,
D-30521 Hannover,
Germany
www.domotex.de

HEIMTEXTIL
Organiser:
Messe Frankfurt GmbH
Ludwig-Erhard-Anlage 1,
60327 Frankfurt a. M.
Germany
www.heimtextil.messefrankfurt.com

JEC COMPOSITES SHOW
19 Boulevard de Courcelles,
75008, Paris
France
Tel: 0158361509
Fax: 0158361513
www.globalcomposites.com

INDEX 05
Organiser:
Fondation Orgexpo
P.O Box 112,
CH-1218 Grand-Saconnex,
Geneva,
Switzerland
Tel: +41-22-761-1111
Fax: +41-22-798-0100
jp.thiemard@palexpo.ch
www.palexpo.ch

CONVENTION INTERNATIONALE D'AFFAIRES DES TEXTILES TECHNIQUES (CITEX EUROPE)
Organiser:
Adhesion Group
Direction du Développement,
71 rue des Tilleuls,
92771 Boulogne Cedex,
France
Tel: 00 33 1 41 86 41 86
Fax: 00 33 1 46 04 57 61
info@adhes.com

AATCC COMMITTEE MEETINGS
Organiser:
American Association of Textile Chemists and Colorists (AATCC)
PO Box 12215,
Research Triangle Park, NC 27709
USA
www.aatcc.org

INTERNATIONAL ISTANBUL TEXTILE & CLOTHING CONFERENCE
Organiser:
Marmara Univeristy
Textile Faculty,
Goztepe 34722,
Istanbul,
Turkey
Tel: +90 (216) 336 57 70 / 117
Fax: +90 (216) 337 89 87
www.tef.marmara.edu.tr

EATP EXECUTIVE SEMINAR
Organiser:
European Association for Textile Polyolefins (EATP)
Avenue E Van Nieuwenhyss 4,
B-1160 Brussels,
Belgium
Tel: +32-2-676-7472
Fax: +32-2-676-7474
info@eatp.org
www.eatp.org

THE FIBER SOCIETY CONFERENCE
Organiser:
ETH Zurich
Eidgenössische Technische Hochschule Zürich
ETH Zentrum, HG
Rämistrasse 101,
CH-8092 Zürich,
Switzerland
Tel: +41 44 632 11 11
Fax: +41 44 632 10 77
www.ethz.ch

SHANGHAITEX
Organiser:
Adsale Exhibition Services Ltd
Unit 1101-1106
11/F, Island Place Tower,
510 King's Road,
North Point,
Hong Kong
Tel: +852-2516-3354
Fax: +852-2516-5024
textile@adsaleexh.com
www.adsaleexh.com

AVANTEX
Organiser:
Messe Frankfurt GmbH
Ludwig-Erhard-Anlage 1,
60327 Frankfurt a. M.,
Germany
Tel: +49-69-7575-6840
Fax: +49-69-7575-6633
avantex@messefrankfurt.com
www.avantex.messefrankfurt.com

TECHTEXTIL
Organiser:
Messe Frankfurt GmbH
Ludwig-Erhard-Anlage 1,
60327 Frankfurt a. M.
Germany
Contact: Katrin Klepsch
Tel: +49-69-7575-6738/5822
Fax: +49-69-7575-6950
katrin.klepsch@messefrankfurt.com
www.messefrankfurt.com

REINFORCED PLASTICS CONFERENCE AND EXHIBITION (RPASIA)
Organiser:
Elsevier Advanced Technology
The Boulevard,
Langford Lane,
Kidlington,
Oxford OX5 1GB,
UK
Tel: +44-(0)1865-843208
Fax: +44-(0)1865-843973
a.jacob@elsevier.com
www.rpasia.com

TECHTEXTIL ROSSIJA
Organiser:
Messe Frankfurt GmbH
Ludwig-Erhard-Anlage 1,
60327 Frankfurt a. M.
Germany
lisa.schuertzmann@russia.messefrankfurt.com
www.techtextil.com

SHANGHAI INTERNATIONAL NONWOVENS CONFERENCE AND EXHIBITION
Organiser:
CMP Asia Trade Fairs Pte Ltd
CMP Asia Ltd - New York Office,
600 Community Drive,
Manhanet, NY 11030,
USA
Tel: +65-6735-3366
Fax: +65-6738-9644
since@cmpasia.com.sg
www.since2005.com

OUTLOOK
Organiser:
EDANA
Avenue Eugène Plasky, 157,
B-1030 Brussels,
Belgium
Tel: +32 2 734 93 10
Fax: +32 2 733 35 18
philip.preest@edana.org
www.edana.org

ITMA ASIA
Organiser:
Cematex
c/o British Textile Machinery Association
Glazebrook Lane,
Glazebrook,
Warrington WA3 5BN,
UK
Tel: +33-1-4717-6345
Fax: +33-1-4717-6348
info@ucmtf.com
www.itmaasia2005.com

IFAI EXPO
Organiser:
IFAI
1801 County Road B W,
Roseville, MN 55113,
USA
Tel: 651/222-2508, 800/225-4324
Fax: 651/631-9334
generalinfo@ifai.com
www.ifai.com

FILTRATION
Organiser:
Association of the Nonwoven Fabrics Industry (INDA)
PO Box 1288,
Cary, NC 27512
USA
Tel: +1-919-233-1210 x123
Fax: +1-919-233-1282
gthien@inda.org
www.inda.org

INTERNATIONAL EXHIBITION OF FINISHING AND KNITTING (IKME)
Organiser:
Fieria Milano International
Largo Domodossola, 1 Palazzina F,
MI 20145 Milano,
Italy
Tel: (39) 02.48550.1
segreteria.ikme@fmi.it
www.ikme.it

TECHTEXTIL SOUTH AMERICA
Organiser:
Guazzelli Messe Frankfurt
Messe Frankfurt GmbH
Ludwig-Erhard-Anlage 1,
60327 Frankfurt a. M.
Germany
Contact: Peter Thomaschewski
Tel: +55-11-4688-6019
Fax: +55-114688-6050
www.guazzelli.com.br

INTC
Organiser:
Association of the Nonwoven Fabrics Industry (INDA)
PO Box 1288,
Cary, NC 27512,
USA
Tel: (919) 233 1210
www.inda.org

VISION
Organiser:
Association of the Nonwoven Fabrics Industry (INDA)
PO Box 1288,
Cary, NC 27512,
USA
Tel: (919) 233 1210
www.inda.org

TECHTEXTIL NORTH AMERICA
Organiser:
Messe Frankfurt GmbH
Ludwig-Erhard-Anlage 1,
60327 Frankfurt a. M.
Germany
Tel: +1-770-984-8016
Fax: +1-770-984-8023
ttna@usa.messefrankfurt.com
www.techtextil.com

IMB – WORLD FAIR FOR APPAREL PRODUCTION TECHNOLOGY AND TEXTILE PROCESSING
Organiser:
Koelnmesse GmbH
Messeplatz 1
50679 Köln,
Germany
Tel: + 49 221 821 -0
Fax: + 49 221 821 -2574
info@koelnmesse.de
u.deitersen@koelnmesse.de
www.imb.de

AMERICAN TEXTILE MACHINERY EXHIBITION INTERNATIONAL (ATME-I)
Organiser:
ATMA
201 Park Washington Court
Falls Church
Virginia 22046
USA
Tel: +1-770-434-7331
Fax: +1-770-438-6172
kevin@atme-i06.com
www.atme-i06.com

CHINA INTERNATIONAL TEXTILE MACHINERY EXHIBITION (CITME)
Organiser:
Sub-Council of Textile Industry,
CCPIT No.12 East Chang An Street
Beijing 100742
China
Tel: 86-10-85229436 / 85229474
Fax: 86-10-85229480 / 65121732
citme@ccpittex.com
www.citme.com.cn

EXPO DETERGO
Organiser:
EXPO DETERGO S.R.L.
Via Angelo Masina 9
I - 20158 Milano
Italy
Tel: +39 02 39314120
Fax: +39 02 39315160
expodetergo@libero.it

IDEA
Organiser:
Association of the Nonwoven Fabrics Industry (INDA)
PO Box 1288
Cary, NC 27512-1288
USA
Tel: (919)233-1210
Fax: (919)233-1282
www.inda.org

INDEX
Organiser:
dmg world media Dubai Ltd.,
PO Box 33817,
Dubai
UAE
Tel. +971 4 331 9688,
Fax +971 4 331 9480 E.
dmg@emirates.net.ae
www.dmgdubai.com

JIAM
Organiser:
Japan Sewing Machinery Manufacturers Association
Fuji Building,
1-14-12 Nishi-Shinbashi
Minato-ku
Tokyo 105-0003
Japan
Tel:+81-3-3597-0470
Fax:+81-3-3597-0477
info@jasma.or.jp
www.fair.or.jp/jiam

SEAMS
Organiser:
SEAMS
4921-C Broad River Road
Columbia
South Carolina
USA
www.seams.org

ATIRA
Organiser:
Ahmedabad Textile Industry's Research Association
P.O. Ambawadi Vistar,
Ahmedabad - 380 015
India
Tel: (079) 26307921, 26307922, 26307923, 26305131, 26305132, 26305133
Fax: (079) 26302874, 26301969
www.atira-tex-rnd.org

BELTWIDE COTTON
Organiser:
National Cotton Council of America
1918 N Parkway
Memphis TN 38112-5000
USA
Tel: (901) 274-9030
Fax: (901) 725-0510
www.cotton.org/beltwide

BTRA
Organiser:
The Bombay Textile Research Association
Lal Bahadur Shastri Marg
Ghatkopar(W)
Mumbai - 400 086
India
Tel: 2500 3651 / 2652 / 2117 / 1119
Fax: 91 - 22 - 2500 0459
www.btraindia.com

CIRCOT
Organiser:
Central Institute for Research on Cotton Technology
(Indian Council of Agricultural Research)
Adenwala Road,
Matunga, Mumbai 400 019
India
Tel: 91-022-24127273,76
Fax: 91-022-24130835/24157239
circot@vsnl.com
www.circot.res.in

COTTECH
Organiser:
Cotton Incorporated
World Headquarters
6399 Weston Parkway
Cary, North Carolina 27513
USA
Tel: (919) 678-2220
Fax: (919)678-2230
www.cottoninc.com

DWI
Organiser:
Deutsches Wollforschungsinstitut an der RWTH Aachen e.V.
Pauwelsstraße 8
52056 Aachen
Germany
Tel. ++49 241 80-233-00
Fax ++49 241 80-233-01
www.dwi.rwth-aachen.de

IAF
Organiser:
International Apparel Federation
5 Portland Place
London W1B 1PW
UK
Tel: +44(0)20-7636 7788
Fax: +44(0)20-7436 8261
www.iafnet.org

IFVTCC
Organiser:
VDTF e.V.
Weinbrenner Str. 4,
D-68723 Schwetzingen,
Germany
www.vdtf.de

INTERNATIONAL MAN-MADE FIBRE
Organiser:
CIRFS
Avenue E. Van Nieuwenhuyse 4,
B-1160 Brussels,
Belgium
Tel. +32 (2) 676 74 55
Fax: +32 (2) 676 74 54
www.cirfs.org

Major Conferences

INTERNATIONAL SILK ASSOCIATION
Organiser:
The Silk Association of Great Britain
5 Portland Place,
London W1N 3AA
UK
Tel: 0171 636 7788
Fax: 0171 636 7515

INTERNATIONAL WOOL RESEARCH
Organiser:
University of Leeds
11th International Wool Research Conference
Department of Colour and Polymer Chemistry,
University of Leeds,
Leeds LS2 9JT
UK
Tel: 0044 (0)113 343 2776
Fax: 0044 (0)113 343 2947
www.colour.leeds.ac.uk/wool2005

ITMF
Organiser:
International Textile Manufacturers Federation
Am Schanzengraben 29,
Postfach,
CH-8039 Zürich
Switzerland
Tel: +41 -1 283 63 80
Fax: +41 -1 283 63 89
www.itmf.org

MEDICAL TEXTILES
Organiser:
The University of Bolton,
Deane Road,
Bolton, BL3 5AB
UK
www.bolton.ac.uk

NITRA
Organiser:
Northern India Textile Research Association
Sector 23,
Rajnagar,
Ghazibad 201002
India
Tel: 0120 2786451
www.nitratextile.org

SAMPE
Organiser:
SAMPE
1161 Park View Drive,
Covina, CA 91724-3751,
USA
Tel: +1 626-331-0616
Toll Free: +1 800 562-7360 (U.S. only)
Fax: 626-332-8929
sampeibo@sampe.org
www.sampe.org

SDC
Organiser:
Society of Dyers and Colourists
PO Box 244,
Perkin House, 82 Grattan Road,
Bradford, BD1 2JB
UK
www.sdc.org.uk

SITRA
Organiser:
SITRA
P.B.No.: 3205,
Coimbatore Aerodrome Post,
Coimbatore - 641 014,
Tamil nadu,
India
Tel: 91 - 422 - 2574367-9,
Fax: 91 - 422 – 2571896
www.sitra.org.in

TIFCON
Organiser:
Textile Institute
1st Floor, St. James's Buildings,
Oxford Street,
Manchester, M1 6FQ
UK
Tel: +44(0)161 237 1188.
Fax: +44(0)161 236 1991
www.texi.org

TIWC
Organiser:
Textile Institute
1st Floor, St. James's Buildings,
Oxford Street,
Manchester, M1 6FQ
UK
Tel: +44(0)161 237 1188.
Fax: +44(0)161 236 1991
www.texi.org

SASMIRA
Organiser:
The Synthetic and Art Silk Mills' Researcg Association
Sasmira Marg,
Worli,
Mumbai – 400025,
India

GARMENTECH
Organiser:
Zak Group
3a Crown Court, 3rd Floor,
34 (Old No 128) Cathedral Road,
Chennai, 600 086
India
Tel: + 91 44 28114578
www.zakgroup.com

TANDEC NONWOVENS CONFERENCE

Organiser:
TANDEC
1321 White Ave.,
Knoxville, TN 37996-1950,
USA
Tel: 865-974-6298
Fax: 865-974-3580
tandec@utk.edu

ITAA MEETING

Organiser:
ITAA
ITAA Offices,
1401 Wilson Boulevard,
Suite 1100,
Arlington, VA 22209
USA
Tel:(703)522-5055